Crop Genetics and Breeding

Crop Genetics and Breeding

Edited by Harold Salazar

Larsen & Keller
www.larsen-keller.com

Crop Genetics and Breeding
Edited by Harold Salazar
ISBN: 979-8-88836-046-0 (Hardback)

Larsen & Keller
Published by Larsen and Keller Education,
5 Penn Plaza,
19th Floor,
New York, NY 10001, USA

Cataloging-in-Publication Data

Crop genetics and breeding / edited by Harold Salazar.
p. cm.
Includes bibliographical references and index.
ISBN 979-8-88836-046-0
1. Genetics. 2. Genomics. 3. Plant breeding. 4. Crops. I. Salazar, Harold.
QH430 .G46 2023
576.5--dc23

For more information regarding Larsen and Keller Education and its products, please visit the publisher's website www.larsen-keller.com

Table of Contents

Preface

Crop breeding refers to the science of altering the traits of plants for human benefit. It can be classified into several types, including hybrid breeding, inbreeding, mutation breeding, backcrossing and gene editing. It can be used to grow superior crop varieties by utilizing biotechnological and genetic advancements. Crop genetics is a branch of genomics that investigates the complete DNA sequence of a crop, including all of its chromosomes. The knowledge of DNA sequence helps to make decisions regarding various problems, such as determining which plant genes should be selected for enhanced variety of crops. Crop genetics facilitates breeding crops with particular traits, like biotic and abiotic tolerance, and superior quality. The ultimate goal is to forecast which plants will thrive and under what situations. This book aims to shed light on some of the unexplored aspects of crop genetics and breeding. It is appropriate for students seeking detailed information in this area of agricultural science as well as for experts.

Significant researches are present in this book. Intensive efforts have been employed by authors to make this book an outstanding discourse. This book contains the enlightening chapters which have been written on the basis of significant researches done by the experts.

Finally, I would also like to thank all the members involved in this book for being a team and meeting all the deadlines for the submission of their respective works. I would also like to thank my friends and family for being supportive in my efforts.

Editor

1

Allelic Differentiation at the *E1/Ghd7* Locus has Allowed Expansion of Rice Cultivation Area

Hiroki Saito [1,2,*], Yutaka Okumoto [1], Takuji Tsukiyama [1,3], Chong Xu [1,4], Masayoshi Teraishi [1] and Takatoshi Tanisaka [1,4]

1 Graduate School of Agriculture, Kyoto University, Kyoto, Kyoto 606-8502, Japan; okumoto.yutaka.4w@kyoto-u.ac.jp (Y.O.); tsukiyama@nara.kindai.ac.jp (T.T.); xuchong@kiui.ac.jp (C.X.); temple@kais.kyoto-u.ac.jp (M.T.); t_tanisa@kiui.ac.jp (T.T.)
2 Tropical Agriculture Research Front, Japan International Research Center of Agricultural Science, Ishigaki, Okinawa 907-0002, Japan
3 Faculty of Agriculture, Kindai University, Nara, Nara 631-8505, Japan
4 School of Agriculture, Kibi International University, Minami-Awaji 656-0484, Japan
* Correspondence: hirokisaito@affrc.go.jp.

Abstract: The photoperiod-insensitivity allele *e1* is known to be essential for the extremely low photoperiod sensitivity of rice, and thereby enabled rice cultivation in high latitudes (42–53° north (N)). The *E1* locus regulating photoperiod-sensitivity was identified on chromosome 7 using a cross between T65 and its near-isogenic line T65w. Sequence analyses confirmed that the *E1* and the *Ghd7* are the same locus, and haplotype analysis showed that the *e1/ghd7-0a* is a pioneer allele that enabled rice production in Hokkaido (42–45° N). Further, we detected two novel alleles, *e1-ret/ghd7-0ret* and *E1-r/Ghd7-r*, each harboring mutations in the promoter region. These mutant alleles alter the respective expression profiles, leading to marked alteration of flowering time. Moreover, *e1-ret/ghd7-0ret*, as well as *e1/ghd7-0a*, was found to have contributed to the establishment of Hokkaido varieties through the marked reduction effect on photoperiod sensitivity, whereas *E1-r/Ghd7-r* showed a higher expression than the *E1/Ghd7* due to the nucleotide substitutions in the *cis* elements. The haplotype analysis showed that two photoperiod-insensitivity alleles *e1/ghd7-0a* and *e1-ret/ghd7-0ret*, originated independently from two sources. These results indicate that naturally occurring allelic variation at the *E1/Ghd7* locus allowed expansion of the rice cultivation area through diversification and fine-tuning of flowering time.

Keywords: rice; flowering time; photoperiod sensitivity; allelic variation; fine-tuning

---

1. Introduction

Rice is a major cereal extensively cultivated in a wide range of latitudes from 55° N to 35° S. Because rice is formerly a facilitative short-day (SD) plant well adapted to warm climate, photoperiodic control of flowering time is a key factor in the regional and seasonal adaptability of rice varieties [1]. In high latitudes (>ca. 40° N), rice cultivation had been impracticable due to the short summer and long-day (LD) more than 15 h during the summer, until early flowering varieties with extremely weak photoperiod sensitivity were raised [2–4]. It was during 1900 to 1930 that such varieties were first released and planted in the northernmost rice cultivation area, Hokkaido, in Japan (42–45° N) [5]. The varieties raised for Hokkaido also enabled rice cultivation even in Hei Long Jiang province (43–53° N) of China [6].

Also in low latitudes (ca. 20° S–20° N), a recent rice breeding program aims to produce varieties with weak photoperiod sensitivity (PS), though a long basic vegetative growth period is necessary at the same time, because such a combination of the two traits for heading will permit almost constant and adequate vegetative growth periods under SD (less than 13.5 h) [7–9]. In addition, in middle latitudes (30–40° N), there is a close relation between the photoperiod sensitivity of varieties and the latitude of their cultivation area [3,4]. Thus, understanding of the genetic factors responsible for photoperiod sensitivity, as well as basic vegetative growth, will be essential for not only guaranteeing stable rice production but also allowing further expansion of rice cultivation area.

Genetic studies on rice flowering (heading) time started in 1915 [10]. Since then, many flowering time loci were reported: among them, *E1* [11–14], *Photosensitivity 1* (*Se1*) [15], and *Earliness 1* (*Ef1*) [16], have been intensively studied about their genetic characteristics, such as allelic variation, response to photoperiod, geographical distribution, and interaction with other loci. The geographical studies showed that these three loci play especially important roles in regional adaptabilities of Japanese and Taiwanese *japonica* rice varieties and *japonica*/*indica* cross varieties in Korea [4–6,15,17–23].

The Committee on Gene Symbolization, Nomenclature and Linkage Groups of the Rice Genetics Cooperative made a rule that the gene symbols which have been commonly used by many workers in the past should be retained [24], and recommended to categorize flowering time genes into three types, earliness and lateness (gene symbol: *E*), photoperiod sensitivity (gene symbol: *Se*), and basic vegetative growth (gene symbol: *Ef*). With the advance of quantitative trait locus (QTL) analysis, however, it has become difficult to categorize newly found QTLs into three types because they are detected only from flowering time data. Since then, the gene symbols, *E*, *Se*, and *Ef*, did not come to be retained. Recent molecular genetic analyses identified three key flowering time loci, *Heading date 1* (*Hd1*) [25], *Early heading date 1* (*Ehd1*) [26], and *Grain number, plant height, and heading date 7* (*Ghd7*) [27], all of which were named regardless of the rule, and subsequent studies on these three loci provided new numerous molecular-based knowledges of rice flowering. Similarly, about the *E1*, *Se1* and *Ef1* loci, the information useful for rice breeding has been accumulated with enormous numbers until now. Therefore, it is significant to clarify the relationships of the three loci, *E1*, *Se1* and *Ef1*, to the loci named regardless of the rule. To date, the *Se1* and *Ef1* loci proved to be identical with the *Hd1* [21,28] and the *Ehd1* loci [7], respectively.

The *E1* was first identified as a late flowering time locus: the functional allele *E1* is completely dominant over the nonfunctional allele *e1* [11,12]. This locus was also involved in plant height. Later, this locus proved to control PS [13], and its functional allele *E1* was shown essentially important in rice varieties for temperate areas in Japan (30–40° N) [17,18] because of firmly inhibiting the panicle primordial differentiation under LD until it becomes SD conditions, and thereby ensuring normal vegetative growth and stable yields. In contrast, a photoperiod-insensitivity allele *e1* was found to be essential for the varieties commercially cultivated in Hokkaido (42–45° N), because of its marked reducing effect on photoperiod sensitivity: use of *e1* enabled rice cultivation in high latitudes where LD conditions continue during the summer [5,21]. The *E1* locus was found to be located on chromosome 7, linked to the *rfs* (rolled fine strip) and *slg* (slender glume) loci with recombination values of 16.3% and 9.1%, respectively [29]. This locus has been well investigated for its effects on photoperiod sensitivity and regional adaptabilities of rice plants [5,11–19], but little is known about the relationship with the loci which were identified by molecular genetic analysis and named regardless of the rule. Recently, the *Ghd7* locus was precisely mapped on chromosome 7, and this locus exert major effects on not only heading date but also number of grains per panicle and plant height [27]. In addition, subsequent molecular analyses of the *Ghd7* locus demonstrated that a loss-of-function allele of *Ghd7* is essential for the extremely early flowering of Hokkaido varieties [30–32]. These reports make a conjecture that the *E1* and *Ghd7* are the same locus.

In the present study, we first analyzed the effects of three alleles at the *E1* locus on photoperiod sensitivity using the Taiwanese *japonica* rice variety "Taichung 65 (T65)" harboring *E1* [33], its isogenic line T65m harboring *e1* [16,33–35], and T65w that harbors a chromosome segment of *O. rufipogon* Griff. including the *E1* locus in the genetic background of T65 [36]. Subsequently, we attempted to determine the precise chromosomal location of the *E1* locus using the progenies from T65 × T65w, and then conducted sequence analysis to learn the sequences of the three alleles, also to investigate the relationship between the two loci, *E1* and *Ghd7*. We finally applied a haplotype analysis of the chromosomal region surrounding the *E1/Ghd7* locus to 44 Hokkaido and 50 Japanese-core-collection varieties in order to prove correctness of the findings by Okumoto et al. (1996) [5] and Ichitani et al. (1998) [21] that *e1* is the key allele for establishing the varieties for the northernmost rice cultivation area, and its history and origin.

2. Results

2.1. Photoperiod Sensitivities of T65, T65w, and T65m

Days to heading (DH) of T65, T65m, and T65w under a SD were 84.7, 81.4, and 82.0 respectively, while those under a LD were 95.6, 90.4, and 118.0, respectively (Figure 1). Thus, the photoperiod sensitivities of T65, T65m, and T65w were estimated at 11.1, 9.0, and 36.0, respectively. Since T65m is an isogenic line of T65 for the *E1* locus, the weaker PS of T65m was attributable to the photoperiod-insensitivity allele e1 at the *E1* locus. T65w showed far stronger photoperiod sensitivity than T65 and T65m. The genotypic difference between T65w and T65 is only in the chromosome region including the *E1* locus, where only T65w harbors a chromosome segment induced from *O. rufipogon* Griff. Since any other photoperiod sensitivity genes have not yet been reported in this region, we conclude that the chromosome segment introduced from *O. rufipogon* Griff. in T65w certainly harbors a strong photoperiod-sensitivity allele, probably at the *E1* locus.

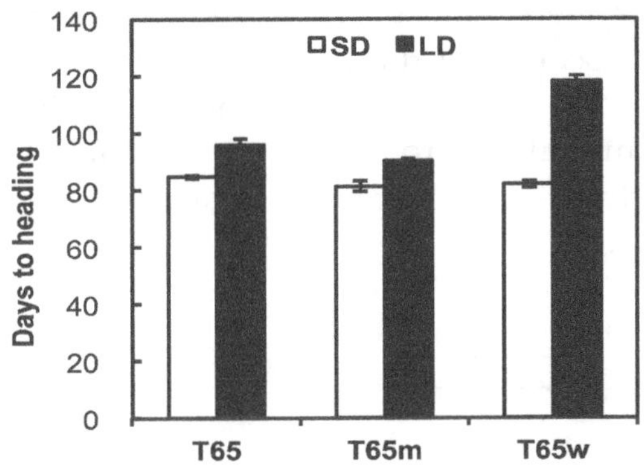

Figure 1. Days to heading of T65, T65m, and T65w under short-day (SD, white bar) and long-day (LD, black bar) conditions.

2.2. Chromosomal Location of the E1 Locus

The F_2 population from the cross between T65 and T65w, comprising 205 plants, showed a continuous distribution of DH within the parental ranges (Figure 2a). We conducted a progeny test using 38 F_3 lines, which were derived from randomly selected F_2 plants. In the test, all the F_3 lines were clearly classified into three groups. The ratio of [T65-type]:[segregating-type]:[T65w-type] lines was 13:14:11, which fitted the 1:2:1 ratio expected for one-locus segregation ($\chi^2 = 8.904$, $P > 0.05$) (Table S1). In contrast, the F_2 population from the cross between T65w and T65m showed a bimodal distribution of DH within the parental ranges, with a clear breakpoint dividing the population into early (T65m-type) and late (T65w-type) groups (Figure 2b). The ratio of early type (34 plants): late type (91 plants) fitted the 1:3 ratio expected for one-locus segregation ($\chi^2 = 0.570$, $P > 0.05$). In the progeny test, all the 40 F_3 lines were clearly classified into three groups. The ratio of [T65m type]:[segregating type]:[T65w type] lines fitted the 1:2:1 ratio expected for one-locus segregation ($\chi^2 = 0.150$, $P > 0.05$) (Table S2). T65m is an isogenic line of T65 harboring a recessive allele *e1* at the *E1* locus. We accordingly inferred that T65w harbors a novel allele at the *E1* locus, whose heading-date delaying effect was stronger than *E1* in T65. We designated this allele *E1-r* (a novel photoperiod-sensitivity allele at the *E1* locus).

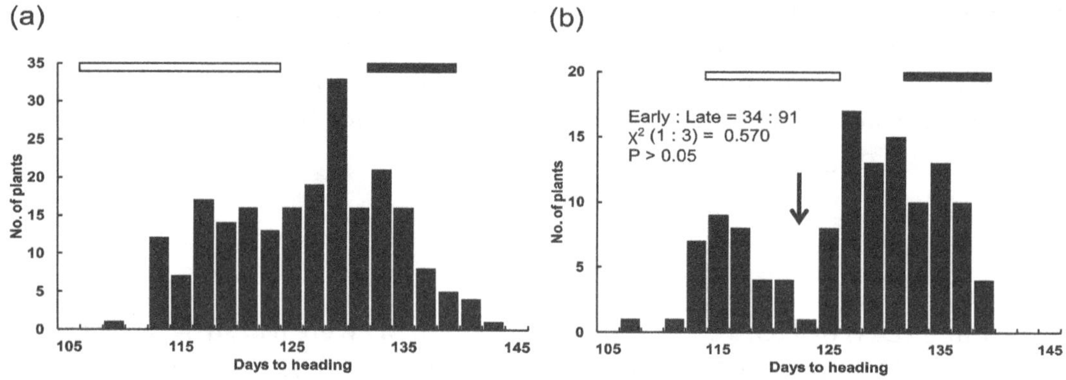

Figure 2. Distributions of days to heading in two F_2 populations from crosses between (**a**) T65 × T65w and (**b**) T65w × T65m. The black bar indicates the range of days to heading of T65w. The white bar indicates the ranges of days to heading of (**a**) T65 and (**b**) T65m. The arrow indicates the breakpoint between early and late heading groups.

Using 546 F_3 plants from the cross between T65 and T65w, we tried to identify the chromosomal location of the *E1* locus. The result showed that the *E1* locus was present in the region with a physical distance of 4.11 Mb between two simple sequence repeat (SSR) markers, RM1253 and RM3635, on chromosome 7 (Figure 3). Subsequently, we attempted to narrow down the candidate region of the *E1* locus, using 1263 F_4 progenies derived from several F_3 recombinants between RM1253 and RM3635; consequently, the chromosomal location of the *E1* locus was narrowed down to the region with a physical length of approximately 228.1-kb between RM5436 and RM21341 (Figure 3). In this region, 11 genes are reported in Rice Annotation Program Database [37]. Among them, we proposed that Os07g0261200, which was reported as *Ghd7*, a repressor of flowering time under LD conditions [27], was likely to be a candidate of *E1*.

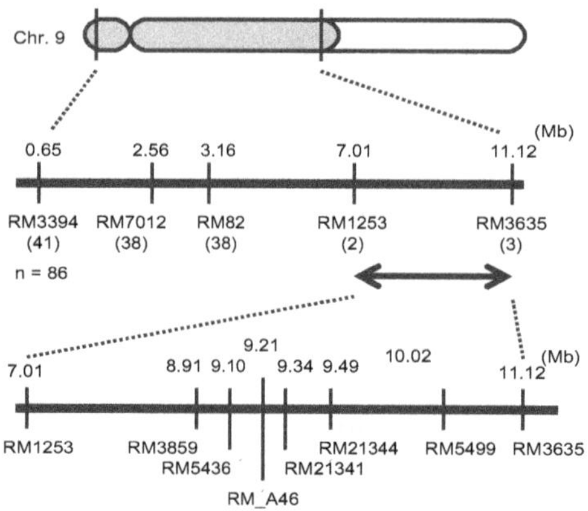

RM1253	RM3859	RM5436	RM_A46	RM21341	RM21344	RM5499	RM3635	No. of plants	Ave. of DH	Plant type
T	T	T	T	T	T	T	T	233	103.0	T
T	T	T	T	T	T	T	H	1	104.0	T
T	T	T	T	T	T	H	H	2	102.5	T
T	T	T	T	H	H	H	H	1	104.0	T
T	T	T	H	H	H	H	H	1	125.0	W
T	T	H	H	H	H	H	H	120	115.6	W
T	H	H	H	H	H	H	H	251	115.1	W
T	H	H	H	H	H	H	T	1	121.0	W
T	H	H	H	H	H	T	T	1	122.0	W
T	H	H	H	H	T	T	T	1	114.0	W

Figure 3. Map-based cloning and graphical genotypes of the candidate region of the E1 locus. "T" and "H" at each marker indicate T65 homozygous and heterozygous, respectively. "T" and "W" at plant type indicate T65-type (early heading) and T65w-type (late heading), respectively.

Sequence analyses showed that the sequences of the alleles at the *Ghd7* locus in T65 and T65m were completely consistent with a functional allele *Ghd7-2* [38] and a nonfunctional allele *ghd7-0a* [27], respectively (Figure 4a,b). Since the genotypic difference in flowering time between T65 and T65m is only at the *E1* locus, this suggests that *E1* and *Ghd7* are the same locus (hereafter we tentatively designate *E1* (=*Ghd7*) as *E1/Ghd7*), and that T65m flowered earlier than T65 because the former harbors a loss-of-function allele *e1/ghd7-0a*. In contrast, the allele of T65w at the *E1* locus harbored four nonsynonymous substitutions and two nucleotide substitutions in the promoter region (Figure 4a,b). Among the substitutions, two in the promoter region were in the transcriptional signal motifs (cis elements): low temperature response element (LTRE) core actor (located at −284) and the TATA box (located at −564). Thus, the two nucleotide substitutions were considered to modify the expression of *E1/Ghd7*. Subsequent expression analysis of *E1/Ghd7* showed that the expression of T65w was higher than that of T65 (Figure 4c). This suggests that the late flowering of T65w is caused by high expression of *E1/Ghd7* due to the nucleotide substitutions in the cis elements.

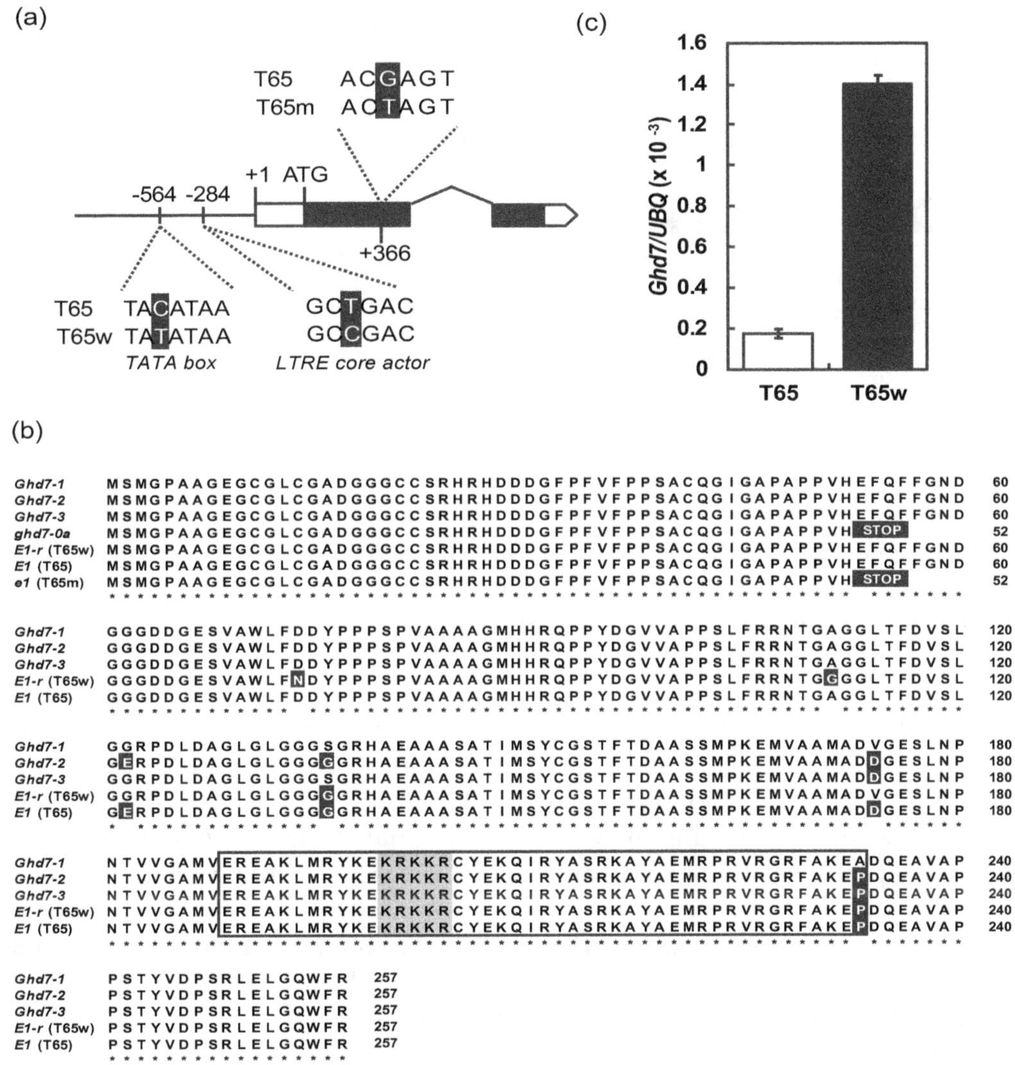

Figure 4. (**a**) Schematic diagrams of the alleles at the *E1/Ghd7* focused on nucleotide substitutions among T65 T65m, and T65w. (**b**) Alignments of amino acid sequences of the alleles at the *E1/Ghd7* locus. The white and black characters with black and gray cells indicate amino acid substitutions and CONSTANS, CO-like, and TOC1 (CCT)-motif, respectively. The box indicates the CCT-motif region. *Ghd7-1*, *Ghd7-2* and *Ghd7-3* were functional alleles [29]. (**c**) Comparison of the expression level of the allele at the *E1/Ghd7* locus between T65 and T65w.

2.3. A Novel Nonfunctional Allele at the E1/Ghd7 Locus

Okumoto et al. (1996) [5] showed that nine Hokkaido varieties tested all harbored a nonfunctional (photoperiod-insensitivity) allele *e1* at the *E1* locus thorough a conventional genetic analysis, and assumed that this allele has played an essential role in the establishment of rice varieties for the Hokkaido district. To confirm this assertion, we analyzed the presence of the nucleotide substitution from GAG (Glu) to TAG (stop codon) in exon 1 at the *E1/Ghd7* locus (Figure 4a) of 44 Hokkaido varieties using a cleaved amplified polymorphic sequence (CAPS) marker. The result showed that 37 varieties harbored the *e1/ghd7-0a* allele, and 7 varieties did not (Table S3). This single nucleotide substitution was not observed in EG5 (Aikoku), which is one of the tester lines for the *E1*, *E2* and *E3* loci involved in the flowering time, and which harbors *e1* allele at the *E1* locus [11–13]. Sequence analysis for the EG5 revealed that a Ty1-copia like retrotransposon (TE) was inserted in the promoter region of the *E1/Ghd7* allele (Figure 5a). The seven varieties, which did not harbor the *e1/ghd7-0a* allele, also harbored the same TE insertion. We named this novel nonfunctional allele *e1-ret/ghd7-0ret*. The expression of the *e1-ret/ghd7-0ret* allele was far lower than the *E1/Ghd7-2* allele in the Japanese variety "Nipponbare" with the reference genome (Figure 5b), indicating that *e1-ret/ghd7-0ret* confers extremely weak photoperiod sensitivity by losing the normal function of the promoter.

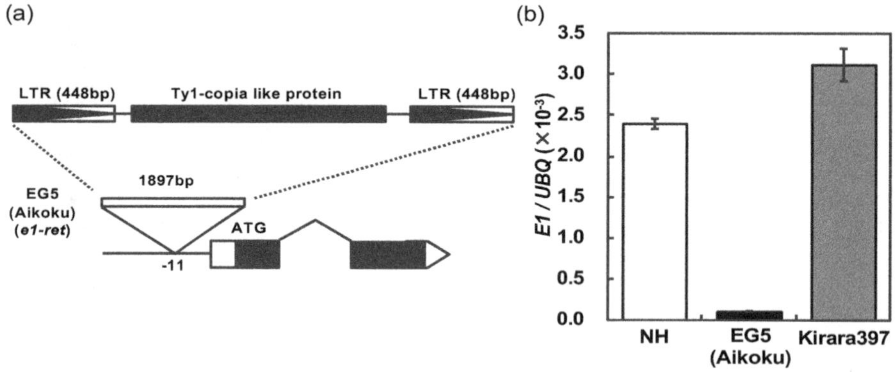

Figure 5. (**a**) Schematic diagram of the allele at the *E1/Ghd7* locus in "EG5 (Aikoku)". (**b**) Comparison of the expression level of the allele at the *E1/Ghd7* locus among three Japanes varieties "Nipponbare" (NH), "EG5 (Aikoku)" and "Kirara397". "EG5 (Aikoku)" and "Kirara397" are a tester line for the *E1*, *E2* and *E3* locus and an elite Hokkaido variety, respectively.

2.4. Haplotype Patterns of the Chromosomal Region Surrounding E1/Ghd7 Locus

We surveyed DNA polymorphisms between EG5 (Aikoku) (*e1-ret/ghd7-0ret*) and Kirara397 (*e1/ghd7-0a*) around *E1/Ghd7* locus. Subsequently, we found three polymorphisms (two SNPs and a 20-bp deletion) other than the nucleotide substitution from GAG (Glu) to TAG (stop codon) and the TE insertion. To know the origins of two nonfunctional alleles *e1/ghd7-0a* and *e1-ret/ghd7-0ret*, we investigated the haplotypes of Hokkaido and Japanese-core collection varieties using five markers surrounding the *E1/Ghd7* locus (three single nucleotide polymorphisms (SNPs), a 20-bp deletion, and a TE insertion). The Japanese-core-collection varieties were classified into at least four haplotypes, Hap2, Hap3, Hap4, and Hap5 (Figure 6 and Table S4). In contrast, Hokkaido varieties were classified into two distinct haplotypes, Hap1 and Hap3. This suggests that Hap1 was derived from Hap2 via nucleotide substitution from GAG (Glu) to TAG (stop codon) in exon 1, whereas Hap3 was derived from Hap4 via the TE insertion in the promoter region. These results indicate that two independent mutational events contributed to the occurrence of the two nonfunctional alleles *e1/ghd7-0a* and *e1-ret/ghd7-0ret*. Interestingly, although Hap1 was found only in Hokkaido varieties, Hap3 was found not only in Hokkaido varieties but also in some Japanese varieties, particularly in the Aikoku-related varieties (Figure 6 and Table S4). Further, the varieties of the Hap3 group, except for Hokkaido varieties, flowered about 20 days later than the Hokkaido varieties, implying that such varieties do not adapt to Hokkaido where autumn comes early (Figure 6 and Table S4). These findings indicate that other genetic factor (s) were involved in the early flowering of Hokkaido varieties belonging to the Hap3 group. We accordingly investigated allelic variations in the *Se1/Hd1* and another major photoperiod-sensitivity gene, *Hd5*, which is known to be involved in the PS in the Hokkaido varieties [30–32]. The results showed that varieties with *e1/ghd7-0a* flowered early regardless of harboring a functional allele(s) (photoperiod-sensitivity allele) at the *Se1/Hd1* and/or *Hd5* locus, whereas varieties with *e1-ret/ghd7-0ret* flowered early only when harboring a nonfunctional allele at either of the *Se1/Hd1* or *Hd5* locus (Figure 7). These results indicate that coexistence of *e1-ret/ghd7-0ret* with a photoperiod-insensitivity allele either at the *Se1/Hd1* or at the *Hd5* locus is necessary to promote flowering under LD conditions.

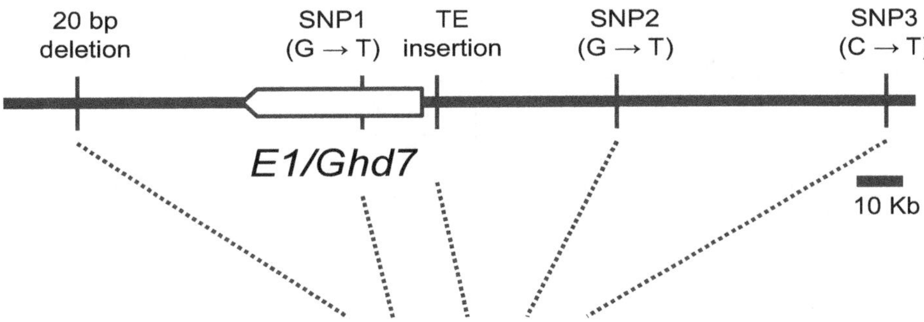

	Haplotype	Variety name	20 bp deletion	SNP1	TE insert	SNP2	SNP3
Hokkaido varieties	Hap 3	Matsumae	-	G	+	G	C
		Tomoemasari	-	G	+	G	C
		Eiko	-	G	+	G	C
		Shimahikari	-	G	+	G	C
		Tomoenishiki	-	G	+	G	C
		Fukoku	-	G	+	G	C
		Sorachi	-	G	+	G	C
	Hap 1	Hayakogane	▲20	T	-	T	T
		Ishikari	▲20	T	-	T	T
		Kitahikari	▲20	T	-	T	T
		Bozu	▲20	T	-	T	T
		Iburiwase	▲20	T	-	T	T
		Akage	▲20	T	-	T	T
		Hashiribozu	▲20	T	-	T	T
		Bozu 6	▲20	T	-	T	T
		Wasefukoku	▲20	T	-	T	T
		Megurosakaemochi	▲20	T	-	T	T
		Ishikarishiroke	▲20	T	-	T	T
		Kiyokaze	▲20	T	-	T	T
		Shiokari	▲20	T	-	T	T
		Sasahonami	▲20	T	-	T	T
		Fukuyuki	▲20	T	-	T	T
		Kitaake	▲20	T	-	T	T
		Dohoku 5	▲20	T	-	T	T
		Kitakogane	▲20	T	-	T	T
		Kirara 397	▲20	T	-	T	T
		Hokuto	▲20	T	-	T	T
		Joiku 272	▲20	T	-	T	T
		Kyowa	▲20	T	-	T	T
		Sakigake	▲20	T	-	T	T
		Norin 11	▲20	T	-	T	T
		Joiku 100	▲20	T	-	T	T
		Wasebozu	▲20	T	-	T	T
		Norin 15	▲20	T	-	T	T
		Norin 33	▲20	T	-	T	T
		Shinei	▲20	T	-	T	T
		Akatsuki	▲20	T	-	T	T
		Hayakaze	▲20	T	-	T	T
		Hokkai 126	▲20	T	-	T	T
		Hokkai 242	▲20	T	-	T	T
		Hokkai 244	▲20	T	-	T	T
		Hayamasari	▲20	T	-	T	T
		Tokachikuroge	▲20	T	-	T	T
Japanese rice core collection	Hap 3	Aikoku	-	G	+	G	C
	Hap 2	Ishijiro	▲20	G	-	T	T
		Shinyamadaho 2	▲20	G	-	T	T
		Akamai	▲20	G	-	T	T
		Akamai	▲20	G	-	T	T
		Himenomochi	▲20	G	-	T	T
		Shinsyu	▲20	G	-	T	T
	Hap 4	Hinode	-	G	-	G	C
		Sensyo	-	G	-	G	C
		Yamadabake	-	G	-	G	C
		Kaneko	-	G	-	G	C
		Irumanishiki	-	G	-	G	C
		Okkamodoshi	-	G	-	G	C
		Hirayama	-	G	-	G	C
		Oiran	-	G	-	G	C
		Bozumochi	-	G	-	G	C
		Meguromochi	-	G	-	G	C
		Hassokuho	-	G	-	G	C
		Wataribune	-	G	-	G	C
		Hosokan	-	G	-	G	C
		Akamai	-	G	-	G	C
		Mansaku	-	G	-	G	C
		Joshu	-	G	-	G	C
		Dango	-	G	-	G	C
		Ginbozu	-	G	-	G	C
		Shinrikimochi	-	G	-	G	C
		Shichimenchomochi	-	G	-	G	C
		Moritawase	-	G	-	G	C
		Kameji	-	G	-	G	C
		Omachi	-	G	-	G	C
		Shinriki	-	G	-	G	C
		Kyotoasahi	-	G	-	G	C
		Kabashiko	-	G	-	G	C
		Sekiyama	-	G	-	G	C
		Kemomi	-	G	-	G	C
		Karahoshi	-	G	-	G	C
		Akamai	-	G	-	G	C
		Karahoushi	-	G	-	G	C
		Okabo	-	G	-	G	C
		Hakamuri	-	G	-	G	C
		Aichiasahi	-	G	-	G	C
		Raiden	-	G	-	G	C
		Houmanshindenine	-	G	-	G	C
	Hap5	Nagoyashiro	-	G	-	G	T

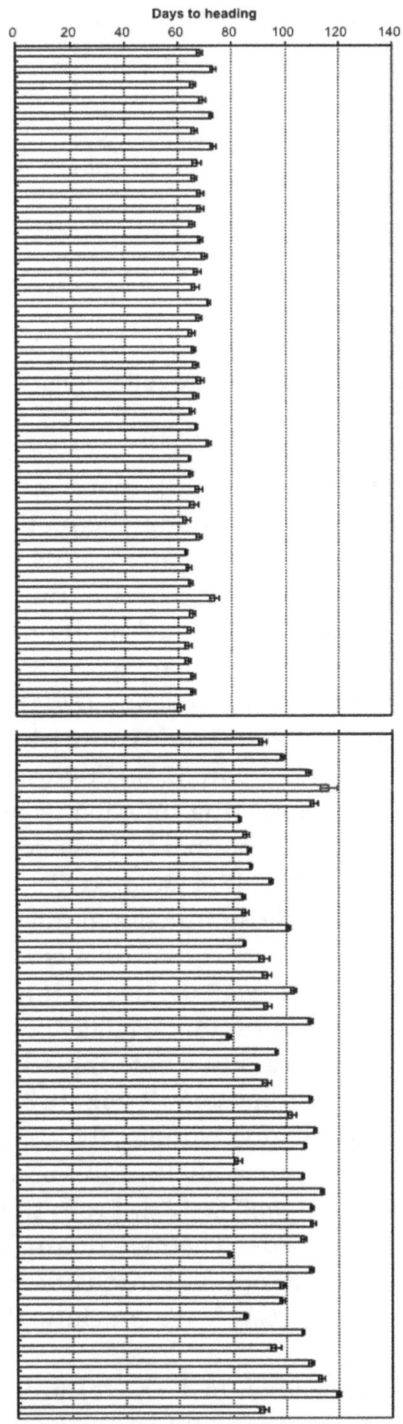

Figure 6. Haplotypes around the *E1/Ghd7* locus and days to heading of Hokkaido and Japanese core collection varieties.

Variety name	Genotype		
	E1	Se1	Hd5
Matsumae	e1-ret	se1	Hd5
Tomomasari	e1-ret	se1	Hd5
Eiko	e1-ret	se1	Hd5
Tomonishiki	e1-ret	se1	Hd5
Fukoku	e1-ret	se1	Hd5
Sorachi	e1-ret	Se1	hd5
Shimahikari	e1-ret	se1	hd5
Shinnriki	e1-ret	Se1	Hd5
Ouu20	e1-ret	Se1	Hd5
Aikoku 1)	e1-ret	Se1	Hd5
Kirara397	e1	Se1	Hd5
Ishikari	e1	Se1	Hd5
Hayamasari	e1	Se1	hd5
Hokkai244	e1	Se1	hd5
Akage 2)	e1	se1	Hd5
Bozu	e1	se1	Hd5
Bozu6	e1	se1	hd5
Norin11	e1	se1	hd5

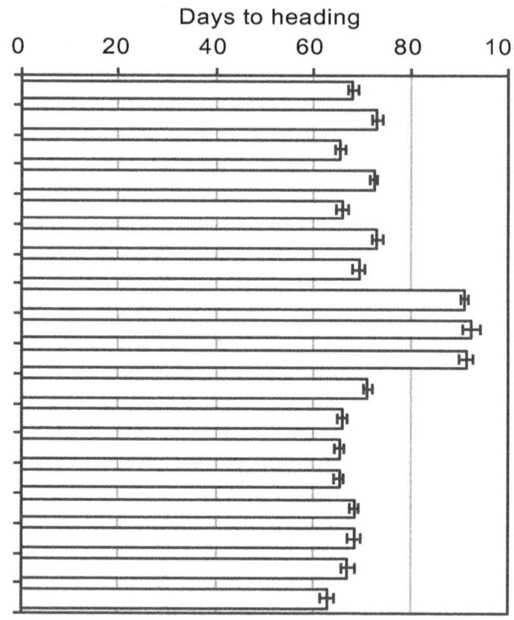

Figure 7. Gene combinations for the *E1*, *Se1*, and *Hd5* loci and days to heading. 1) This "Aikoku" variety belongs to the Japanese core collections. 2) This "Akage" variety belongs to the Hokkaido varieties.

3. Discussion

Since Hoshino (1915) [10], many genes (loci) controlling flowering time have been reported (reviewed in [39–41]). Among them, the *E1* is an important locus closely associated with the regional adaptability of rice varieties: its photoperiod insensitivity allele *e1* enabled rice cultivation even in Hokkaido, one of the northernmost rice cultivation area (42–45° N) [5,13,21]. A dominant photoperiod-sensitivity allele *E1* at the *E1* locus widely deployed among Japanese varieties for all regions other than Hokkaido in Japan [17,18,21]. Ichitani et al. (1998) [21] reported that the *E1* locus was identical to the *Heading date 4* (*Hd4*) locus, which was identified by Quantitative Trait Locus (QTL) analysis of flowering time using progenies from the cross of the *indica* variety Kasalath and the *japonica* variety Nipponbare [25]. Fujino and Sekiguchi (2005) [30] identified two QTLs, *qDTH-7-1* and *qDTH-7-2*, for flowering time using progenies from the cross between two Hokkaido varieties, Hoshinoyume and Nipponbare. They concluded that *qDTH-7-1* is the same locus as the *E1* (*Hd4*). Later, Xue et al. (2008) [27] isolated *the grain number, plant height, and heading date 7* (*Ghd7*) locus on chromosome 7, whose functional allele *Ghd7* encodes a *CONSTANS*, *CO-like*, and *TOC1* (CCT) domain-containing protein that delays flowering under LD conditions. They also reported that a nonfunctional allele *ghd7-0a*, harboring a premature termination in the predicted coding region, deployed among varieties commercially cultivated in Hei Long Jiang province, China (43–53° N). In the present study, we successfully determined the chromosomal location of the *E1* locus within a 228.1-kb physical region on chromosome 7 (Figure 3). According to the rice public database RAP-DB [38], 11 loci (genes) exist in this region. Among the genes, only Os07g0261200 (=*Ghd7*) showed a SNP in exon 1 of a photoperiod insensitive allele *e1* in T65m (Figure 4). This substitution was the same as that of the nonfunctional allele *ghd7-0a* [27]. From these results, we concluded that the *E1*, *Hd4*, *qDTH-7-1* and *Ghd7* are the same locus. Then we finally designate this locus as *E1/Ghd7*. It is noteworthy that we identified a novel photoperiod-insensitivity allele *e1-ret/ghd7-0ret* that harbored a TE insertion in the promoter region of *E1/Ghd7*. We conclude that this TE-insertional mutation causes the loss of function of the *E1/Ghd7* allele.

In addition, we identified a novel strong photoperiod-sensitivity allele *E1-r/Ghd7-r*, which harbors three nonsynonymous substitutions in the coding sequence (CDS) and two SNPs in the promoter region: one is in the LTRE core actor (CCGAC), and the other is in the TATA-box (Figure 5a). The LTRE core actor was identified in the regulatory regions of all cold-induced genes in Arabidopsis [42]. The CCGAC core motif, also known as C-repeat / drought response element (CRT/DRE), is essential for transcriptional activation in response to cold, drought, and/or high-salt treatments [43]. Kim et al. (2002) [44] showed that light signaling mediated by phytochrome activates cold-induced gene expression through CRT/DRE. The expression of *E1-r/Ghd7-r* was higher than that of a PS allele of *E1/Ghd7-2* in Nipponbare (Figure 5b). The expression level of *E1/Ghd7* is regulated by red light signal and correlated well with its LD specific activity [45,46]. These suggest that the mutations in the promoter region modify the *E1-r/Ghd7-r* expression, which delays flowering under LD conditions. Elucidation of the influences of the three amino acid substitutions in the CDS of *E1-r/Ghd7-r* awaits further study.

Haplotype analysis showed that two photoperiod-insensitive alleles, *e1/ghd7-0a* and *e1-ret/ghd7-0ret*, originated independently from two sources (Figure 6). *e1/ghd7-0a* widely deployed among improved and landrace varieties in Hokkaido, including "Akage", which was one of the pioneers of the Hokkaido rice varieties in the late 1800's [47]. This indicates that *e1/ghd7-0a* is a pioneer allele, leading to raising extremely early heading varieties with extremely weak photoperiod sensitivity during 1900–1930 (see Introduction). In contrast, *e1-ret/ghd7-0ret* was detected in "Fukoku", one of the past leading varieties in Hokkaido. "Fukoku" was bred from the cross between the Japanese warm region variety "Nakate-Aikoku", and the Hokkaido variety "Bozu 6". "Nakate-Aikoku" harbors *e1-ret/ghd7-0ret*, whereas "Bozu 6" does not. This indicates that *e1-ret/ghd7-0ret* of "Fukoku" was derived from "Nakate-Aikoku" (Figure S1). Interestingly, although most of the Aikoku-related varieties harbor *e1-ret/ghd7-0ret* (Table S4), none of them flowered as early as "Fukoku". This indicates that some genetic factor(s) other than *e1-ret/ghd7-0ret* is also responsible for the early flowering of "Fukoku". The varieties with *e1-ret/ghd7-0ret* flowered extremely early when harboring a nonfunctional allele either at the *Se1/Hd1* locus or at the *Hd5* locus (Figure 7). Therefore, we conclude that a nonfunctional allele either at the *Se1/Hd1* locus or at the *Hd5* locus is necessary for early flowering of the Hokkaido varieties with *e1-ret/ghd7-0ret*. Fujino et al. (2013) [32] reported that the nonfunctional allele at the *Hd5* locus was a spontaneous mutant gene that occurred in the Hokkaido local landrace "Bozu," and that this allele contributed to the expansion of rice cultivation to the northern area of Hokkaido. In contrast, the nonfunctional allele at the *Se1/Hd1* locus widely deployed among the varieties for the Tohoku-Hokuriku region (37–40° N) [15,18]. In the early rice breeding in Hokkaido, many Aikoku-related varieties, which were chiefly cultivated in the Tohoku-Hokuriku region, Japan, were frequently used as cross-parents to increase the genetic diversity and improve grain quality [47]. This suggests that *e1-ret/ghd7-0ret* was introduced from the Aikoku-related varieties, and the combination with the nonfunctional allele at the *Hd1* or *Hd5* locus made *e1-ret/ghd7-0ret* available for rice breeding programs in Hokkaido. It is also suggested that mutations in the promoter region often make functional differentiations of alleles at the *E1/Ghd7* locus, bringing about flowering time diversification in rice.

Recent molecular genetic studies revealed that there are large natural allelic variations in key loci controlling flowering time, such as *Ghd7*, *Hd1*, *DTH2*, and *DTH7*, which contribute to the diversity of flowering time, and to the regional adaptability by adjusting flowering time to distinct environmental conditions [27,32,48–51]. Zhang et al. (2015) [52] and Zheng et al. (2016) [53] classified *Ghd7* alleles into three (strong function, weak function and non-function) and two (function and non-function) groups based on the sequence polymorphisms in the coding region, respectively. On the other hand, Lu et al. (2012) [38] analyzed 104 varieties (*O. sativa*) and three wild rice accessions (*O. rufipogon*) and found that 76 SNPs and six insertions and deletions within a 3932-bp DNA fragment of *Ghd7*. Among them, the functional C/T mutation in the promoter region was related to plant height probably by altering gene expression. In this study, we identified two novel alleles with functional differentiations in the promoter region of the *E1/Ghd7* allele. The interaction of these alleles with other flowering time genes except *Hd1* and *Hd5* have not yet been elucidated, additional detailed analysis of their effects on flowering time will contribute to fine-tuning of the flowering time well adapting to the climatic conditions in each area.

Photoperiod sensitivity is an important trait responsible for regional and seasonal adaptability of rice varieties. In this study, we detected two novel alleles at the *E1/Ghd7* locus. It is known that many other loci are involved in the photoperiod sensitivity pathway of flowering in rice. Recent studies showed that photoperiod sensitivity loci, *Se1/Hd1*, *OsPRR37* and *Ghd8*, each have a large allelic variation [49,54–57]. Therefore, analyzing the functional and inter-locus interactions should be advanced, which will lead to practice the fine-tuning of flowering time in rice breeding programs. In addition, we elucidated that the *E1* and *Ghd7* are the same locus. Until now, numerous genetic information has been accumulated for each locus (*E1*, *Ghd7*). For further genetic analyses of the *E1/Ghd7* locus, however, all information about *E1* and *Ghd7* should be available.

4. Materials and Methods

4.1. Photoperiod Sensitivities of T65, T65m, and T65

The Taiwanese *japonica* rice variety "Taichung 65 (T65)", and its isogenic line T65m and its near-isogenic line T65w were used. T65 harbors a photoperiod sensitivity allele *E1* at the *E1* locus, while T65m harbors a photoperiod-insensitivity allele *e1*, which is derived from the cross with Bozu5 [16,19,34,35]. T65w is a chromosome substitution line which was developed by introducing the *E1* region of *O. rufipogon* (W107) into the genetic background of T65 [36]. Five seeds were sown on field soil in a 3.6 L pot and covered with granulated soils. Seedlings were thinned to one plant per pot 14 days after sowing, and were grown under two photoperiod conditions, SD (12-h light/12-h dark) and LD (14-h light/10-h dark). Photoperiod treatments were conducted using two growth cabinets without temperature control. In addition to natural daylight (8:00–18:00), supplementary artificial light was used for the 12-h and 14-h light conditions. The degree of photoperiod sensitivity of each line was expressed as a difference of DH between SD and LD. The experiment was conducted by using five plants per line with three replications from 1 May to October 16. Heading date was recorded for each plant when the first panicle emerged from the sheath of the flag leaf.

4.2. Chromosomal Location of the E1 Locus

Two F_2 populations from crosses of T65 × T65w, comprising 205 plants, and T65w × T65m, comprising 125 plants, were subjected to genetic analysis of heading date. They were grown in a paddy field of Kyoto University, Kyoto, Japan (35°01′ N). Seeds were sown on April 25 in 2007, and seedlings were transplanted on May 16 in 2007. The progeny test was conducted with 40 F_3 lines (sowing on April 30 in 2008 and transplanting on May 21 in 2008). Each F_3 line with approximately 25 plants was the progeny of an F_2 plant randomly selected from the F_2 population. To narrow down the candidate region of the *E1* locus, the F_4 progenies derived from several F_3 recombinants between RM1253 and RM3635 were cultivated (sowing on May 1 in 2009 and transplanting on May 22 in 2009).

4.3. Expression Analysis of E1/Ghd7

Plants of T65, T65m, and T65w were grown in a growth cabinet with a temperature controller under a LD condition (14.5-h light, 30 °C/9.5-h dark, 25 °C) at 70% relative humidity. Seedlings were grown on sand in 3.6 L pots (two plants/pot) with additional liquid fertilizer (Kimura's B Culture Solution, Nippon Medical and Chemical Instruments Co., Ltd., Osaka, Japan). Thirty days before flowering, leaves were collected at 4-h intervals during that day. Total RNAs were extracted with Trizol reagent (Life Technologies Inc., Gaithersburg, Maryland, USA) according to the manufacturer's protocols. Total RNA was subjected to DNA digestion by treatment with RNase-free DNase I (Takara Bio Inc.). The transcriptor first-strand cDNA synthesis kit (Roche Applied Science, Indianapolis, Indiana, USA) was used to reverse-transcribe cDNA from 1 μg of RNA using anchored-oligo (dT)18 primers. Real-time PCR analysis was performed by TaqMan PCR using a LightCycler 1.5 (Roche Applied Science) according to the manufacturer's instructions. The primer sets of *Ghd7* and *UBQ* genes and Universal Probe Library probes of each gene were designed with ProbeFinder version 2.45 (Roche; https://www.roche-applied-science.com/). Primer and probe sequences are shown in Table S5. Expression analysis using the standard curve method was performed to determine the expression level of each gene. The relative expression level of each gene was calculated using *UBQ* gene. The RNA gene standards for the seven genes were applied to their plasmids prepared by the pGEM-T Easy Vector System (Promega Corp., Madison, Wisconsin, USA) using PCR amplicons from the total RNA of T65.

4.4. Identification of the Genotype at the E1/Ghd7 Locus

To identify the *E1/Ghd7* genotype, we developed two DNA markers based on an SNP and a TE insertion. CAPS marker analysis was based on a nucleotide substitution from GAG (Glu) to TAG (stop codon) in exon 1, producing an additional restriction enzyme (*Spe* I) site. A pair of PCR primers (Ghd7_CAPS_F: 5′-CCAACTTGCCCTGTCTTCTT-3′, Ghd7_CAPS_R: 5′-AGCTGCTGCAAGCCAGTAAT-3′) was designed to amplify the 950 bp. PCR was performed with a 20 μL reaction mixture containing 2 μL template DNA, 10× PCR buffer, 25 mM $MgCl_2$, 2 mM of each deoxyriboside-triphosphate (dNTP), 0.2 μL *Taq* DNA polymerase (5 U/μL), 4 μL of a 2.5 mM solution of each primer, and 3.2 μL H_2O. PCR conditions were as follows: 94 °C for 5 min, followed by 35 cycles (1 min at 94 °C, 1 min at 60 °C, and 2 min at 72 °C) with a final extension of 7 min at 72 °C. The amplified products were digested with *Spe* I at 37 °C for 6 h. After digestion, the nonfunctional allele produced three fragments (81, 287, and 582 bp), whereas the functional allele produced two fragments (81 and 869 bp). Amplicons and digested amplicons were separated on 1% agarose gel. After electrophoresis, the gel was stained with ethidium bromide, and the DNA fragments were visualized under UV light. Insertion and deletion (INDEL) marker analysis was based on an 1897-bp copia-like TE insertion. A pair of PCR primers (Ghd7_INDEL_F: 5′-CGTTTCAGCAATAGCATTATGG-3′, Ghd7_INDEL_R: 5′-GCGGGTAGTCATCGAACAG-3′) was designed to amplify the 824 bp in the wild type and the 2721 bp in the insertion type. PCR was performed with a 20 μL reaction mixture containing 2 μL template DNA, 10× PCR buffer, 25 mM $MgCl_2$, 2 mM of each dNTP, 0.2 μL *Taq* DNA polymerase (5 U/μL), 4 μL of a 2.5 mM solution of each primer, and 3.2 μL H_2O. PCR conditions were as follows: 94 °C for 5 min, followed by 35 cycles (1 min at 94 °C, 1 min at 60 °C, and 2 min at 72 °C) with a final extension of 7 min at 72 °C. Amplicons were separated on 1% agarose gel. After electrophoresis, the gel was stained with ethidium bromide, and the DNA fragments were visualized under UV light.

4.5. Haplotype Patterns of the Chromosomal Region around the E1/Ghd7 Locus

Three DNA polymorphisms (two single-nucleotide substitutions and a 20 bp deletion) within the 100-kb chromosomal region surrounding *Ghd7* existed among 44 Hokkaido, 50 Japanese-core-collection [58] and 71 Aikoku-related varieties, which are considered to the derivatives of "Aikoku" variety based on their names. The Japanese rice core collection is a limited set of accessions representing, with a minimum set repetitiveness, the genetic diversity among Japanese rice varieties [58]. The two substitutions (Hap_SNP1 and Hap_SNP2) were detected by CAPS marker analyses. Hap_SNP1 harbors a G-to-T substitution, resulting in changing the recognition site of the restriction enzyme *Hpy188* I. Hap_SNP2 harbors a C-to-T substitution, resulting in changing the recognition site of *Hpy188* I. Primer pairs for Hap_SNP1 and Hap_SNP2 were designed to amplify 623 and 295 bp, respectively (Table S6). In addition, two polymorphism surveys of the *copia*-like TE insertion and the SNP in the first exon of *Ghd7* were performed. The CAPS and INDEL marker analyses to detect the insertion and SNP were described above.

Supplementary Materials:
Figure S1: Pedigree of the varieties, which harbor *e1-ret/ghd7-0ret* allele. Bold with underline indicates that the varieties harbor *e1-ret/ghd7-0ret* allele. Gray indicates varieties whose allele is unknown. Other indicates that the varieties harbor *e1/ghd7-0a* allele., Table S1: Frequency distributions of days to heading in F_3 lines (T65 × T65w), Table S2: Frequency distributions of days to heading in F_3 lines (T65m × T65w), Table S3: Lists of Hokkaido and Japanese core collection varieties used in this study, Table S4: Lists of Aikoku-related varieties used in this study, Table S5: Primer and probe sequences for expression analysis, Table S6 Primer sequences for haplotype analysis.

Author Contributions: Conceptualization, H.S., Y.O. and T.T. (Takatoshi Tanisaka); Methodology, H.S., Y.O. and T.T. (Takatoshi Tanisaka); Formal analysis, H.S., T.T. (Takuji Tsukiyama) and M.T.; Investigation, H.S. and C.X.; Writing—original draft preparation, H.S.; Writing—review and editing, H.S., Y.O. and T.T. (Takatoshi Tanisaka); Project administration, H.S. and T.T. (Takatoshi Tanisaka).

Acknowledgments: We thank the genetic resources center, NARO (National Agriculture and Food Research Organization) for providing the seeds of Hokkaido, Japanese-core-collection and Aikoku-related varieties. T65m was kindly provided from Shigetoshi Sato in Ryukyu University and Kuo Hai Tsai in National Chung-Hsing University. T65w was kindly provided from Yoshio Sano in Hokkaido University. We also would like to thank Syo Asano and Shinsuke Hamada for their supports of our experiments.

References

1. Khush, G.S. Green revolution: The way forward. *Nat. Rev. Genet.* **2001**, *2*, 815–820. [CrossRef] [PubMed]
2. Wada, E. Studies on the response of heading to daylength and temperature in rice plants. : II. Response of varieties and the relation to their geographical distribution in Japan. *Jpn. J. Breed.* **1952**, *2*, 55–62. [CrossRef]
3. Hosoi, T. Studies on meteorological fluctuation in the growth of rice plants. V. Regional references of thermo-sensitivity, photo-sensitivity, basic vegetative growth and factors determining the growth duration of Japanese varieties. *Jpn. J. Breed.* **1981**, *31*, 239–250. [CrossRef]
4. Tanisaka, T.; Inoue, H.; Uozu, S.; Yamagata, H. Basic vegetative growth and photoperiod sensitivity of heading-time mutants induced in rice. *Jpn. J. Breed.* **1992**, *42*, 657–668. [CrossRef]
5. Okumoto, Y.; Ichitani, K.; Inoue, H.; Tanisaka, T. Photoperiod insensitivity gene essential to the varieties grown in the northern limit region of paddy rice (*Oryza sativa* L.) cultivation. *Euphytica.* **1996**, *92*, 63–66. [CrossRef]
6. Zhang, Y. An Economic Analysis on Growth and Technological Progress of Rice Production in Three Northeastern Provinces of China. *J. Rural Problem* **2002**, *38*, 1–12. [CrossRef]
7. Saito, H.; Yuan, Q.; Okumoto, Y.; Doi, K.; Yoshimura, A.; Inoue, H.; Teraishi, M.; Tsukiyama, T.; Tanisaka, T. Multiple alleles at *Early flowering 1* locus making variation in the basic vegetative growth period in rice (*Oryza sativa* L.). *Theor. Appl. Genet.* **2009**, *119*, 315–323. [CrossRef]
8. Saito, H.; Ogiso-Tanaka, E.; Okumoto, Y.; Yoshitake, Y.; Izumi, H.; Yokoo, T.; Matsubara, K.; Hori, K.; Yano, M.; Inoue, H.; et al. *Ef7* encodes an ELF3-like protein and promotes rice flowering by negatively regulating the floral repressor gene *Ghd7* under both short-and long-day conditions. *Plant Cell Physiol.* **2012**, *53*, 717–728. [CrossRef]
9. Yuan, Q.; Saito, H.; Okumoto, Y.; Inoue, H.; Nishida, H.; Tsukiyama, T.; Teraishi, M.; Tanisaka, T. Identification of a novel gene *ef7* conferring an extremely long basic vegetative growth phase in rice. *Theor. Appl. Genet.* **2009**, *119*, 675–684. [CrossRef]
10. Hoshino, Y. On the inheritance of the flowering time in peas and rice. *J. Col. Agric.* **1915**, *6*, 229–288.
11. Syakudo, K.; Kawase, T. Studies on the quantitative inheritance (11): A Rice (*Oryza sativa* L.) (d) Inheritance of the heading duration and the quantitative function of the causal genes in its determination. : (1) On the quantitative function of the genes E_1, E_2 and D_1. *Jpn. J. Breed.* **1953**, *3*, 6–12. [CrossRef]
12. Syakudo, K.; Kawase, T.; Yoshino, K. Studies on the quantitative inheritance (13): A Rice (*Oryza sativa* L.) (d) Inheritance of the heading duration and the quantitative function of the causal genes in its determination. : (2) On the quantitative function of the genes E_3, E_4 and E_5. *Jpn. J. Breed.* **1954**, *4*, 83–91. [CrossRef]
13. Yamagata, H.; Okumoto, Y.; Tanisaka, T. Analysis of genes controlling heading time in Japanese rice. *Rice Genet.* **1986**, *1*, 351–359.
14. Okumoto, Y.; Tanisaka, T.; Yamagata, H. Genotypic difference in response to light interruption in Japanese rice varieties. *Rice Genet.* **1991**, *2*, 778–780.
15. Yokoo, M.; Kikuchi, F.; Nakane, A.; Fujimaki, H. Genetical analysis of heading time by aid of close linkage with blast resistance in rice. *Bull. Nat. Inst. Agric. Sci. Ser.* **1980**, *D31*, 95–126.
16. Tsai, K.H. Studies on earliness genes in rice, with special reference to analysis of isoalleles at the *E* locus. *Jpn. J. Genet.* **1976**, *51*, 115–128. [CrossRef]
17. Okumoto, Y.; Tanisaka, T.; Yamagata, H. Heading-time genes of the rice varieties grown in south-west-warm region in Japan. *Jpn. J. Breed.* **1991**, *41*, 135–152. [CrossRef]

18. Okumoto, T.; Tanisaka, T.; Yamagata, H. Heading-time genes of the rice varieties grown in the Tohoku-Hokuriku region in Japan. *Jpn. J. Breed.* **1992**, *42*, 121–135. [CrossRef]
19. Okumoto, Y.; Yoshimura, A.; Tanisaka, T.; Yamagata, H. Analysis of a rice variety Taichung 65 and its isogenic early-heading lines for late-heading genes *E1*, *E2* and *E3*. *Jpn. J. Breed.* **1992**, *42*, 415–429. [CrossRef]
20. Ichitani, K.; Okumoto, Y.; Tanisaka, T. Photoperiod sensitivity gene of *Se-1* locus found in photoperiod insensitive rice cultivars of the northern limit region of rice cultivation. *Breed. Sci.* **1997**, *47*, 145–152.
21. Ichitani, K.; Okumoto, Y.; Tanisaka, T. Genetic analysis of low photoperiod sensitivity of rice cultivars for the northernmost regions of Japan. *Plant Breed.* **1998**, *117*, 543–547. [CrossRef]
22. Ichitani, K.; Inoue, H.; Nishida, H.; Okumoto, Y.; Tanisaka, T. Interactive effects of two heading-time loci, *Se1* and *Ef1*, on pre-flowering developmental phases in rice (*Oryza sativa* L.). *Euphytica.* **2002**, *12*, 227–234. [CrossRef]
23. Saito, H.; Okumoto, Y.; Teranishi, T.; Yuan, Q.; Nakazaki, T.; Tanisaka, T. Heading time genes responsible for the regional adaptability of 'Tongil-type short-culmed rice cultivars' developed in Korea. *Breed. Sci.* **2007**, *57*, 135–143. [CrossRef]
24. Kinoshita, T. Report of the committee on gene symbolization, nomenclature and linkage groups. *Rice Genet. Let.* **1986**, *3*, 3.
25. Yano, M.; Sasaki, T. Genetic and molecular dissection of quantitative traits in rice. *Plant Mol. Biol.* **1997**, *35*, 145–153. [CrossRef]
26. Doi, K.; Izawa, T.; Fuse, T.; Yamanouchi, U.; Kubo, T.; Shimatani, Z.; Yano, M.; Yoshimura, A. *Ehd1*, a B-type response regulator in rice, confers short-day promotion of flowering and controls *FT*-like gene expression independently of *Hd1*. *Gene Dev.* **2004**, *18*, 926–936. [CrossRef]
27. Xue, W.; Xing, Y.; Weng, X.; Zhao, Y.; Tang, W.; Wang, L.; Zhou, H.; Yu, S.; Xu, C.; Li, X.; et al. Natural variation in *Ghd7* is an important regulator of heading date and yield potential in rice. *Nat. Genet.* **2008**, *40*, 761–767. [CrossRef]
28. Yano, K.; Yamamoto, E.; Aya, K.; Takeuchi, H.; Lo, P.C.; Hu, L.; Yamasaki, M.; Yoshida, S.; Kitano, H.; Matsuoka, M. Genome-wide association study using whole genome sequencing rapidly identifies new genes influencing agronomic traits in rice. *Nat. Genet.* **2016**, *48*, 927–934. [CrossRef]
29. Okumoto, Y.; Tanisaka, T. Trisomic analysis of a strong photoperiod-sensitivity gene *E1* in rice (*Oryza sativa* L.). *Euphytica* **1997**, *95*, 301–307. [CrossRef]
30. Fujino, K.; Sekiguchi, H. Mapping of QTLs conferring extremely early heading in rice (*Oryza sativa* L.). *Theor. Appl. Genet.* **2005**, *111*, 393–398. [CrossRef]
31. Nonoue, Y.; Fujino, K.; Hirayama, Y.; Yamanouchi, U.; Lin, S.Y.; Yano, M. Detection of quantitative trait loci controlling extremely early heading in rice. *Theor. Appl. Genet.* **2008**, *116*, 715–722. [CrossRef] [PubMed]
32. Fujino, K.; Yamanouchi, U.; Yano, M. Roles of the *Hd5* gene controlling heading date for adaptation to the northern limits of rice cultivation. *Theor. Appl. Genet.* **2013**, *126*, 611–618. [CrossRef] [PubMed]
33. Inoue, H.; Nishida, H.; Okumoto, Y.; Tanisaka, T. Identification of an early heading time gene found in the Taiwanese rice cultivar Taichung 65. *Breed. Sci.* **1998**, *48*, 103–108. [CrossRef]
34. Tsai, K.H.; Oka, H. Genetic studies of yielding capacity and adaptability in crop plants. 4. Effects on an earliness gene, m^b in the genetic background of a rice variety, Taichung 65. *Bot. Bull. Acad. Sinica.* **1970**, *11*, 16–26.
35. Itoh, Y.; Sano, Y. Phyllochron dynamics under controlled environments in rice (*Oryza sativa* L.). *Euphytica* **2006**, *150*, 87–95. [CrossRef]
36. Eiguchi, M.; Sano, Y. Evolutionary significance of chromosome 7 in an annual type of wild rice. *Rice Genet. Newsl.* **1995**, *12*, 187.
37. The Rice Annotation Project Database. Available online: http://rapdb.dna.affrc.go.jp (accessed on 29 August 2019).
38. Lu, L.; Yan, W.H.; Shao, D.; Xing, Y.Z. Evolution and association analysis of *Ghd7* in rice. *PLoS ONE* **2012**, *7*, e34021. [CrossRef]
39. Vergara, B.S.; Chang, T.T. *The Flowering Response of the Rice Plant to Photoperiod: A Review of the Literature*, 2nd ed.; IRRI: Manila, Philippine, 1985; pp. 1–65.
40. Izawa, T. Adaptation of flowering-time by natural and artificial selection in Arabidopsis and rice. *J. Exp. Bot.* **2011**, *58*, 3091–3097. [CrossRef]

41. Tsuji, H.; Taoka, K.; Shimamoto, K. Regulation of flowering in rice: Two florigen genes, a complex gene network, and natural variation. *Curr. Opin. Plant Biol.* **2011**, *14*, 45–52. [CrossRef]
42. Baker, S.S.; Wilhelm, K.S.; Thomashow, M.F. The 5′-region of *Arabidopsis thaliana cor15a* has cis-acting elements that confer cold-, drought- and ABA-regulated gene expression. *Plant Mol. Biol.* **1994**, *24*, 701–713. [CrossRef]
43. Yamaguchi-Shinozaki, K.; Shinozaki, K. A novel cis-acting element in an Arabidopsis gene is involved in responsiveness to drought, low-temperature, or high-salt stress. *Plant Cell* **1994**, *6*, 251–264.
44. Kim, H.J.; Kim, Y.K.; Park, J.Y.; Kim, J. Light signaling mediated by phytochrome plays an important role in cold-induced gene expression through the C-repeat/dehydration responsive element (C/DRE) in Arabidopsis thaliana. *Plant J.* **2002**, *29*, 693–704. [CrossRef] [PubMed]
45. Itoh, H.; Izawa, T. The coincidence of critical day length recognition for florigen gene expression and floral transition under long-day conditions in rice. *Mol. Plant* **2013**, *6*, 635–649. [CrossRef] [PubMed]
46. Osugi, A.; Itoh, H.; Ikeda-Kawakatsu, K.; Takano, M.; Izawa, T. Molecular dissection of the roles of phytochrome in photoperiodic flowering in rice. *Plant Physiol.* **2011**, *157*, 1128–1137. [CrossRef] [PubMed]
47. Shinada, H.; Yamamoto, T.; Yamamoto, E.; Hori, K.; Yonemaru, J.; Matsuba, S.; Fujino, K. Historical changes in population structure during rice breeding programs in the northern limits of rice cultivation. *Theor. Appl. Genet.* **2014**, *127*, 995–1004. [CrossRef]
48. Yano, M.; Katayose, Y.; Ashikari, M.; Yamanouchi, U.; Monna, L.; Fuse, T.; Baba, T.; Yamamoto, T.; Umehara, Y.; Nagamura, Y.; et al. *Hd1*, a major photoperiod sensitivity quantitative trait locus in rice, is closely related to the *Arabidopsis* flowering time gene *CONSTANS*. *Plant Cell* **2001**, *12*, 2473–2484. [CrossRef]
49. Fujino, K.; Wu, J.; Sekiguchi, H.; Ito, T.; Izawa, T.; Matsumoto, T. Multiple introgression events surrounding the *Hd1* flowering-time gene in cultivated rice, *Oryza sativa* L. *Mol. Gen. Genom.* **2010**, *284*, 137–146. [CrossRef]
50. Wu, W.; Zheng, X.M.; Lu, G.; Zhong, Z.; Gao, H.; Chen, L.; Wu, C.; Wang, H.J.; Wang, Q.; Zhou, K.; et al. Association of functional nucleotide polymorphisms at *DTH2* with the northward expansion of rice cultivation in Asia. *Proc. Natl. Acad. Sci. USA* **2013**, *110*, 2775–2780. [CrossRef]
51. Gao, H.; Jin, M.; Zheng, X.M.; Chen, J.; Yuan, D.; Xin, Y.; Wang, M.; Huang, D.; Zhang, Z.; Zhou, K.; et al. *Days to heading 7*, a major quantitative locus determining photoperiod sensitivity and regional adaptation in rice. *Proc. Natl. Acad. Sci. USA* **2014**, *111*, 16337–16342. [CrossRef]
52. Zhang, J.; Zhou, X.; Yan, W.; Zhang, Z.; Lu, L.; Han, Z.; Zhao, H.; Liu, H.; Song, P.; Hu, Y.; et al. Combinations of the *Ghd7*, *Ghd8* and *Hd1* genes largely define the ecogeographical adaptation and yield potential of cultivated rice. *New Phytol.* **2015**, *208*, 1056–1066. [CrossRef]
53. Zheng, X.M.; Feng, L.; Wang, J.; Qiao, W.; Zhang, L.; Cheng, Y.; Yang, Q. Nonfunctional alleles of long-day suppressor genes independently regulate flowering time. *J. Int. Plant Biol.* **2016**, *58*, 540–548. [CrossRef] [PubMed]
54. Takahashi, Y.; Teshima, K.M.; Yokoi, S.; Innan, H.; Shimamoto, K. Variations in Hd1 proteins, *Hd3a* promoters, and *Ehd1* expression levels contribute to diversity of flowering time in cultivated rice. *Proc. Natl. Acad. Sci. USA* **2009**, *106*, 4555–4560. [CrossRef] [PubMed]
55. Wei, X.; Xu, J.; Guo, H.; Jiang, L.; Chen, S.; Yu, C.; Zhou, Z.; Hu, P.; Zhai, H.; Wan, J. *DTH8* suppresses flowering in rice, influencing plant height and yield potential simultaneously. *Plant Physiol.* **2010**, *153*, 1747–1758. [CrossRef] [PubMed]
56. Ebana, K.; Shibaya, T.; Wu, J.; Matsubara, K.; Kanamori, H.; Yamane, H.; Yamanouchi, U.; Mizubayashi, T.; Kono, I.; Shomura, A.; et al. Uncovering of major genetic factors generating naturally occurring variation in heading date among Asian rice cultivars. *Theor. Appl. Genet.* **2011**, *122*, 1199–1210. [CrossRef]
57. Koo, B.H.; Yoo, S.C.; Park, J.W.; Kwon, C.T.; Lee, B.D.; An, G.; Zhang, Z.; Li, J.; Li, Z.; Paek, N.C. Natural variation in *OsPRR37* regulates heading date and contributes to rice cultivation at a wide range of latitudes. *Mol. Plant* **2013**, *6*, 1877–1888. [CrossRef]
58. Ebana, K.; Kojima, K.; Fukuoka, S.; Nagamine, T.; Kawase, M. Development of mini core collection of Japanese rice landrace. *Breed Sci.* **2008**, *58*, 281–291. [CrossRef]

2

Effects of Genotype, Growing Season and Nitrogen Level on Gluten Protein Assembly of Durum Wheat Grown under Mediterranean Conditions

Anna Gagliardi [1], Federica Carucci [1], Stefania Masci [2], Zina Flagella [1], Giuseppe Gatta [1,*] and Marcella Michela Giuliani [1,*]

1 Department of Agricultural Food and Environmental Science, University of Foggia, 71122 Foggia FG, Italy; anna.gagliardi@unifg.it (A.G.); federica.carucci@unifg.it (F.C.); zina.flagella@unifg.it (Z.F.)
2 Department of Agricultural and Forest Sciences, University of Tuscia, 01100 Viterbo, Italy; masci@unitus.it
* Correspondence: giuseppe.gatta@unifg.it (G.G.); marcella.giuliani@unifg.it (M.M.G.)

Abstract: Water deficit and high temperatures are the main environmental factors which affect both wheat yield and technological quality in the Mediterranean climate. The aim of the study was to evaluate the variation in the gluten protein assembly of four durum wheat genotypes in relation to growing seasons and different nitrogen levels. The genotypes, Marco Aurelio, Quadrato, Pietrafitta and Redidenari, were grown under three nitrogen levels (36, 90 and 120 kg ha^{-1}) during two growing seasons in Southern Italy. Significant lower yield and a higher protein concentration were observed in the year characterized by a higher temperature at the end of the crop cycle. The effect of the high temperatures on protein assembly was different for the genotypes in relation to their earliness. Based on PCA, in the warmer year, only the medium-early genotype Quadrato showed positive values along the "protein polymerization degree" factor, while the medium and medium-late genotypes, Marco Aurelio and Pietrafitta showed negative values along the "proteins assembly" factor. No clear separation along the two factors was observed for the early genotype Redidenari. The variation in gluten protein assembly observed in the four genotypes in relation to the growing season might help breeding programs to select genotypes suitable for facing the ongoing climate changes in Mediterranean area.

Keywords: durum wheat; glutenin polymers; gluten quality; high temperature; nitrogen fertilization

1. Introduction

Durum wheat (*Triticum turgidum* L., subsp. *durum* Desf.) is the most widespread cereal crop in Mediterranean countries and is grown in various climatic conditions [1].

Water deficit and high temperatures are the main environmental factors which affect both wheat yield and technological quality in the Mediterranean climate [2,3]. According to studies performed by the Intergovernmental Panel on Climate Change (IPCC), further increase in temperatures is predicted in Europe, especially in the Southern and Central parts [4,5]. In this context, the maintenance of adequate yield and quality standards is of particular interest, since the annual variability of product quality cannot be acceptable, especially for dry pasta production [6].

The wheat grain quality mainly depends on the quantity and type of gluten proteins, as well as on their aggregation/polymerization level [7,8]. In particular, gliadins, which are monomeric proteins, are mainly responsible for the viscous nature of the dough, and interact mostly via non-covalent links, while glutenin, which are polymeric proteins stabilized by disulphide bonds, determine its elasticity [9–12].

In the literature [13–22], conflicting results on the effect of high temperatures on the quality of the gluten proteins have been reported. Studies made on bread wheat suggest that when high temperatures occur in the middle of grain filling, they positively affect dough strength [13], while very high temperatures near physiological maturity can have a negative effect [14]. Ciaffi et al. [15] reported that in bread wheat, high temperatures increased the accumulation of glutenins compared to gliadins. On the contrary, O'Leary et al. [16] reported that water or thermal stress conditions throughout the grain filling period determine a delay in the synthesis of glutenins while the synthesis of gliadins is not altered. Furthermore, for common wheat, it is reported that short periods of very high temperatures can significantly reduce the proportion of SDS-insoluble polymers (UPP) [15,17], which in bread wheat (*Triticum aestivum* L.) have been positively correlated with dough viscoelasticity [7,8]. On the contrary, some authors have reported that short periods of very high temperatures can lead to an increase in the size of glutenin polymers in both soft and durum wheat [18,19]. While numerous are the studies available in the literature on the effect of high temperatures on gluten protein concentration, composition and on polymeric proteins size and distribution in common wheat [20–22], very few are the studies relative to durum wheat and to its pasta-making quality [8]. Moreover, pasta-making quality in durum wheat is mostly determined by low-molecular-weight glutenin subunits (LMW-GS), especially the B-type [23], whereas in bread wheat high molecular weight glutenin subunits (HMW-GS) play the major role in determining dough technological properties [24].

In the Mediterranean areas, after climate conditions, the nitrogen (N) availability represents the main constraint in obtaining adequate yield and quality in durum wheat [25]. Some studies on bread wheat have suggested that high doses of N tend to increase the amount of monomer proteins [26,27] and to reduce the percentage of UPP causing an increase in the extensibility of the dough [28–31]. Moreover, some authors have highlighted that the effect of nitrogen on gluten proteins composition and on polymers organization may vary according to the genotype [26,30,32]. Finally, for the same parameters, significant effect of the interaction between the high temperatures and N availability has been reported [29,33]. Malik et al. [33] highlighted that the combinations of cultivars, nitrogen and temperature were needed to explain the variation in the quantity and size distribution of the polymer proteins and their effects on the quality of the end-product. To the best of our knowledge, for durum wheat, this type of information is still lacking.

Thus, the aim of the present study was to evaluate the variation in gluten proteins quality, in terms of their capacity to assembly in a visco-elastic structure, of four durum wheat genotypes in relation to the growing season and different nitrogen levels, including a low input rate.

2. Materials and Methods

2.1. Field Trials

Four durum wheat cultivars, Marco Aurelio, Quadrato, Pietrafitta and Redidenari, that are used in an important Italian pasta supply chain, (Table 1), were grown in two rain-fed field experiments carried out at Foggia (latitude 41°46′ N and longitude 15°54′ E, 74 m a.s.l.) during two growing seasons (2016–2017 and 2017–2018, hereafter indicated as 2017 and 2018, respectively) in a clay loam soil.

Table 1. Main characteristics of the genotypes under study.

Genotype	Year of Release	Pedigree	Earliness
Pietrafitta	1999	Grazia x Isa	medium-late
Quadrato	1999	Creso x Trinakria	medium early
Marco Aurelio	2010	Orobel//Arcobaleno/Svevo	medium
Redidenari	2015	Kofa x N185	early

The main chemical and physical soil characteristics in the two experimental year, 2017 and 2018, are reported in Table 2.

Table 2. Soil physical and chemical characteristics in the two experimental years.

Soil Characteristics		2017	2018
Sand	%	21.5	25.2
Silt	%	39.8	36.2
Clay	%	38.7	38.6
pH		8.1	8.2
Organic Matter *	%	1.9	1.9
Total Nitrogen **	‰	1.3	1.3
Assimilable Phosphorus √	mg kg^{-1}	80	64
Exchangeable Potassium ◊	mg kg^{-1}	461	422
Field Capacity (−0.03 MPa)	%	37.3	33.13
Wilting Point (−1.5 MPa)	%	19.7	18.5
Bulk Density	Mgm^3	1.15	1.10

* Walkley-Black method; ** Kjeldhal method; √ Olsen method; ◊Ammonium acetate method.

The four cultivars were sown on November 17 in 2016 and November 25 in 2017, at a seeding rate of 240 kg ha^{-1}. In both years, the experiment was in a field where the previous crop was durum wheat.

Three different nitrogen levels were adopted corresponding to 36, 90 and 120 kg ha^{-1} (N36, N90 and N120, respectively). The fertilizers used were Yara Mila Supersemina (18% nitrogen) at pre-sowing fertilization and Yara Bela Sulfan (24% nitrogen) at tillering, stem elongation and inflorescence emergence fertilization.

Each year, the experiment was arranged in a split-plot design with two factors (genotype in plots and nitrogen levels in sub-plots) and three replications; each sub-plot was 20.4 m^2.

The grain harvest was carried out at physiological maturity on 13 June 2017 and on 22 June 2018. During the experimental period, the daily climatic parameters of rainfall and temperature were recorded by a weather station near the experimental area.

2.2. *Yield and Technological Quality Parameters*

At harvest, grain yield (t ha^{-1}) and thousand kernel weight (TKW) were determined. Moreover, grain protein content (GPC) was performed by NIR System Infratec 1241 Analyzer (Foss, Hillerod, Denmark).

Semolina flours have been obtained from kernels milled by Bona mill 4 cylinders (sieve 180 μm).

The gluten index (GI), an indicator of the gluten strength, was determined on semolina samples using the Glutomatic system according to ICC standard 155 [34].

2.3. *Calculation of %UPP and Analysis of Gluten Protein Molecular Size Distribution*

The percentage of Unextractable Polymeric Proteins (%UPP) was measured trough the SE-HPLC procedure according to the method reported in Tosi et al. [35] with minor modifications. The SDS-soluble fraction was obtained by adding to the semolina a solution consisting of 0.5% (w/v) SDS in 0.05 M sodium phosphate buffer, pH 6.9 to a final concentration of 10 mg/mL (0.3 g semolina on 30 mL buffer). The mixture was stirred for 30 min at room temperature and then centrifuged at 20.000 g for 20 min at 15 °C. The supernatant was filtered through 0.45 μm PVDF filters and 20 μl were injected into a Biobasic Thermo Scientific SEC-300 Columns (300 mm × 7.8 mm; flow rate: 0.7 mL/min) and run for 40 min, with an eluent consisting of 0.05 M sodium phosphate buffer pH 6.9, containing 0.08 M NaCl and 0.1 % (w/v) SDS, using the UHPLC Ultimate 3000 Thermo scientific. Detection was at 214 nm. The SDS soluble fraction profiles were divided into four areas, corresponding to HPLC fractions F1, F2, F3 and F4 (Figure S2a). The first two areas correspond to large and medium size polymers, with both being enriched in HMW-GS (mainly F1) and B-type LMW-GS (mainly F2) of glutenin. F3 corresponds to ω-gliadins and small oligomers enriched in C-type and D-type LMW-GS subunits [23], while F4 corresponds to monomeric gliadins (α-type and β-type) and non-gluten proteins [35].

The SDS-insoluble fraction was obtained from the residue of the centrifugation step. The pellet was resuspended in 30 mL of the same extraction buffer and sonicated in a probe type sonicator (SONICS Vibracell Model VCX 130 -max output power 130 W at a frequency 20 KHz) for 30 s at 45% power setting. After centrifugation at 20.000 *g* for 20 min at 15 °C, the supernatant was filtered through 0.45 μm PVDF filters and 20 μL were injected into column in the same condition described above. The SDS-insoluble fraction profile (Figure S2b) showed only one peak (F1*) containing the largest glutenin polymers, insoluble in SDS solution alone, but rendered soluble by sonication.

Samples were extracted in duplicate and two replicate separations for each extraction were performed. The proportions of each peak (%F1* and%F1–%F4) were calculated as percentages of the total areas of the two chromatograms (SDS-insoluble and SDS-soluble fractions). The amount of monomeric over polymeric proteins (mon/pol) was calculated as the ratio between the sum of F3 and F4 areas and the sum of F1*, F1 and F2 areas. %UPP was determined as the ratio between F1* area and the sum of F1 and F1* areas (*100).

2.4. Statistical Analysis

The dataset was tested according to the basic assumptions of analysis of variance (ANOVA). The normal distribution of the experimental error and the common variance of the experimental error were verified through Shapiro–Wilk and Bartlett's tests, respectively. When required, Box-Cox transformations [36] were applied prior to analysis. The ANOVA procedure was performed according to a split-plot design with three replicates. Three-way ANOVA procedure was performed considering the factors (growing season, genotype and nitrogen level) as fixed factors. The statistical significance of the difference among the means was determined using Tukey's honest significance difference post hoc test at the 5% probability level. A principal component analysis (PCA) was performed on the correlation matrix of technological and SE-HPLC parameters. We obtained Principal Components (PCs) on centered and scaled variables, through diagonalization of the correlation matrix and extraction of the associated eigenvectors and eigenvalues. Grain protein content, gluten index, and SE-HPLC parameters were set as quantitative variables and used to define PCs, while genotype, N level and growing season were used as categorical variables, not considered in the computation of PCs. The coordinates of the categorical variables were calculated in order to enhance the interpretation of data and were represented as barycenter in the Principal Component biplot. The number of factors needed to adequately describe the data was determined on the basis of the eigenvalues and of the percentage of the total variance accounted by the different factors. The results of PCA were graphically represented in two-dimensional plot, using the SigmaPlot software (Systat Software, Chicago, IL, USA). ANOVA and PCA analyses were performed using the JMP software package, version 14.3 (SAS Institute Inc., Cary, NC, USA).

3. Results

3.1. Weather Condition

The climatic data related to the two growing seasons are reported in Table 3, while the rainfall distribution and maximum and minimum daily mean temperatures of the 2017 (a) and 2018 (b) crop seasons are reported in Figure S1 (Supplementary File).

The first growing season was characterized by lower rainfall compared to the second year (about 340 mm vs. 401 mm). Moreover, in the first experimental year the rain distribution was not regular, with the most intense rainfall occurred in the second decade of January, the third decade of February, the second decade of April and the first decade of May. As for the second growing season, rainfall was observed throughout the crop cycle, especially during the grain filling period, in the first ten days of May and June. In addition to rainfall, the two years differed also for the maximum temperatures during the grain filling period showing the second year the highest values. Moreover, during 2018, more days with temperatures between 30 and 35 °C and three days with temperatures higher than 35 °C, compared to 2017, occurred.

Table 3. Climatic data related to the two growing seasons.

		2017	2018
Crop cycle duration	d	209	210
Crop cycle rainfall	mm	339.9	401.4
From seeding to heading rainfall	mm	204.2	198.6
Grain filling rainfall	mm	135.7	202.5
Crop cycle Mean T	°C	12.3	13.1
Grain filling Mean T	°C	18.3	21.7
Grain filling Mean T max	°C	25.5	29.1
30 °C < T < 35 °C	d	15	23
T > 35 °C	d	-	3

3.2. *Yield and Technological Parameters*

The analysis of variance (ANOVA) generally showed a significant effect of year (Y), genotype (G) and nitrogen (N) on the parameters considered (Table S1). The two growing seasons differently influenced the yield and the technological parameters considered. In the second growing season (Table 4), a significant lower yield, a thousand kernel weight and gluten index were observed with respect to the first one. On the contrary, grain protein content was higher in 2018 than in 2017. Relative to the nitrogen level (Table 4), a significant positive effect on grain yield was evident only under N90, while for protein content the highest value was observed under N120. Finally, the gluten index values decreased with N level increasing.

Table 4. Effect of the year, nitrogen level and genotype on grain yield, thousand kernel weight, grain protein content and gluten index.

Experimental Factors	Grain Yield (t ha^{-1})	Thousand Kernel Weight (g)	Grain Protein Content (%)	Gluten Index (-)
Year				
2017	6.66 a	60.91 a	14.53 b	64.44 a
2018	5.91 b	50.21 b	16.00 a	58.50 b
Nitrogen level				
N36	6.20 b	55.16 a	14.25 c	63.83 a
N90	6.36 a	55.90 a	15.33 b	62.71 a
N120	6.28 ab	55.62 a	16.23 a	57.88 b
Genotype				
Marco Aurelio	7.11 a	50.62 d	15.74 b	57.72 bc
Pietrafitta	5.75 c	64.47 a	15.29 c	56.50 c
Quadrato	6.42 b	54.56 b	14.08 d	61.39 b
Redidenari	5.85 c	52.60 c	15.97 a	70.28 a

For each experimental factor, values in column followed by different letters are significantly different at $P \leq 0.05$ according to Tukey's test.

Among the genotypes (Table 4), Marco Aurelio showed the highest yield value even if associated with lower thousand kernel weight. Instead, Redidenari was the genotype with the best technological quality performance showing the highest protein content and gluten index values. However, the behavior of the genotypes changed in relation to growing seasons (Table 5) and nitrogen levels adopted (Table 6). In particular, the yield decrease observed in the second year was different among the genotypes (Table 5); it was 5% and 9% for Marco Aurelio and Redidenari, and 14% and 17% for Pietrafitta and Quadrato, respectively. Moreover, Marco Aurelio in addition to presenting lower yield decrease in the second year also showed an increase in the protein content that was double compared to the other genotypes (3.1% vs. 0.4%, 1.36% and 1.07% for Pietrafitta, Quadrato and Redidenari, respectively). Finally, as for gluten index, Marco Aurelio and Redidenari showed a significant decrease in the second year, more marked for Redidenari (Table 5).

Table 5. Effect of the year x genotype interaction on grain yield, thousand kernel weight, grain protein content and gluten index.

	2017				2018			
	Marco Aurelio	**Pietrafitta**	**Quadrato**	**Redidenari**	**Marco Aurelio**	**Pietrafitta**	**Quadrato**	**Redidenari**
Grain yield (t ha^{-1})	7.29 a	6.20 c	7.02 ab	6.12 c	6.92 b	5.31 e	5.82 d	5.57 de
Thousand kernel weight (g)	54.82 c	71.21 a	59.25 b	58.39 b	46.42 e	57.74 b	49.88 d	46.80 e
Grain protein content (%)	14.19 f	15.11 d	13.40 g	15.43 c	17.29 a	15.47 c	14.76 e	16.50 b
Gluten index (-)	60.78 b	56.44 b	60.89 b	79.67 a	54.67 c	56.56 b	61.89 b	60.89 b

In each row, values followed by different letters are significantly different at $P \leq 0.05$ according to Tukey's test.

Table 6. Effect of the genotype x nitrogen interaction on grain yield, thousand kernel weight, grain protein content and gluten index.

		Marco Aurelio	**Pietrafitta**	**Quadrato**	**Redidenari**
Grain yield (t ha^{-1})					
	N36	7.16 a	5.6 f	6.46 bc	5.61 ef
	N90	7.00 a	5.99 de	6.21 cd	6.25 bcd
	N120	7.17 a	5.68 ef	6.60 b	5.68 ef
Thousand kernel weight (g)					
	N36	50.18 e	64.67 a	53.80 bcd	51.99 cde
	N90	50.76 de	65.38 a	54.22 bc	53.26 bcde
	N120	50.91 de	63.37 a	55.67 b	52.53 bcde
Grain protein content (%)					
	N36	15.33 c	14.1 e	12.8 f	14.75 d
	N90	15.60 c	15.33 c	14.03 e	16.33 b
	N120	16.28 b	16.43 ab	15.40 c	16.82 a
Gluten index (-)					
	N36	57.67 bcde	56. 00 cde	67.33 ab	74.33 a
	N90	54.33 de	61.17 bcd	66.83 abc	68.50 ab
	N120	61.17 bcd	52.33 de	50. 00 e	68. 00 ab

For each parameter, values in each row and column followed by different letters are significantly different at $P \leq 0.05$ according to Tukey's test.

The nitrogen fertilization did not significantly affect the grain yield response in Marco Aurelio, while for both Pietrafitta and Redidenari, the highest values were observed under N90 level; for Quadrato the highest value was observed under N120 even if not significantly different from N36 (Table 6). On the contrary, for all genotypes a positive effect of the nitrogen level on grain protein content was evident with the highest values observed under N120. The effect of nitrogen fertilization on gluten index was not clear; only Quadrato showed a significant decrease under N120 level (Table 6).

3.3. Measurement of %UPP and Analysis of Gluten Protein Molecular Size Distribution

SE-HPLC was used to compare the molecular size distribution of the semolina proteins by a quantitative comparison of elution profiles.

The analysis of variance performed on the percentage of SDS-insoluble protein fraction (F1*), SDS-soluble protein fraction (F1–F4), monomeric/polymeric ratio (mon/pol) and proportion of unextractable polymeric protein (%UPP) showed a general significant effect of the year (Y), genotype (G), nitrogen level (N) and their interactions (Table S2). A significant decrease of F1* and %UPP was observed in 2018 compared to 2017. Moreover, in 2018 a significant increment of the polymeric fraction, due to an increase of both F1 and F2 was observed. On the contrary, in the same year, a decrease of the monomeric fraction, due to a decrease of F4 was evident, determining also a lower mon/pol ratio with respect to 2017 (Table 7). As for the nitrogen levels, a general positive effect of N90 compared with N36 was observed for F1*, %UPP and for the monomeric fraction, while there have never been significant differences between N36 and N120 (Table 7). Finally, as for genotypes, Marco Aurelio showed higher values of %UPP and polymeric fraction, due to higher values of F1* and F2, and lower value of mon/pol ratio. On the contrary Redidenari and Pietrafitta showed lower values of polymeric fraction (again mainly due to lower F1* and F2 values) and higher values of monomeric fraction and mon/pol ratio (Table 7). Finally, Quadrato showed intermediate values for all the fraction considered. The behavior of the genotypes changed in relation to growing seasons (Table 8). A significant decrease of F1* in the second year was evident for Marco Aurelio and Pietrafitta, more marked for the former. As consequence also %UPP significantly decrease in 2018 for Marco Aurelio (13.7%) and Pietrafitta (4.2%). On the contrary, a significant increase of F1* and %UPP was observed in the second year for Quadrato. All genotypes showed the increase of F1 values in the second year and only Marco Aurelio and Pietrafitta the increase of F2 values. Also for the polymeric and monomeric fraction the effect of the growing season was observed only for Quadrato and Redidenari. In particular, in 2018 these two genotypes showed higher polymeric and lower monomeric fraction values than 2017. The increase in polymeric fraction was due mainly to the significant increase in 2018 of both F1* and F1 for Quadrato, and of F1 for Redidenari, while the decrease of the monomeric fraction was due mainly to the F4 decrease.

Table 7. Effect of the year, genotype and nitrogen level on SDS insoluble (F1*) and soluble protein fraction (F1–F4) separated by SE-HPLC, monomeric/polymeric ratio (mon/pol) and proportion of unextractable polymeric protein (%UPP).

Experimental Factors	F1*	F1	F2	F3	F4	F1*+F1	Polymeric Fraction (F1*+F1+F2)	Monomeric Fraction (F3+F4)	%UPP	mon/pol
						(%)				(-)
Year										
2017	10.66 a	24.16 b	11.52 b	22.99 a	30.68 a	34.82 b	46.34 b	53.66 a	30.20 a	1.17 a
2018	9.71 b	26.99 a	12.07 a	23.24 a	27.99 b	36.70 a	48.77 a	51.23 b	26.26 b	1.06 b
Nitrogen level										
N36	10.07 b	26.22 a	11.75 a	22.59 a	29.38 a	36.29 a	48.04 a	51.96 b	27.66 b	1.09 b
N90	10.68 a	24.69 b	11.76 a	23.46 a	29.42 a	35.37 b	47.13 b	52.87 a	30.04 a	1.13 a
N120	9.80 b	25.81 a	11.88 a	23.30 a	29.21 a	35.61 ab	47.49 ab	52.51 ab	27.00 b	1.12 ab
Genotype										
Marco Aurelio	12.32 a	25.71 a	13.22 a	22.73 b	26.02 d	38.04 a	51.26 a	48.74 c	32.18 a	0.96 c
Pietrafitta	8.46 c	25.82 a	10.91 c	24.31 a	30.50 b	34.28 c	45.20 c	54.80 a	24.64 c	1.21 a
Quadrato	10.28 b	25.38 a	11.92 b	22.97 ab	29.45 c	35.66 b	47.58 b	52.42 b	28.69 b	1.11 b
Redidenari	9.68 b	25.37 a	11.12 c	22.46 b	31.37 a	35.05 bc	46.17 c	53.83 a	27.43 b	1.17 a

For each experimental factors, values in column followed by different letters are significantly different at $P \leq 0.05$ according to Tukey's test.

Table 8. Effect of the year x genotype interaction on SDS insoluble (F1*) and soluble protein fraction (F1–F4) separated by SE-HPLC, monomeric/polymeric ratio (mon/pol) and proportion of unextractable polymeric protein (%UPP).

	2017				2018			
(%)	**Marco Aurelio**	**Pietrafitta**	**Quadrato**	**Redidenari**	**Marco Aurelio**	**Pietrafitta**	**Quadrato**	**Redidenari**
F1*	14.95 a	9.18 c	8.93 cd	9.56 c	9.70 c	7.74 d	11.62 b	9.80 c
F1	23.11 d	24.98 c	24.00 cd	24.53 c	28.31 a	26.67 b	26.76 b	26.21 b
F2	12.76 b	10.32 f	12.09 c	10.90 ef	13.68 a	11.50 cde	11.76 cd	11.34 de
F3	22.50 b	23.72 ab	23.27 ab	22.45 b	22.96 ab	24.89 a	22.66 ab	22.47 b
F4	26.68 d	31.79 a	31.70 a	32.56 a	25.36 e	29.21 c	27.19 d	30.19 b
F1*+F1	38.1 a	34.2 c	32.3 c	34.1 c	38 a	34.4 bc	38.4 a	36 b
Polymeric fraction (F1*+F1+F2)	50.83 a	44.49 c	45.03 c	45.00 c	51.68 a	45.9 bc	50.14 a	47.34 b
Monomeric fraction (F3+F4)	49.17 c	55.51 a	54.97 a	55.00 a	48.32 c	54.1 ab	49.86 c	52.66 b
UPP	39.0 a	26.7 cd	27.1 cd	28.0 bc	25.3 d	22.5 e	30.3 b	26.9 cd
mon/pol (-)	0.97 c	1.25 a	1.22 a	1.22 a	0.94 c	1.18 ab	0.99 c	1.12 b

Values in each row followed by different letters are significantly different at $P \leq 0.05$ according to Tukey's test.

Relative to the effect of the genotype x nitrogen level interaction (Table 9), a significant effect of nitrogen level on F1* was evident for Marco Aurelio and Redidenari; in particular, for the former the F1* values increased with N level increasing, while for Redidenari the highest value was observed under N90. Both of these genotypes showed also highest %UPP values under N90. Moreover, only Redidenari showed a significant effect of the nitrogen level on the polymeric and the monomeric fraction, showing under N120 lower polymeric and higher monomeric fraction values.

Table 9. Effect of the genotype x nitrogen level interaction on SDS insoluble (F1*) and soluble protein fraction (F1–F4) separated by SE-HPLC, monomeric/polymeric ratio (mon/pol) and proportion of unextractable polymeric protein (%UPP).

(%)		Marco Aurelio	Pietrafitta	Quadrato	Redidenari
F1*	N36	11.18 bcd	9.25 efg	10.26 cde	9.58 def
	N90	12.70 ab	8.33 fg	9.80 def	11.91 abc
	N120	13.10 a	7.80 g	10.77 cde	7.55 g
F1	N36	25.86 ab	26.51 ab	25.81 ab	26.72 a
	N90	25.07 b	25.10 b	25.17 b	23.42 c
	N120	26.21 ab	25.86 ab	25.18 b	25.98 ab
F2	N36	13.01 a	10.73 de	11.71 bc	11.54 bcd
	N90	13.45 a	11.02 cde	12.05 b	10.50 e
	N120	13.20 a	10.98 cde	12.02 b	11.32 bcde
F3	N36	23.10 abc	23.38 abc	22.97 abc	20.9 c
	N90	23.28 abc	24.10 ab	23.36 abc	23.09 abc
	N120	21.8 bc	25.45 a	22.57 abc	23.39 abc
F4	N36	26.85 e	30.13 bcd	29.26 d	31.26 ab
	N90	25.51 f	31.45 a	29.63 d	31.09 abc
	N120	25.70 ef	29.91 cd	29.46 d	31.77 a
F1*+F1	N36	37.04 abc	35.76 bcde	36.07 bc	36.29 bc
	N90	37.76 ab	33.43 f	34.96 cdef	35.32 cdef
	N120	39.31 a	33.66 def	35.95 bcd	33.53 ef
Polymeric fraction (F1*+F1+F2)	N36	50.04 ab	46.49 cd	47.77 bc	47.88 bc
	N90	51.22 a	44.46 d	47.01 cd	45.82 cd
	N120	52.51 a	44.64 d	47.97 bc	44.84 d
Monomeric fraction (F3+F4)	N36	49.95 cd	53.51 ab	52.23 bc	52.17 bc
	N90	48.78 d	55.54 a	52.99 ab	54.17 ab
	N120	47.49 d	55.36 a	52.03 bc	55.15 a
UPP	N36	30.2 bc	25.8 def	28.5 cd	26.2 def
	N90	33.8 a	24.9 efg	27.7 cde	33.7 a
	N120	32.5 ab	23.2 fg	29.9 bc	22.4 g
mon/pol (-)	N36	1.00fg	1.15 b-e	1.10 def	1.10 de
	N90	0.96 g	1.25 a	1.14 cde	1.19 abcd
	N120	0.91 g	1.24 ab	1.09 ef	1.23 abc

For each parameter, values in each row and column followed by different letters are significantly different at $P \leq 0.05$ according to Tukey's test.

3.4. PCA Analysis

A principal component analysis (PCA) was performed on the correlation matrix. The results of PCA allowed two factors to be identified explaining 51% and 20.9% of total variance, respectively (Table 10). The first factor (PC1) was highly and positively associated with the largest insoluble polymers (F1*), the medium size soluble polymers (F2), the largest glutenin polymers (both insoluble and soluble; F1*+F1) and with the polymeric fraction (F1*+F1+F2). Moreover, it was highly and negatively related with the small oligomers fraction (F3), the monomeric gliadin fraction (F4), the total monomeric fraction (F3+F4) and mon/pol ratio. Thus, PC1 could be considered a factor linked to the degree of polymerization, mostly depending on the capacity to form covalent bonds. The second factor (PC2) was positively associated with gluten index (depending on the interactions among gluten proteins, both gliadins and glutenins), with the largest insoluble polymers (F1*) and with %UPP (depending on glutenin polymers size and amount) and negatively related with grain protein content (that can affect mostly gliadin accumulation) and the large size soluble polymers (F1) (that affect negatively %UPP). Thus, PC2 could be considered as a "gluten proteins assembly" factor, including the different interactions occurring in the gluten network. Both the factors linked to the degree of polymerization and the gluten proteins aggregation are major determinants of technological quality.

Table 10. Loading matrix values for the first two principal components (PC1 and PC2), considering the original variables. The corresponding percentages of accounted variation are also reported.

Original Variables	Loading Matrix Values	
	PC1	PC2
Grain protein content	0.09	−0.57
Gluten index	−0.21	0.47
F1* (%)	0.72	0.64
F1 (%)	0.29	−0.80
F2 (%)	0.54	−0.35
F3 (%)	−0.57	−0.26
F4 (%)	−0.81	0.29
F1*+F1 (%)	0.94	0.04
F1*+F1+F2 (%)	0.99	−0.09
F3+F4 (%)	−0.99	0.09
UPP (%)	0.55	0.77
mon/pol	−0.99	0.10
Percentage explained variation	**51**	**20.9**
Percentage cumulative variation	**71.9**	

In Figure 1, the biplot relative to the principal component analysis is reported. Based on the barycenter of the categorical variables (Figure 1, yellow marks), the nitrogen level did not show a clear separation along the two factors considered. On the contrary, the separation between the two years was observed mainly along the "gluten proteins assembly" factor (PC2) with the 2018 in the lower part. However, the separation between the crop seasons has to be interpreted also considering the genotype behaviors. Only for Quadrato the two years were separated mainly along the PC1 (polymerization degree factor), with the 2018 showing the positive and higher values. No clear separation was observed for the early maturing genotype Redidenari along the two PC factors. On the other hand, Marco Aurelio and Pietrafitta showed a clear separation of the two years only along the PC2, more marked for Marco Aurelio, with the 2018 showing the lower values. Finally, only the two genotypes, Marco Aurelio and Pietrafitta were clearly separated along PC1, presenting Marco Aurelio positive values and RDD negative values.

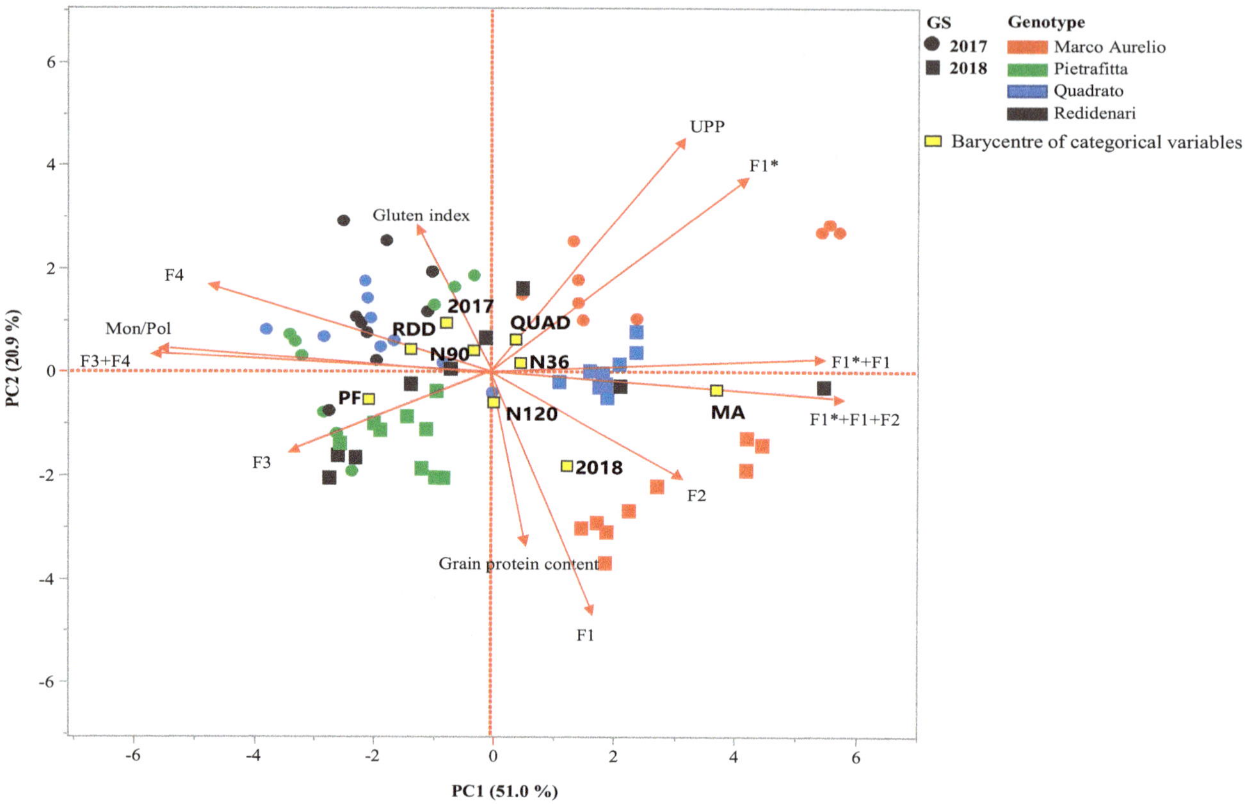

Figure 1. Biplot relative to the principal component analysis performed on grain protein content, gluten index, SDS insoluble (F1*) and soluble protein fraction (F1–F4) separated by SE-HPLC, monomeric/polymeric ratio (mon/pol) and proportion of unextractable polymeric protein. In yellow, the barycenter of the categorical variables, growing season (2017 and 2018), genotype (MA, Marco Aurelio; PF, Pietrafitta; QUAD, Quadrato; RDD, Redidenari) and nitrogen level (N36, N90 and N120) are shown.

4. Discussion

In the Mediterranean climate, the rainfall variability together with the frequency of high temperature during the grain filling period, may cause large fluctuations in durum wheat grain yield and technological quality aspects [3,37]. In semi-arid regions, a further increase in temperatures together with reduced rainfall are expected following the ongoing climate change [38,39]. This trend will influence also the crop responses to nitrogen fertilization, which depend on rainfall amount and distribution during the crop cycle, to the amount and timing of nitrogen applications as well as to the initial soil nitrogen levels [40,41]. Moreover, Malik et al. [33] highlighted that the combinations of cultivars, nitrogen and temperature are needed to explain the variation in the quantity and size distribution of the polymer proteins and their effect on the quality of the end-product. To the best of our knowledge, for durum wheat, this type of information is still lacking. The results obtained in this study represent a tile of the complex mosaic depicting the interactions among environment, fertilization and genotype.

Glutenin polymers are among the major determinants of wheat quality. Polymers are formed by different types of subunits that are functionally divided into chain terminators, chain extenders, and chain branches, according to their possibility to form one, two, or three (or more) intermolecular bonds, respectively (reviewed in [23]). The combination of these three functional glutenin classes gives rise to a range of glutenin polymers with different sizes and structures, that contributes to dough rheological properties. In general, the higher the size and amount of glutenin polymers, the better dough strength, that can be predicted by the %UPP value [7].

In our experimental condition, the two growing seasons showed a different climatic trend in terms of rainfall distribution and temperatures. Significant lower yield and thousand kernel weight, together with higher protein concentration were observed for all the genotypes in 2018, characterized by higher temperatures during the grain filling with respect to the first growing season. Moderate high temperature during grain filling, between 25 °C and 35 °C, and short periods of very high temperature (>35 °C) at the end of grain filling phase, as those we observed in the second growing season, are frequently associated with a decrease in grain yield and an increase in grain protein concentration [8,42]. However, the genotypes Marco Aurelio and Redidenari (released in 2010 and 2015, respectively) were less influenced by the growing season with respect to Quadrato and Pietrafitta (both released in 1999). The positive effect of nitrogen fertilization was clearer for the protein content than for grain yield as also reported in literature under Mediterranean climate [43–46]. However, the high yield response observed for Redidenari under N90 level was particularly interesting, indicating the possibility of limiting nitrogen inputs by adopting genotypes capable to optimize the use of nitrogen.

The growing season differently affected the gluten index, an indicator of gluten strength for durum wheat, in relation to the genotypes, showing only Marco Aurelio and Redidenari lower values in the warmer year. In bread and soft wheat, dough strength has been often positively correlated with the proportion of UPP [15,18,47–49]. As for durum wheat, the relation between %UPP and gluten index has been less investigated. In our experimental condition, this relation was genotype dependent, since only Marco Aurelio and Redidenari showed simultaneous decrease of gluten index and %UPP in the second year.

The composition and functionality of storage proteins have been significantly affected by growing season and genotype, while the effect of N fertilization level was rather small (Table S2) [50] as also resulted by PCA analysis. Several studies reported an increase in the proportions of the monomeric gliadins with increasing N availability [26,27]. In our experimental conditions, this was true only for the genotype Redidenari due to an increase of F4 component represented mainly by α/β type gliadin. An interesting result was the increase of %UPP for both Marco Aurelio and Redidenari under N90 level due to the increase of the F1*. The significant decrease of the larger insoluble polymers fraction (F1*) and %UPP observed in the second growing season for Marco Aurelio and Pietrafitta has to be discussed in relation to their earliness. Indeed, the very high temperature recorded at the end of the crop cycle (3 days with $T > 35$ °C) could have negatively influenced these two genotypes that are medium and medium-late maturing genotypes. This result is probably due to the fact that the assembly of the storage proteins takes place at the end of the grain filling phase [10,51,52]. Shewry et al. [53] proposed that at the end of the cycle, the loss of water favors the polymer chains contact inducing the assembly through disulphide crosslinking or through inter-chain hydrogen bonding. The effect of the temperatures on gluten protein assembly, have been studied mostly in bread wheat and only few studies are available for durum wheat. In common wheat, several research studies suggested that moderate high temperature or few days of very high temperature resulted in a significant reduction in the proportion of the SDS-insoluble protein fraction [15,17,47]. Other studies showed that the size of the glutenin polymers increased in response to short periods of very high temperature [18]. Ferreira et al. [8], in durum wheat, reported also a positive effect of the high temperature during the whole grain filling period on gluten protein assembly. Thus, the relationship between the gluten protein assembly and high temperatures is still not clear and needs more investigation. In our experimental conditions, in the second growing season, the two late maturing genotypes (Marco Aurelio and Pietrafitta), together with the decrease in F1* and %UPP showed an increase of both F2 and F1 fraction, the latter together with the other genotypes, confirming that the synthesis of the SDS soluble polymers continued also under high temperature condition [14,47]. Due to the concurrent decrease in F1* and increase in F1 and F2 fractions, Marco Aurelio and Pietrafitta did not significantly change their polymeric fraction between the two years. The increase of both %UPP and polymeric fraction observed in Quadrato and only of polymeric fraction observed in Redidenari in the second growing seasons is also linked to their earliness. Indeed, it seems like that on these genotypes, which are medium-early and early maturing, respectively, only the moderately high temperatures occurring during the grain filling acted, but not the extreme ones recorded at the end of the crop cycle. Indeed, also the results of the PCA highlighted the negative effect of the extreme temperatures on the gluten proteins assembly properties (PC2) only for Marco Aurelio and Pietrafitta, while for Quadrato a separation of the values only along the polymerization degree factor (PC1) was observed, with the warmer year showing the positive and higher values.

Because %UPP depends on protein distributions among the four areas typically used for its calculation, with the chain branchers and extenders mostly present in the fractions F1 (in particular F1*) and F2, it is important not only to select durum wheat varieties with proper glutenin compositions able to give rise to polymers of adequate size and amounts, but also that are synthetized in periods less susceptible to environmental changes, such it has occurred here for the medium early and early maturing varieties.

5. Conclusions

In the two growing seasons, the four durum wheat genotypes showed different capacities of the gluten proteins to assembly in a visco-elastic structure in relation to their earliness. In particular, in the second warmer year the late maturing genotype, Marco Aurelio and Pietrafitta showed a significant decrease of larger insoluble polymers fraction (F1*) and %UPP with a negative effect on their protein assembly level, despite Marco Aurelio always showed higher degree of polymerization. On the contrary, the medium-early and early maturing genotypes Quadrato and Redidenari, probably due to their earliness, did not change their "protein assembly level" in relation to the growing season.

The effect of N fertilization on the gluten protein polymerization and assembly was rather small, but among the N levels utilized the increase of F1*, %UPP and monomeric fraction under N90 was observed. Moreover, also the highest yield and gluten index values were obtained under N90. This was true especially for Redidenari.

In general, the effect of the growing season on the parameters evaluated was more evident than those of genotype and nitrogen level.

The results obtained in this study regarding four durum wheat genotypes clearly indicate different patterns of protein assembly in relation to the growing season, a factor that has a great influence on quality characteristics, thus contributing to the rational selection of the durum wheat genotypes, in particular those to include in supply chains, suitable for facing the ongoing climate changes in Mediterranean area.

Supplementary Materials:
Table S1. Mean square of effects (year, Y; genotype, G; nitrogen level, N) resulting from analysis of variance (ANOVA) performed on yield and technological parameters. Table S2: Mean square of effects (year, Y; genotype, G; nitrogen level, N) resulting from analysis of variance (ANOVA) performed on sonicated protein fraction (F1*) and SDS-soluble protein fraction (F1–F4) separated by SE-HPLC, monomeric/polymeric ratio (mon/pol) and proportion of unextractable polymeric protein (UPP). Figure S1. Rainfall distribution and maximum and minimum mean temperatures for the two growing seasons 2017 (a) and 2018 (b). Figure S2. SE-HPLC chromatograms of SDS-extractable protein fraction (a) and of SDS-unextractable protein fraction (b).

Author Contributions: Conceptualization, M.M.G. Methodology, M.M.G. and G.G. Validation, M.M.G., G.G., S.M., Z.F. Formal analysis, A.G., G.G. and M.M.G.. Investigation, A.G., F.C. Writing—original draft preparation, A.G., F.C. Writing—review and editing, M.M.G., G.G., S.M. and Z.F. Visualization, A.G. and F.C. Supervision, M.M.G., G.G. Project administration, M.M.G. and G.G. All authors have read and agreed to the published version of the manuscript.

Acknowledgments: We would like to show our gratitude to Luigi Toriaco (Syngenta Italia) and to Damiana Tozzi for their skilful technical assistance during the experimental trials and in the SE-HPLC analysis, respectively.

References

1. Ayadi, S.; Karmous, C.; Hammami, Z.; Trifa, Y.; Rezgui, S. Variation of durum wheat yield and nitrogen use efficiency under Mediterranean rainfed environment. *Int. J. Agric. Crop Sci.* **2016**, *7*, 693–699. [CrossRef]
2. Fagnano, M.; Fiorentino, N.; D'Egidio, M.G.; Quaranta, F.; Ritieni, A.; Ferracane, R.; Raimondi, G. Durum wheat in conventional and organic farming: Yield amount and pasta quality in southern Italy. *Sci. World J.* **2012**. [CrossRef] [PubMed]
3. Ercoli, L.; Lulli, L.; Mariotti, M.; Masoni, A.; Arduini, I. Post-anthesis dry matter and nitrogen supply and soil water availability. *Eur. J. Agron.* **2008**, *28*, 138–147. [CrossRef]
4. ICCP, Climatic Change 2011. IPCC Special Report on Renewable Energy Sources and Climate Change Mitigation. Summary for Policymakers. In Proceedings of the 11th Session of Working Group III of the IPCC, Abu Dhabi, UAE, 5–8 May 2011.
5. Bita, C.; Gerats, T. Plant tolerance to high temperature in a changing environment: Scientific fundamentals and production of heat stress-tolerant crops. *Front. Plant Sci.* **2013**, *4*, 273. [CrossRef] [PubMed]
6. Marchylo, B.A.; Dexter, J.E.; Clarke, F.R.; Clarke, J.M.; Preston, K.R. Relationships among bread-making quality, gluten strength, physical dough properties, and pasta cooking quality for some Canadian durum wheat genotypes. *Can. J. Plant Sci.* **2011**, *81*, 611–620. [CrossRef]
7. Gupta, R.B.; Popineau, Y.; Lefebvre, J.; Cornec, M.; Lawrence, G.J.; MacRitchie, F. Biochemical basis of flour properties in bread wheats. II. Changes in polymeric protein formation and dough/gluten properties associated with the loss of low Mr or high Mr glutenin subunits. *J. Cereal Sci.* **1995**, *21*, 103–116. [CrossRef]
8. Ferreira, M.S.L.; Martre, P.; Mangavel, C.; Girousse, C.; Rosa, N.N.; Samson, M.F.; Morel, M.H. Physicochemical control of durum wheat grain filling and glutenin polymer assembly under different temperature regimes. *J. Cereal Sci.* **2012**, *56*, 58–66. [CrossRef]
9. Gras, P.W.; Anderssen, R.S.; Keentok, M.; Békés, F.; Appels, R. Gluten protein functionality in wheat flour processing: A review. *Aust. J. Agric. Res.* **2001**, *52*, 1311–1323. [CrossRef]
10. Shewry, P.; Halford, N.G. Cereal seed storage proteins: Structure, properties and role in grain utilization. *J. Exp. Bot.* **2002**, *53*, 947–958. [CrossRef]
11. Triboï, E.; Martre, P.; Triboi-Blondel, A. Environmentally-induced changes in protein composition in developing grains of wheat are related to changes in total protein content. *J. Exp. Bot.* **2003**, *54*, 1731–1742. [CrossRef]
12. Anjum, F.M.; Khan, M.R.; Din, A.; Saeed, M.; Pasha, I.; Arshad, M.U. Wheat gluten: High molecular weight glutenin subunits-structure, genetics, and relation to dough elasticity. *J. Food Sci.* **2007**, *72*, 56–63. [CrossRef] [PubMed]
13. Stone, P.J.; Gras, P.W.; Nicolas, M.E. The influence of recovery temperature on the effects of a brief heat shock on wheat. III. Grain protein composition and dough properties. *J. Cereal Sci.* **1997**, *25*, 129–141. [CrossRef]
14. Stone, P.J.; Nicolas, M.E. Effect of timing of heat stress during grain filling on two wheat varieties differing in heat tolerance. II. Fractional protein accumulation. *Aust. J. Plant Physiol.* **1996**, *23*, 739–749. [CrossRef]
15. Ciaffi, M.; Tozzi, L.; Borghi, B.; Corbellini, M.; Lafiandra, D. Effect of heat shock during grain filling on the gluten protein composition of bread wheat. *J. Cereal Sci.* **1996**, *24*, 91–100. [CrossRef]
16. O'Leary, G.; Walker, C.; Panozzo, J.; Thayalakumaran, T.; McCaskill, M.; Nuttall, J.; Barlow, K.; Christy, B.; Asseng, S. Linking wheat grain quality to environmental processes. In Proceedings of the 2019 Agronomy Australia Conference, Wagga Wagga, Australia, 25–29 August 2019. Available online: www.agronomyaustralia.org/conference-proceedings (accessed on 24 February 2020).
17. Corbellini, M.; Canevar, M.G.; Mazza, L.; Ciaffi, M.; Lafiandra, D.; Borghi, B. Effect of the duration and intensity of heat shock during grain filling on dry matter and protein accumulation, technological quality and protein composition in bread and durum wheat. *Aust. J. Plant Physiol.* **1997**, *24*, 245–260. [CrossRef]
18. Don, C.; Lookhart, G.; Naeem, H.; MacRitchie, F.; Hamer, R.J. Heat stress and genotype affect the glutenin particles of the glutenin macropolymer-gel fraction. *J. Cereal Sci.* **2005**, *42*, 69–80. [CrossRef]
19. Flagella, Z.; Giuliani, M.M.; Giuzio, L.; Volpi, C.; Masci, S. Influence of water deficit on durum wheat storage protein composition and technological quality. *Eur. J. Agron.* **2010**, *33*, 197–207. [CrossRef]
20. Edwards, N.M.; Gianibelli, M.C.; McCaig, T.N.; Clarke, J.M.; Ames, N.P.; Larroque, O.R.; Dexter, J.E. Relationships between dough strength, polymeric protein quantity and composition for diverse durum wheat genotypes. *J. Cereal Sci.* **2007**, *45*, 140–149. [CrossRef]

21. Ohm, J.B.; Hareland, G.; Simsek, S.; Seabourn, B. Size-exclusion HPLC of protein using a narrow-bore column for evaluation of breadmaking quality of hard spring wheat flours. *Cereal Chem.* **2009**, *86*, 463–469. [CrossRef]
22. Malalgoda, M.; Ohm, J.B.; Meinhardt, S.; Simsek, S. Association between gluten protein composition and breadmaking quality characteristics in historical and modern spring wheat. *Cereal Chem.* **2018**, *95*, 226–238. [CrossRef]
23. D'Ovidio, R.; Masci, S. The low-molecular-weight glutenin subunits of wheat gluten. *J. Cereal Sci.* **2004**, *39*, 321–339. [CrossRef]
24. Shewry, P.R.; Gilbert, S.M.; Savage, A.W.J.; Tatham, S.; Wan, Y.F.; Belton, P.S.; Wellner, N.; D'Ovidio, R.; Bekes, F.; Halford, N.G. Sequence and properties of HMW subunit 1Bx20 from pasta wheat (*Triticum durum*) which is associated with poor end use properties. *Theor. Appl. Genet.* **2003**, *106*, 744–750. [CrossRef] [PubMed]
25. Ryan, J.; Masri, S.; Pala, M.; Singh, M. Nutrient Dynamics in a Long-Term Cereal-Based Rotation Trial in a Mediterranean Environment: Nitrogen Forms. *Commun. Soil Sci. Plant* **2009**, *40*, 931–946. [CrossRef]
26. Zhu, J.; Khan, K. Effects of genotype and environment on glutenin polymers and breadmaking quality. *Cereal Chem.* **2001**, *78*, 125–130. [CrossRef]
27. Kindred, D.R.; Verhoeven, T.M.O.; Weightman, R.M.; Swanston, J.S.; Agu, R.C.; Brosnan, J.M.; Sylvester-Bradley, R. Effects of variety and fertiliser nitrogen on alcohol yield, grain yield, starch and protein content, and protein composition of winter wheat. *J. Cereal Sci.* **2008**, *48*, 46–57. [CrossRef]
28. Johansson, E.; Prieto-Linde, M.L.; Svensson, G. Influence of nitrogen applicationrate and timing on grain protein composition and gluten strength in Swedishwheat cultivars. *J. Plant Nutr. Soil Sci.* **2004**, *167*, 345–350. [CrossRef]
29. Johansson, E.; Kuktaite, R.; Andersson, A.; Prieto-Linde, M.L. Protein polymer build-up during wheat grain development: Influences of temperature andnitrogen timing. *J. Sci. Food Agric.* **2005**, *85*, 473–479. [CrossRef]
30. Godfrey, D.; Hawkesford, M.; Powers, S.; Millar, S.; Shewry, P.R. Effects of Crop Nutrition on Wheat Grain Composition and End Use Quality. *J. Agric. Food Chem.* **2010**, *58*, 3012–3021. [CrossRef]
31. Ferrise, R.; Bindi, M.; Martre, P. Grain filling duration and glutenin polymerization under variable nitrogen supply and environmental conditions for durum wheat. *Field Crops Res.* **2015**, *171*, 23–31. [CrossRef]
32. Peckanek, U.; Karger, A.; Groger, S.; Charvat, B.; Schoggl, G.; Lelley, T. Effect of nitrogen fertilization on quantity of flour protein components, dough properties and breadmaking quality of wheat. *Cereal Chem.* **1997**, *74*, 800–805. [CrossRef]
33. Malik, A.H.; Prieto-Linde, M.L.; Kuktaite, R.; Andersson, A.; Johansson, E. Individual and interactive effects of cultivar maturation time, nitrogen regime and temperature level on accumulation of wheat grain proteins. *J. Sci. Food Agric.* **2011**, *91*, 2192–2200. [CrossRef] [PubMed]
34. International Association of Cereal Chemistry (ICC). *Standard Methods of the ICC*; ICC: Vienna, Austria, 1986.
35. Tosi, P.; Masci, S.; Giovangrossi, A.; D'Ovidio, R.; Bekes, F.; Larroque, O.; Napier, J.; Shewry, P. Modification of the low molecular weight (LMW) glutenin composition of transgenic durum wheat: Effects on glutenin polymer size and gluten functionality. *Mol Breed.* **2005**, *16*, 113–126. [CrossRef]
36. Box, G.; Cox, D. An analysis of transformations. *J. R. Stati. Soc. Ser. B Methodol.* **1964**, *26*, 211–252. [CrossRef]
37. Ierna, A.; Lombardo, G.M.; Mauromicale, G. Yield, nitrogen use efficiency and grain quality in durum wheat as affected by nitrogen fertilization under a Mediterranean environment. *Expl. Agric.* **2016**, *52*, 314–329. [CrossRef]
38. ICCP, Climatic Change 2012. *Managing the Risks of Extreme Events and Disasters to Advance Climate Change Adaptation*; A Special Report of Working Groups I and II of the Intergovernmental Panel on Climate Change; Field, C.B., V. Barros, T.F., Stocker, D., Qin, D.J., Dokken, K.L., Ebi, M.D., Mastrandrea, K.J., Mach, G.K., Plattner, S.K., Allen, M., Eds.; Cambridge University Press: Cambridge, UK, 2012.
39. Nuttall, J.G.; O'Leary, G.J.; Panozzo, J.F.; Walker, C.K.; Barlow, K.M.; Fitzgerald, G.J. Models of grain quality in wheat—A review. *Field Crop Res.* **2015**, *202*, 136–145. [CrossRef]
40. Cossani, C.M.; Slafer, G.A.; Savina, R. Nitrogen and water use efficiencies of wheat and barley under a Mediterranean environment in Catalonia. *Field Crops Res.* **2012**, *128*, 109–118. [CrossRef]
41. López-Bellido, L.; Muˇnoz-Romero, V.; Benítez-Vega, J.; Fernández-García, P.; Redondo, R.; López-Bellido, R.J. Wheat response to nitrogen splitting applied to a vertisols in different tillage systems and cropping rotations under typical Mediterranean climatic conditions. *Eur. J. Agron.* **2012**, *43*, 24–32. [CrossRef]

42. Triboï, E.; Martre, P.; Girousse, C.; Ravel, C.; Triboï-Blondel, A.M. Unravelling environmental and genetic relationships between grain yield and nitrogen concentration for wheat. *Eur. J. Agron.* **2006**, *25*, 108–118. [CrossRef]
43. Giuliani, M.M.; Giuzio, L.; De Caro, A.; Flagella, Z. Relationships between Nitrogen Utilization and Grain Technological Quality in Durum Wheat: II. grain yield and quality. *Agron J.* **2011**, *103*, 1668–1675. [CrossRef]
44. Li, Y.; Yin, Y.P.; Zhao, Q.; Wang, Z.L. Changes of glutenin subunits due to water-nitrogen interaction influence size and distribution of glutenin macropolymer particles and flour quality. *Crop Sci.* **2011**, *51*, 2809–2819. [CrossRef]
45. Malik, A.H.; Kuktaite, R.; Johansson, E. Combined effect of genetic and environmental factors on the accumulation of proteins in the wheat grain and their relationship to bread-making quality. *J. Cereal Sci.* **2013**, *57*, 170–174. [CrossRef]
46. Bouacha, O.D.; Rhazi, L.; Aussenac, T.; Rezgui, S.; Nouaigui, S. Molecular characterization of storage proteins for selected durum wheat varieties grown in different environments. *J. Cereal Sci.* **2015**, *61*, 97–104. [CrossRef]
47. Labuschagne, M.T.; Elago, O.; Koen, E. Influence of extreme temperatures during grain filling on protein fractions, and its relationship to some quality characteristics in bread, biscuit, and durum wheat. *Cereal Chem.* **2009**, *86*, 61–66. [CrossRef]
48. Corbellini, M.; Mazza, L.; Ciaffi, M.; Lafiandra, D.; Borghi, B. Effect of heat shock during grain filling on protein composition and technological quality of wheats. In *Wheat, Prospects for Global Improvement, Proceedings of the 5th International Wheat Conference, Ankara, Turkey, 10–14 June 1996*; Springer: Berlin/Heidelberg, Germany, 1998.
49. Wardlaw, I.F.; Blumenthal, C.S.; Larroqu, O.; Wrigley, C.W. Contrasting effects of chronic heat stress and heat shock on kernel weight and flour quality in wheat. *Funct. Plant Biol.* **2002**, *29*, 25–34. [CrossRef]
50. Rossmann, A.; Buchner, P.; Savill, G.P.; Hawkesford, M.J.; Scherf, K.A.; Mühling, K.H. Foliar N application at anthesis alters grain protein composition and enhances baking quality in winter wheat only under a low N fertiliser regimen. *Eur. J. Agron.* **2019**, *109*. [CrossRef]
51. Loussert, C.; Popineau, Y.; Mangavel, C. Protein bodies ontogeny and localization of prolamin components in the developing endosperm of wheat caryopses. *J. Cereal Sci.* **2008**, *47*, 445–456. [CrossRef]
52. Koga, S.; Aamot, H.U.; Uhlen, A.K.; Seehusen, T.; Veiseth-Kent, E.; Hofgaard, I.S.; Moldestad, A.; Böcker, U. Environmental factors associated with glutenin polymer assembly during grain maturation. *J. Cereal Sci.* **2020**, *91*. [CrossRef]
53. Shewry, P.R.; Toscano Underwood, C.; Wan, Y.; Lovegrove, A.; Bhandari, D.; Toole, G.; Mills, E.N.; Denyer, K.; Mitchell, R. Storage product synthesis and accumulation in developing grains of wheat. *J. Cereal Sci.* **2009**, *50*, 106–112. [CrossRef]

3

Segregation Distortion Observed in the Progeny of Crosses between *Oryza sativa* and *O. meridionalis* Caused by Abortion during Seed Development

Daiki Toyomoto [1], Masato Uemura [2], Satoru Taura [3], Tadashi Sato [4], Robert Henry [5], Ryuji Ishikawa [6] and Katsuyuki Ichitani [1,2,*]

1 United Graduate School of Agricultural Sciences, Kagoshima University, 1-21-24 Korimoto, Kagoshima, Kagoshima 890-0065, Japan
2 Faculty of Agriculture, Kagoshima University, 1-21-24 Korimoto, Kagoshima, Kagoshima 890-0065, Japan
3 Institute of Gene Research, Kagoshima University, 1-21-24 Korimoto, Kagoshima, Kagoshima 890-0065, Japan
4 Graduate School of Agricultural Science, Tohoku University, Sendai, Miyagi 980-8577, Japan
5 Queensland Alliance for Agriculture and Food Innovation, University of Queensland, Brisbane, Queensland 4072, Australia
6 Faculty of Agriculture and Life Science, Hirosaki University, Hirosaki, Aomori 036-8561, Japan
* Correspondence: ichitani@agri.kagoshima-u.ac.jp.

Abstract: Wild rice relatives having the same AA genome as domesticated rice (*Oryza sativa*) comprise the primary gene pool for rice genetic improvement. Among them, *O. meridionalis* and *O. rufipogon* are found in the northern part of Australia. Three Australian wild rice strains, Jpn1 (*O. rufipogon*), Jpn2, and W1297 (*O. meridionalis*), and one cultivated rice cultivar Taichung 65 (T65) were used in this study. A recurrent backcrossing strategy was adopted to produce chromosomal segment substitution lines (CSSLs) carrying chromosomal segments from wild relatives and used for trait evaluation and genetic analysis. The segregation of the DNA marker RM136 locus on chromosome 6 was found to be highly distorted, and a recessive lethal gene causing abortion at the seed developmental stage was shown to be located between two DNA markers, KGC6_10.09 and KGC6_22.19 on chromosome 6 of W1297. We name this gene as *SEED DEVELOPMENT 1* (gene symbol: *SDV1*). *O. sativa* is thought to share the functional dominant allele *Sdv1-s* (s for *sativa*), and *O. meridionalis* is thought to share the recessive abortive allele *sdv1-m* (m for *meridionalis*). Though carrying the *sdv1-m* allele, the *O. meridionalis* accessions can self-fertilize and bear seeds. We speculate that the *SDV1* gene may have been duplicated before the divergence between *O. meridionalis* and the other AA genome *Oryza* species, and that *O. meridionalis* has lost the function of the *SDV1* gene and has kept the function of another putative gene named *SDV2*.

Keywords: reproductive barrier; segregation distortion; abortion; wild rice; *O. meridionalis*; *O. sativa*; gene duplication

---

1. Introduction

Rice (*Oryza sativa*) is one of the most important staple crops in the world. It feeds about one-third of the world population. Wild rice relatives having the same AA genome as domesticated rice comprise the primary gene pool for rice genetic improvement and include the following species; *O. rufipogon*, *O. meridionalis*, *O. glumaepatula*, *O. barthii*, *O. longistaminata*. Another domesticated *Oryza* species *O. glaberrima* (African rice) also has an AA gemome, and contributes to rice improvement. Though there are several reproductive barriers among these species as described below, transfer of useful genes such as disease resistance gene from AA genome *Oryza* species to rice has been successful via hybridization.

O. meridionalis and *O. rufipogon* are found in the northern part of Australia [1]. *O. rufipogon* is inferred to be the direct progenitor of *O. sativa* [2], and widely distributed not only in Australia but also in South and South East Asia and New Guinea. On the other hand, the distribution of *O. meridionalis* is confined to the northern parts of Australia and Irian Jaya, Indonesia [1]. Molecular data provides support for the divergence of *O. meridionalis* from the other AA genome *Oryza* species [3–7]. This is reflected by low pollen fertility of the hybrids between *O. meridionalis* and the other AA genome species [8,9], with almost no progeny being produced from the selfing of the hybrids. To utilize the rice breeding potential of wild relatives of rice, a recurrent backcrossing strategy has been adopted to produce chromosomal segment substitution lines (CSSLs) carrying chromosomal segments from wild relatives of rice in the genetic background of cultivated rice [10–13]. Subsequent backcrossing with *O. sativa* as pollen parent was successful, because the F_1 plants between *O. sativa* and its wild relatives retained female fertility.

To elucidate the genetic potential for the improvement of cultivated rice using these wild species, we produced three kinds of CSSLs with different Australian wild rice strains in the same genetic background. As a model agronomic trait, we selected late-heading, because the wild rice strains in this study head later than the recurrent parent Taichung 65 by about 50 days, and heading-time is easily scored. We have succeeded in mapping the late-heading time genes from these wild rice strains (see below) and found a new genetic distortion phenomenon in the *Oryza*. In this study, we report the genetic mechanism of the new distortion phenomenon.

2. Results

2.1. Mapping of Photoperiod Sensitivity Gene

Three wild rice strains, Jpn1, Jpn2, and W1297, and one cultivated rice cultivar Taichung 65 (T65) were used in this study. We bred various CSSLs in a T65 genetic background incorporating the three Australian wild rice strains, W1297, Jpn1, and Jpn2, chromosomal segments by recurrent backcrossing (see Material and Methods). Hereafter, the backcrossing populations using Jpn1, Jpn2, and W1297 as donor parent are described as BCnFm (Jpn1), BCnFm (Jpn2), and BCnFm (W1297), respectively. "n" and "m" represent numbers of backcrossing and selfing, respectively. The frequency distributions of days to heading of the three BC_3F_2 populations are shown in Figure 1. All populations showed bimodal distributions. A total of 94 DNA markers covering the whole 12 chromosomes and showing polymorphism between T65 and the three wild rice strains were subjected to preliminary linkage analysis using bulked DNA from the three BC_3F_2 populations. Only one marker RM136, located 568 kbp away from a photoperiod sensitivity gene *HD1* [14], showed heterozygosity in all the bulk DNAs. Chi square values for the independence between genotypes of RM136 and days to heading (early and late heading divided by the dotted line in Figure 1) were 26.880, 81.073, and 86.693 for Jpn1, Jpn2, and W1297, respectively, all highly significant ($P < 0.0001$). These results suggest that the three strains from Australia carry photoperiod sensitive alleles of the *HD1* locus, because heterozygotes and homozygotes of these strains at the RM136 locus headed much later than the homozygotes of T65. This cultivar proved to carry a photoperiod insensitive allele at the *HD1*(= *Se1*) locus [15], which behaved as an early heading-time allele in a usual cropping season in Japan [16–18].

In the BC_3F_2 (W1297), the segregation of the RM136 locus was highly distorted: very few homozygotes of W1297 appeared. To check if this phenomenon was specific to the cross with W1297 as donor parent, and to evaluate this phenomenon more clearly under a more uniform genetic background, BC_4F_2 populations with the same cross combinations were subject to further study. As for W1297, the BC_3F_1 plants producing the BC_3F_2 population for the analysis was backcrossed again to produce BC_4F_1 plants. Among them, late flowering plants were selected to produce BC_4F_2 populations. As for Jpn1 and Jpn2, different BC_3F_1 plants from that producing the BC_3F_2 population for the above experiment were backcrossed to produce BC_4F_1 plants. Among them, late flowering plants were selected to produce BC_4F_2 populations.

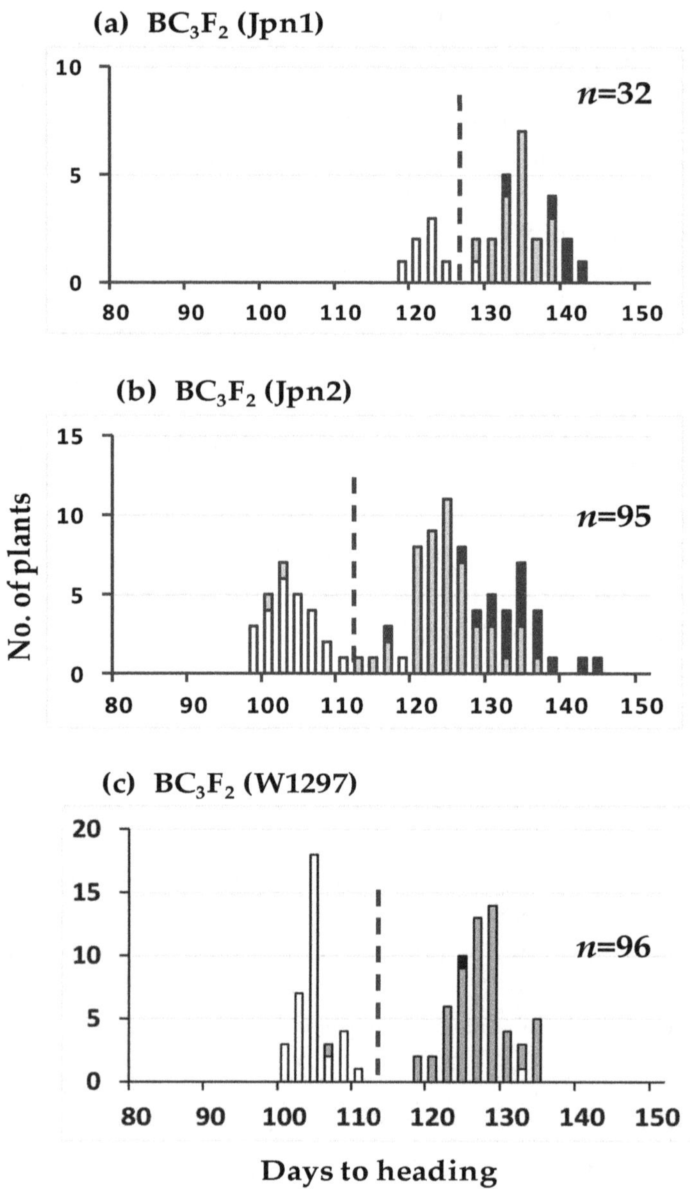

Figure 1. Frequency distributions of days to heading of the three BC_3F_2 populations using T65 as the recurrent parent. Jpn1, Jpn2, and W1297 were respectively used as donor parent in subfigure (**a**), (**b**), and (**c**). Three classified genotypes were assessed for RM136 as indicated: white, homozygous for T65, grey, heterozygous, black, homozygous for wild rice strains, Jpn1 (**a**), Jpn2 (**b**), and W1297 (**c**). Dotted lines dividing each population into early heading and late heading were drawn for chi-square analysis (see text).

2.2. *Mapping of Segregation Distortion Gene*

In the BC_4F_2 (W1297), the genotype of RM136 was distorted again (data not shown). In our preliminary experiment, among the published DNA markers around RM136, RM314 [19] located at 4,845kb, RM276 [19] at 6,231kb, RM7023 [20] at 6972kb, RM3628 [20] at 23,738kb and RM5314 [20] at 24,843kb on the IRGSP 1.0 pseudomolecule for chromosome 6 were fixed for the T65 allele. On the other hand, RM6818 and RM193 (Table 1) were segregating. These results suggest that the cause of segregation distortion is located between RM7023 and RM3628. Because other published DNA markers in our stocks failed in amplification of W1297 or did not distinguish T65 from W1297, we designed new DNA markers (Table 1), and performed linkage analysis. For the five consecutive markers from KGC6_12.02 to KGC6_19.48, only homozygotes of T65 and heterozygotes appeared (Figure 2), and no recombination occurred among the five markers (Table 2). The ratio of 64 homozygotes of T65: 119 heterozygotes fitted very well to 1:2 (χ^2(1:2) = 0.221, P = 0.638).

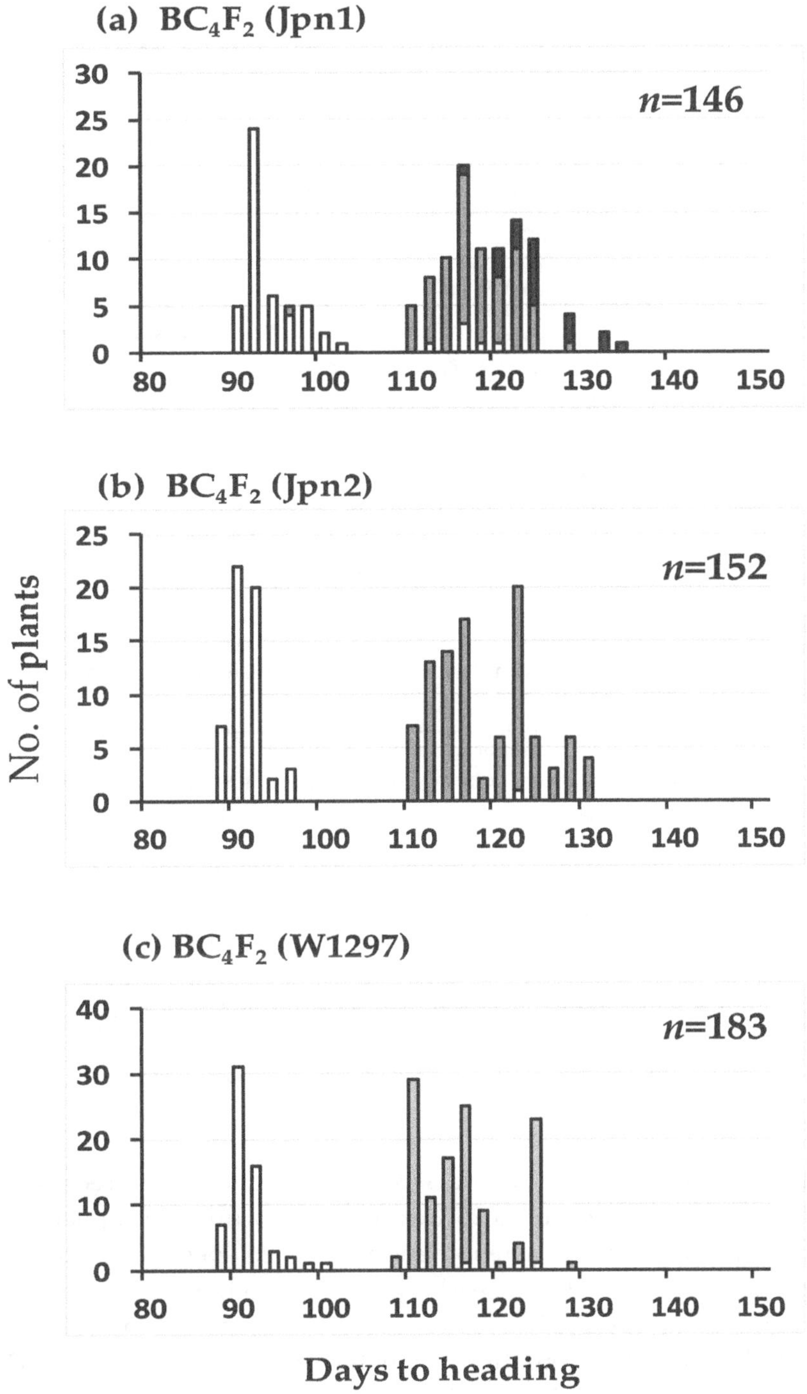

Figure 2. Frequency distributions of days to heading of the three BC_4F_2 populations using T65 as recurrent parent. Jpn1, Jpn2, and W1297 were respectively used as donor parent in subfigure (**a**), (**b**) and (**c**). Three classified genotypes were assessed for KGC6_12.02 as indicated: white, homozygous for T65, grey, heterozygous, black, homozygous for wild rice strains, Jpn1 (**a**), Jpn2 (**b**), and W1297 (**c**).

Table 1. Primer sequences of DNA markers designed or redesigned for linkage analysis of *SDV1* gene.

Marker Name	Kind of DNA Marker		Primer Sequence	Location on IRGSP 1.0 pseudomolecule chromosome 6		
				From	To	Source
KGC6_8.73	Indel	F	GAAGAGGAACATATGTGGTGTAAGC	8731826	8731914	This study
		R	AAAATTTATACTCTTGGTGACGTGA			
RM136	SSR	F	GAGAGCTCAGCTGCTGCCTCTAGC	8752461	8752562	Temnykh et al. [19]
		R	GAGGAGCGCCACGGTGTACGCC			
KGC6_8.82	Indel	F	TCTCTACCACACTCATCATCTGC	8820385	8820484	This study
		R	CCCTCGAGTAATAAACGATCCAG			
KGC6_10.09	Indel	F	TAGTCCTACGAAAACCCCTACTAGA	10090008	10090165	This study
		R	TTCCACGCACTAATACTACTACCTC			
KGC6_12.02	Indel	F	TTGATTTTGGGAAACATCAGGTAGC	12020476	12020625	This study
		R	AGCATGGTAATTTCATCGGATTCAA			
KGC6_13.00	Indel	F	CATTCGCATGGTAGCCTTTTCTTAT	13008017	13008169	This study
		R	CATAGGTGCCACAAGAGAAATCTTC			
RM6818	SSR	F	CGGCGAAGACTTGGAACCT	16582450	16582596	McCouch et al. [20]
		R	CCGTCACAAGGCTCGTCC			redesigned in this study
RM193	SSR	F	CAATCAACCAAACCGCGCTC	18086456	18086578	Temnykh et al. [19]
		R	CGCGGGCTTCTTCTCCTTC			redesigned in this study
KGC6_19.48	Indel	F	GAAGATAGTTAAGGGGTGTAGTGTGA	19483821	19484065	This study
		R	GACCAAAAGTTAAACAACATATTC TTCTAACCTAG			
KGC6_22.19	Indel	F	ACAAAATATGCTTTCTTCGTGCGTA	22191370	22191498	This study
		R	GCACTCAACTGTATCGTCTTTGAAA			

Table 2. Haplotypes around the segregation distortion region on rice chromosome 6 of BC_4F_2 (W1297).

Haplotype	Genotype of DNA Marker [1]								No. of Plants
	KGC6_8.73	KGC6_10.09	KGC6_12.02	KGC6_13.00	RM6818	RM193	KGC6_19.48	KGC6_22.19	
1	H	H	H	H	H	H	H	H	115
2	T	T	T	T	T	T	T	T	59
3	H	H	T	T	T	T	T	T	2
4	T	T	T	T	T	T	T	H	2
5	H	T	T	T	T	T	T	T	1
6	H	H	H	H	H	H	H	H	1
7	W	H	H	H	H	H	H	H	1
8	W	W	H	H	H	H	H	H	1
9	H	H	H	H	H	H	H	W	1
									183

[1] T, H, and W respectively denote homozygotes for T65, heterozygotes, and homozygotes for W1297.

The distorted segregation ratio 1:2:0 can be explained by one pair of recessive lethal genes. If the lethality occurred at the seedling stage, about 25% of seedlings would be expected to die. However, our visual observation did not fit with such a phenomenon. We then speculated that segregation distortion occurred during seed development. If so, seed fertility of the heterozygotes should be lower than that of the T65 homozygotes by about 25%. To test this, we examined the seed fertility of each of the BC_4F_2 plants. In the BC_4F_2 (W1297) population, the heterozygotes for KGC6_12.02 showed lower seed fertility than the homozygotes of the T65 allele (Figure 3). If lower fertility was caused by one recessive gene, many of the sterile seeds were expected to be aborted after fertilization. Therefore, sterile seeds were dehusked to see if sterility occurred before or after fertilization (Figure 4). The proportion of seeds aborted after fertilization for heterozygotes for KGC6_12.02 was higher than that for homozygotes of the T65 allele (Figure 5). These results suggested that homozygotes of the W1297 allele for KGC6_12.02 die at the seed developmental stage.

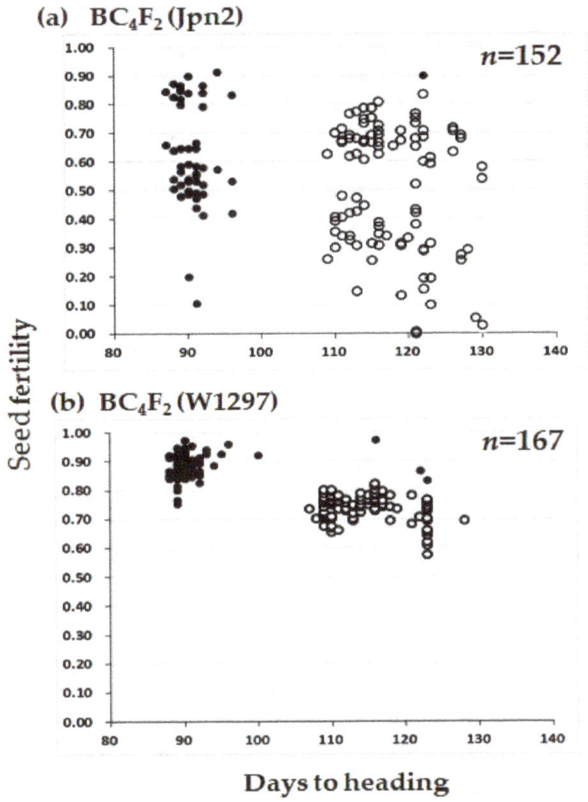

Figure 3. Scatter diagram of days to heading and seed fertility in the two BC_4F_2 populations using T65 as the recurrent parent. Jpn2 and W1297 were respectively used as donor parent in subfigure (**a**) and (**b**). Two classified genotypes were assessed for KGC6_12.02 as indicated: solid circle, homozygous for T65; open circle, heterozygous. In (**b**), plants used for testcross (Table 3) or damaged by birds in Figure 2 were removed in this figure.

Figure 4. Sterile seeds aborted after fertilization (top) and normal fertile seeds (bottom) found in the BC_4F_2 (W1297). One unit of the rightmost scale indicates 1 mm.

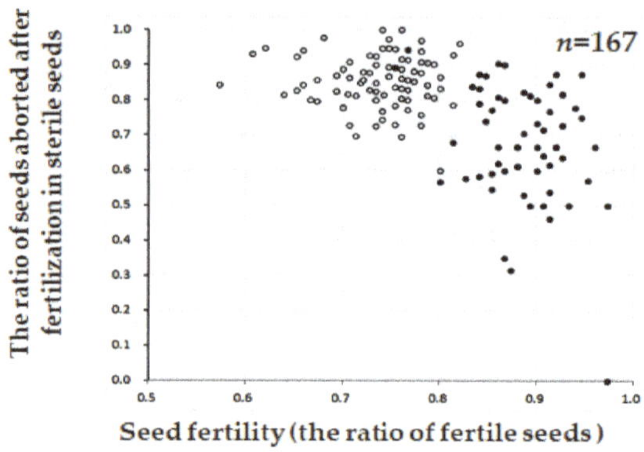

Figure 5. The scatter diagram between seed fertility (the ratio of fertile seeds) (X-axis) and the ratio of seeds aborted after fertilization in sterile seeds (Y-axis) in the BC_4F_2 population (W1297). Two classified genotypes were assessed for KGC6_12.02 as indicated: solid circle, homozygous for T65; open circle, heterozygous. Plants used for testcross (Table 3) or damaged by birds in Figure 2 were removed in this figure.

2.3. Segregation Distortion Caused by Abortion During Seed Development

To confirm this hypothesis, the following experiments were performed. First, pollen fertility was examined for all BC_4F_2 (W1297) plants, with the result that all plants showed more than 90% pollen fertility (data not shown), suggesting that pollen sterility was not the cause of the distorted segregation ratio. Second, reciprocal backcrossing of heterozygotes for KGC6_12.02 to T65 to produce a BC_5F_1 generation was undertaken. The BC_5F_1 from both cross combinations showed the segregation ratio fitted a 1 heterozygote:1 homozygote ratio for the T65 allele (Table 3), indicating that normal gene segregation occurred at both the egg and pollen developmental stage. The BC_4F_2 plants used for backcrossing were also selfed to produce a BC_4F_3 generation. DNA was extracted from the embryo of the fertile seeds. The segregation ratio was largely distorted from 1:2:1 at the KGC6_12.02 locus, and no homozygotes for the W1297 allele for KGC6_12.02 appeared, indicating that segregation distortion occurred during seed development, and was not caused by ungerminated fertile seeds, though the segregation ratio did not fit to a 1:2:0 ratio. (Table 3). The BC_4F_3 plants deriving from selfed seeds of the BC_4F_2 plants heterozygous for the KGC6_12.02 locus also showed distorted segregation, and the ratio fitted a 1: 2: 0 ratio, confirming the other experimental results (Table 3). Taken together, all the experimental results indicated that a recessive lethal gene causing abortion at the seed developmental stage was located between KGC6_10.09 and KGC6_22.19 on chromosome 6 of W1297 (Table 2).

Table 3. Segregation of progeny of BC_4F_2 (W1297) heterozygous for KGC6_12.02 genotype.

BC_4F_2	Genotype of the KGC6_12.02 [1]															
Individual Number	BC_4F_3										BC_5F_1					
	Plant					Embryo of Fertile Seeds					T65 as Pollen Parent			T65 as Egg Donor		
				P					*P*				*P*			*P*
	T	H	W	χ^2 (1:2:1)	χ^2 (1:2:0)	T	H	W	χ^2 (1:2:1)	χ^2 (1:2:0)	T	H	χ^2 (1:1)	T	H	χ^2 (1:1)
1	10	23	0	0.004	0.712	21	26	0	0.000	0.099	10	17	0.178	12	13	0.841
2	12	14	0	0.004	0.166	17	27	0	0.000	0.456	10	17	0.178	8	9	0.808
3	12	20	0	0.004	0.617	16	30	0	0.000	0.835	17	10	0.178	16	7	0.061
4	18	18	0	0.000	0.034	15	33	0	0.000	0.759	12	12	1.000	14	15	0.853
5	10	15	0	0.011	0.480	20	27	0	0.000	0.180	10	14	0.414	15	13	0.705
6	12	18	0	0.005	0.439	21	27	0	0.000	0.126	14	12	0.695	14	9	0.297
7	6	23	0	0.002	0.149	24	24	0	0.000	0.014	8	20	0.023	21	23	0.763
Sum	80	131	0	0.000	0.158	134	194	0	0.000	0.004	81	102	0.121	100	89	0.424

[1] T, H, and W respectively denote homozygotes for T65, heterozygotes, and homozygotes for W1297.

The same segregation distortion was also found in the BC_4F_2 (Jpn2) population (Figure 2). The ratio of 59 homozygotes of T65 allele: 97 heterozygotes at the KGC6_12.02 locus fitted very well to 1:2 (χ^2(1: 2) = 0.556, P = 0.456), and no homozygotes of Jpn2 allele appeared. The seed fertility of heterozygotes of KGC6_12 was lower than that of the homozygote of T65 allele, supporting the view that a recessive lethal gene causing abortion at seed developmental stage was located close to KGC6_12 of Jpn2 (Figure 3). The seed fertility of BC_4F_2 (Jpn2) was highly variable for both homozygotes of the T65 allele and heterozygote at the KGC6_12.02 locus, suggesting that other genetic factor(s) were involved in the large variance of seed fertility. Our preliminary results showed low pollen fertility might be responsible for low seed fertility of some plants (unpublished data). Therefore, the cause of the seed sterility was not investigated further. For Jpn1, both BC_3F_2 (Jpn1) and BC_4F_2(Jpn1) showed that normal gene segregation occurred around the *HD1* locus (Figures 1 and 2).

These results indicated that the two Australian *O. meridionalis* strains, W1297 and Jpn2, carry a recessive lethal gene causing abortion at the seed developmental stage, which was located between the two DNA markers, KGC6_10.09 and KGC6_22.19, spanning 12 Mb on chromosome 6, and that the Australian *O. rufipogon* strain Jpn1 does not carry such a gene.

3. Discussion

There have been many genes conferring hybrid seed sterility, hybrid pollen sterility, and segregation distortion found on *Oryza* chromosome 6 in inter-and intra-specific crosses, most of which *O. sativa* is involved with [21–27]. However, to our knowledge, the segregation distortion caused by seed abortion after fertilization has not been reported in the genus *Oryza*. We name this gene *SEED DEVELOPMENT 1* (gene symbol: *SDV1*), according to the gene nomenclature system for rice [28], because this gene is involved in the early seed developmental stage. In the intraspecific crosses among *O. sativa*, there have been no reports of gene distortion or partial seed sterility phenomena as described above on chromosome 6, though other phenomena have been reported [21–27]. Therefore, all *O. sativa* is thought to share the same functional dominant allele found in T65. This allele was called *Sdv1-s* (s for *sativa*). The homozygotes of the W1297 allele and the Jpn2 allele of this locus do not exist in the T65 genetic background probably because they die at the early seed development stage. W1297 and Jpn2 have originated from different places in Australia: W1297 is from Northern Territory, and Jpn2 is from Queensland. According to Juliano et al. [29], most crosses between Northern Territory and Queensland accessions produced sterile hybrids. Our preliminary results showed the hybrids from the reciprocal crosses between W1297 and Jpn2 were highly sterile (unpublished data). DNA marker-based analyses showed *O. meridionalis* genetic differentiation corresponding to geographic origin [29]. Further, in the CSSL lines of an *O. meridionalis* accession, W1625, chromosomal segments in a T65 genetic background [12], no lines were fixed for the W1625 chromosomal segment on which *SDV1* locus is located (https://shigen.nig.ac.jp/rice/Oryzabase/locale/change?lang=en). The results described above on the whole suggest that all *O. meridionalis* share the recessive abortive allele. This allele was called *sdv1-m* (m for *meridionalis*).

Though carrying the recessive abortive allele in homozygous form at the *SDV1* locus, the *O. meridionalis* accessions can self-fertilize and bear seeds. We speculate that the *SDV1* gene may have been duplicated before the divergence between *O. meridionalis* and the other AA genome *Oryza* species, and that *O. meridionalis* has lost the function of the *SDV1* gene and has kept the function of the other gene while *O. sativa* kept the function of the *SDV1* gene and has lost the function of the other gene (Figure 6). Such duplication and loss of reproductive barrier-related genes has been reported; Yamagata et al. [30] found that the reciprocal loss of duplicated genes encoding mitochondrial ribosomal protein L27, essential for the later stage of pollen development, causes hybrid pollen sterility in F_1 hybrid between *O. sativa* and *O. glumaepatula*. Nguyen et al. [31] reported that the duplication and loss of function of genes encoding RNA polymerase III subunit C4 hybrid causes pollen sterility in F_1 hybrid between *O. sativa* and *O. nivara* (annual form of *O. rufipogon*). Ichitani et al. [32] performed linkage analysis of hybrid chlorosis genes in rice, and found that the causal recessive genes *hca1-1* and *hca2-1* are located on the distal region of the short arm of chromosome 12 and 11, respectively, known to be highly conserved as a duplicated chromosomal segment.

There are other models explaining the hybrid incompatibility (abortion, lethality, or weakness) known as the Bateson–Dobzhansky–Muller (BDM) model (for a review, Bomblies et al. [33]). In the incompatibility caused by the two nonallelic dominant genes, if the one locus is heterozygote or fixed for the incompatibility-causing allele, both heterozygotes and homozygotes of the incompatibility-causing allele of the other locus should show incompatibility. If the one locus is fixed for the normal allele, incompatibility does not occur. In the incompatibility caused by the heterozygote on one locus, only heterozygotes should show incompatibility. Therefore, these models cannot explain the segregation distortion in this study. In the hybrid breakdown model proposed by Oka [34], the combination of the heterozygotes or the homozygotes of recessive alleles at one locus and the homozygotes of recessive alleles at the other locus show incompatibility. This model cannot explain the segregation in this study either. Therefore, the gene duplication model as described above fits the phenomenon in this study best.

As the counterpart of *SDV1*, "the other" putative gene is named *SEED DEVELOPMENT 2* (gene symbol: *SDV2*). *SDV1* and *SDV2* are thought to be derived from duplication. *O. meridionalis* accessions and *O. sativa* accessions should carry the functional allele and the unfunctional allele at the *SDV2* locus, respectively. We name these respective alleles *Sdv2-m* and *sdv2-s*. The presence and chromosomal location of *SDV2* have not been elucidated. We are undertaking the genetic analysis of *SDV2*, tracing back to earlier backcrossing populations.

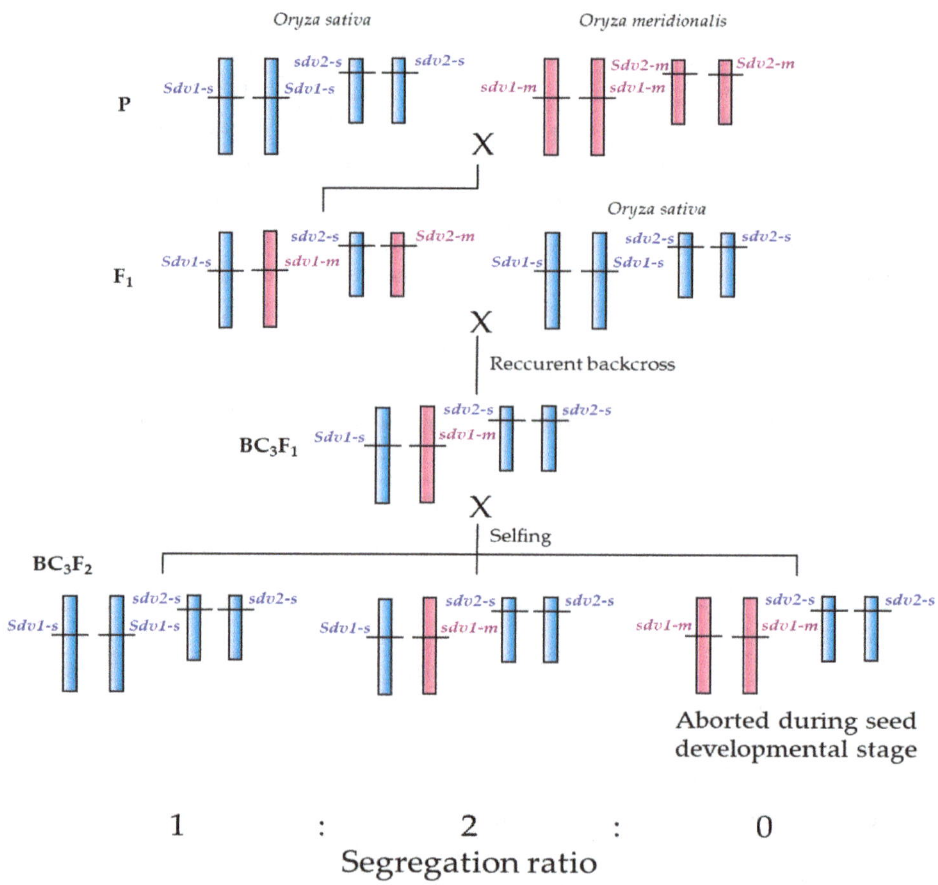

Figure 6. A genetic model that explains segregation distortion and seed sterility assuming gene duplication and loss of gene function for seed development.

If useful genes of *O. meridionalis* for rice genetic improvement are located close to *sdv1-m*, the introgression of these genes into *O. sativa* genetic background should be combined with *Sdv2-m*. Therefore, the chromosomal location of *SDV2* and tightly linked DNA markers to it are urgently needed.

In the frequently cited high-density rice genetic linkage map by Harushima et al. [35], the centromeric region of chromosome six is located between 64.7 cM and 65.7 cM. Some of the DNA marker sequences located on the centromeric region are available in NCBI (https://www.ncbi.nlm.nih.gov). C574 (accession name: D15395) is located at 13,685 kb, and G294 (accession name: D14774) is located at 17,056 kb in Nipponbare genome (Os-Nipponbare-Reference-IRGSP-1.0). Therefore the physical size of the centromeric region is at least 3371 kb. The candidate chromosomal region of *SDV1* encompasses this region (Tables 1 and 2). Recombination events were, in general, highly suppressed around the centromere. Our result is consistent with that. The combination of high resolution linkage analysis with gene expression analysis, gene disruption, and association study will be necessary to identify the *SDV1* gene.

Seed development is dissected into embryogenesis and endosperm development. We are undertaking microscopic observation of seed development of *Sdv1-s sdv1-m* heterozygotes in the T65 background to define the cause of the seed abortion. Several genes required for embryogenesis and endosperm development have been reported [36–39]. Identification of the *SDV1* and *SDV2* genes will contribute to the molecular genetics of seed development.

Direct evidence supporting the gene model was that the DNA from the embryo of aborted seeds deriving from the heterozygote of KGC6_1202 was homozygous for the W1297 allele. In our preliminary experiments, we tried to extract DNA from them, modifying the method below so that the DNA concentration would be higher. A few embryos were homozygous for the W1297 allele. However, PCR failed in most cases. This suggests that embryogenesis stops at an early stage in the homozygotes of *sdv1-m*. One alternative approach might be to extract DNA from developing seeds, not from mature seeds. Combination of microscopic observation of developing embryo and DNA genotyping will contribute to understanding the abortion mechanism caused by the *sdv1-m* gene.

According to the chloroplast genome analyses by Wambugu et al. [40], Yin et al. [41] and Sotowa et al. [42], *O. rufipogon* in Australia carries a chloroplast genome similar to that of *O. meridionalis* rather than that of *O. rufipogon* in Asia and *O. sativa*, probably because of chloroplast capture (introgression). During the process of chloroplast capture, some nuclear genome genes could be shared by *O. rufipogon* in Australia and *O. meridionalis*. However, they might carry distinct alleles on *SDV1* and *SDV2* loci. When the DNA sequences of these alleles of the two loci are uncovered, this information can be applied for the analysis of plants growing in the wild, and possible ongoing hybridization between *O. rufipogon* and *O. meridionalis* can be monitored in the Northern part of Australia, in which the two species are sympatric. Hybrids have been found in the wild and confirmed by molecular analysis [43], but the low frequency of these hybrids and the continued existence of the two distinct AA genome taxa in the northern Australian environment may be explained by these genes that create a reproductive barrier.

4. Materials and Methods

4.1. Plant Material

Three wild rice strains, Jpn1, Jpn2, and W1297, and one cultivated rice cultivar Taichung 65 (T65) were used in this study. W1297 is a strain of *O. meridionalis* collected in Darwin, Northern Territory, Australia, and provided by National Institute of Genetics, Mishima, Japan. Jpn1 and Jpn2 were collected in Australia with the permission from the Queensland government, under the EcoAccess program [42]. Judging from its perennial life history, typical of Australian *O. rufipogon*, and Indel marker genotypes, Jpn1 was classified as *O. rufipogon*. The Australian *O. rufipogon* population at the Jpn1 site has been shown to have a chloroplast similar to that of *O. meridionalis* and a nuclear genome closer to *O. rufipogon* [44] suggesting it may need to be considered as a distinct taxon. Jpn2 was distinct with a short anther, typical of *O. meridionalis*, and perennial life history in its habitat in Queensland, Australia [42]. *O. meridionalis* is now described as including both annual and perennial types [45]. It had Indel marker genotypes that were the same as 18 *O. meridionalis* Core collection accessions. It was treated as a type of *O. meridionalis* based on five Indel DNA markers that reflect varietal differentiation in comparisons, such as Indica–Japonica, temperate Japonica–tropical Japonica with high accuracy [46,47]. Our visual observations indicated that the three wild rice strains each showed a uniform phenotype in the first growing year, suggesting that they had been fixed for at least the loci controlling agronomic traits. Before anthesis, the panicles of the wild rice strains were covered with bags made of glassine paper to force self-fertilization in every generation. The selfed progeny also showed uniform phenotypes. The preliminary analysis of DNA markers covering the whole 12 chromosomes indicated that they were homozygous at all the DNA marker loci. T65 is a Japonica cultivar used frequently in the study of rice genetics, as a recurrent parent of CSSLs, isogenic lines, and the study of induced mutation [12,21,48].

We bred CSSLs in a T65 genetic background incorporating the three Australian wild rice strains, W1297, Jpn1, and Jpn2, chromosomal segments by recurrent backcrossing. First T65 was crossed with W1297, Jpn1, and Jpn2 as pollen parents. One plant per each wild rice strain was used for producing the F_1 generation. Then the F_1 was backcrossed with T65 as a pollen parent in all subsequent backcross generations with some exception described above. A total of 39, 43, and 33 BC_1F_1 plants were obtained using W1297, Jpn1, and Jpn2 as donor parent, respectively. All the BC_1F_1 plants were backcrossed with T65. One BC_2F_1 plant originating from each BC_1F_1 plant was backcrossed with T65 to produce BC_3F_1. One BC_3F_1 plant originating from each BC_2F_1 plant was backcrossed with T65 to produce BC_4F_1. W1297, Jpn1, and Jpn2 have many characters different from T65, such as late heading, red pericarp, long awn, and easy shattering. Some BC_3F_1 plants had such characteristics in T65 genetic background, which was suitable for genetic dissection of these characters. As a model character, late heading was selected. We selected the latest BC_3F_1 plants and collected seeds from these plants to produce a BC_3F_2 generation. As shown above, because the segregation of genes conferring days to heading did not fit the expected Mendelian single gene segregation, we focused on the analysis of the distorted segregation. The BC_4F_2 generations deriving from the late heading BC_4F_1 plants were also examined.

Plant cultivation followed Ichitani et al. [48]. Germinated seeds were sown in nursery beds in a greenhouse. About two weeks after sowing, seedlings were transferred out of the greenhouse. About 30 days after the sowing date, seedlings were planted in a paddy field at the Experimental Farm of Kagoshima University, Kagoshima, Japan. The fertilizers applied were 4, 6, and 5 g/m^2, respectively, for N, K_2O, and P_2O_5. Plant spacing was 15 × 30 cm. Sowing and transplanting were done respectively on May 31 and June 24 in 2015, on May 27and June 28 in 2016, May 25 and July 4 in 2017, respectively. Hybridization was performed as follows: For emasculation, panicles of the egg donor were soaked in hot water at 43 °C for 7 min. For pollination, the upper half of the open spikelets were cut about 30 min after emasculation. All the closed spikelets were cut off. Then pollen of the pollen donor was scattered on them. After pollination, panicles were covered with bags made of glassine paper. At least one panicle was left without pollination to check whether emasculation was complete.

4.2. Trait Evaluation

Heading date was recorded for each plant when the first developing panicle emerged from the leaf sheath of the flag leaf. Heading date was converted into days to heading. Seed fertility was evaluated by collecting 50 seeds from the upper side of each of the three panicles, using a modification of the method of Wan and Ikehashi [49], counting fertile and sterile spikelets on the upper half of 3–4 panicles for each plant. Seeds were scored as fertile or sterile. In the W1297 cross, sterile seeds were dehusked to see if sterility occurred before or after fertilization. The BC_4F_2 plants that produced the BC_5F_1 generation were dug up, and transferred from the paddy field to a glass house a day before pollination. We empirically know that rice plants undergoing such a treatment show lower seed fertility, probably because of root damage. Therefore, we did not evaluate seed fertility of these plants. Panicles of some plants were damaged by birds after heading. This is the reason for the inconsistency in BC_4F_2 plant number among tables and figures.

Pollen fertility of the BC_4F_2 (W1297) population was evaluated using iodine-potassium iodide solution. Panicles were collected about three days after emerging from the leaf sheath of the flag leaf, and dried in paper bags at room temperature. All the anthers in a spikelet collected one day before anthesis were cleaved to gather pollen on a glass slide. Pollen were stained with iodine-potassium iodide solution. More than 200 pollen grains were scored for each individual. Densely stained pollen with a normal size were scored as fertile. The other pollen were scored as sterile.

4.3. DNA Analysis

DNA from leaves and embryo from fertile seeds was extracted according to Ichitani et al. [48] with some modifications: Each leaf tip, 2.5 cm long from a single plant, or embryo from dehusked seeds was put in a well of a 96-deep-well plate. Then 100 μL of extraction buffer (100 mM Tris–HCl (pH 8.0), 1 M KCl, and 10 mM EDTA) was added with a 5-mm-diameter stainless steel ball to the well. After being covered with a hard lid, the plate was shaken hard (ShakeMaster ver. 1.2; BioMedical Science Inc., Tokyo, Japan) for 1 min to grind the leaves or embryos. After centrifuging, the plate was incubated at 70 °C for half an hour, then at room temperature for half an hour. Then 10 μL of the supernatant was recovered and 8 μL of 2-propanol was added. After centrifuging, the supernatant was discarded and the DNA pellet was rinsed with 50 μL of 70% ethanol. The DNA pellet was dried and dissolved in 50 μL of sterilized distilled water. It was very difficult to separate the embryo from the other part of seed completely. However, our preliminary experiment indicated that even if DNA was extracted from the whole dehusked seeds produced by a heterozygote for a DNA marker such as KGC6_12.02 (Table 1), DNA marker segregation was observed, suggesting that DNA from the parts of the dehusked seed other than the embryo was negligible. PCR mixture, cycle, electrophoresis, DNA staining, gel image documentation also followed Ichitani et al. [48].

4.4. DNA Markers

Most published PCR-based DNA markers for *Oryza* are based on an *O. sativa* genome sequence such as Nipponbare (Os-Nipponbare-Reference-IRGSP-1.0, [50]) and 9311 (GCA_0000046551, [51]). However, a preliminary survey comparing the genome of Nipponbare (IRGSP 1.0) and that of *O. meridionalis* accession (GCA_000338895.2. [7]) showed that there were many discrepancies between them, leading to expected failure in amplification from the *O. meridionalis* genome when using *O. sativa* genome-based DNA markers. Our strategy of designing co-dominant DNA markers was that insertion/deletion (indel) polymorphisms ranging from 5 to 100 base pairs were searched for between the Nipponbare and the *O. meridionalis* genomes. Then, the indels found only between *O. meridionalis* and Nipponbare, not between *meridionalis* and two Indica cultivars, 93-11 and HR-12 (GCA_000725085), were selected. The event causing such indels were thought to have occurred in Japonica rice after Japonica-Indica differentiation. T65 is a typical Japonica cultivar. Our preliminary survey showed that T65 shared the banding patterns of Nipponbare in most of the DNA markers examined [52]. Therefore, the indels as described above were expected to show polymorphism between T65 and *O. meridionalis*. The selected indels were screened based on sequence similarity surrounding indels between Nipponbare and the *O. meridionalis* genomes. The primer design followed Busung et al. [53].

Author Contributions: Conceptualization, K.I.; methodology, K.I.; validation, D.T. and M.U.; formal analysis, D.T., M.U., and K.I.; investigation, D.T. and M.U.; resources, S.T., T.S., R.H., R.I., K.I.; data curation, D.T., M.U., K.I.; writing—original draft preparation, D.T. and K.I.; writing—review and editing, D.T., M.U., S.T., T.S., R.H., R.I. and K.I.; visualization, D.T., M.U., and K.I.; supervision, K.I.; project administration, R.I.; funding acquisition, R.I.

Acknowledgments: We are grateful to the National Institute of Genetics for their kind provision of W1297 seeds. We thank Mr. Masaaki Ikenoue, Mr. Nishiobino Tsubasa, Ms. Yoko Nakashima and Ms. Asako Kobai for their technical assistance.

References

1. Henry, R.J.; Rice, N.; Waters, D.L.E.; Kasem, S.; Ishikawa, R.; Hao, Y.; Dillon, S.; Crayn, D.; Wing, R.; Vaughan, D. Australian *Oryza*: Utility and conservation. *Rice* **2010**, *3*, 235–241. [CrossRef]

2. Huang, X.; Kurata, N.; Wei, X.; Wang, Z.-X.; Wang, A.; Zhao, Q.; Zhao, Y.; Liu, K.; Lu, H.; Li, W.; et al. A map of rice genome variation reveals the origin of cultivated rice. *Nature* **2012**, *490*, 497–501. [CrossRef] [PubMed]
3. Ohtsubo, H.; Cheng, C.; Ohsawa, I.; Tsuchimoto, S.; Ohtsubo, E. Rice retroposon *p-SINE1* and origin of cultivated rice. *Breed. Sci.* **2004**, *54*, 1–11. [CrossRef]
4. Cheng, C.; Tsuchimoto, S.; Ohtsubo, H.; Ohtsubo, E. Evolutionary relationships among rice species with AA genome based on SINE insertion analysis. *Genes Genet. Syst.* **2002**, *77*, 323–334. [CrossRef] [PubMed]
5. Xu, J.-H.; Kurata, N.; Akimoto, M.; Ohtsubo, H.; Ohtsubo, E. Identification and characterization of Australian wild rice strains of *Oryza meridionalis* and *Oryza rufipogon* by SINE insertion polymorphism. *Genes Genet. Syst.* **2005**, *80*, 129–134. [CrossRef]
6. Zhang, Q.-J.; Zhu, T.; Xia, E.-H.; Shi, C.; Liu, Y.-L.; Zhang, Y.; Liu, Y.; Jiang, W.-K.; Zhao, Y.-J.; Mao, S.-Y.; et al. Rapid diversification of five *Oryza* AA genomes associated with rice adaptation. *Proc. Natl. Acad. Sci. USA* **2014**, *111*, E4954–E4962. [CrossRef] [PubMed]
7. Stein, J.C.; Yu, Y.; Copetti, D.; Zwickl, D.J.; Zhang, L.; Zhang, C.; Chougule, K.; Gao, D.; Iwata, A.; Goicoechea, J.L.; et al. Genomes of 13 domesticated and wild rice relatives highlight genetic conservation, turnover and innovation across the genus *Oryza*. *Nat. Genet.* **2018**, *50*, 285–296. [CrossRef] [PubMed]
8. Naredo, M.E.B.; Juliano, A.B.; Lu, B.-R.; Jackson, M.T. Hybridization of AA genome rice species from Asia and Australia I. Crosses and development of hybrids. *Genet. Resour. Crop Evol.* **1997**, *44*, 17–23. [CrossRef]
9. Naredo, M.E.B.; Juliano, A.B.; Lu, B.-R.; Jackson, M.T. Taxonomic status of *Oryza glumaepatula* Steud. II. Hybridization between New World diploids and AA genome species from Asia and Australia. *Genet. Resour. Crop Evol.* **1998**, *45*, 205–214. [CrossRef]
10. Shim, R.A.; Angeles, E.R.; Ashikari, M.; Takashi, T. Development and evaluation of *Oryza glaberrima* Steud. chromosome segment substitution lines (CSSLs) in the background of *O. sativa* L. cv. Koshihikari. *Breed. Sci.* **2010**, *60*, 613–619. [CrossRef]
11. Hirabayashi, H.; Sato, H.; Nonoue, Y.; Kuno-Takemoto, Y.; Takeuchi, Y.; Kato, H.; Nemoto, H.; Ogawa, T.; Yano, M.; Imbe, T.; et al. Development of introgression lines derived from *Oryza rufipogon* and *O. glumaepatula* in the genetic background of *japonica* cultivated rice (*O. sativa* L.) and evaluation of resistance to rice blast. *Breed. Sci.* **2010**, *60*, 604–612. [CrossRef]
12. Yoshimura, A.; Nagayama, H.; Sobrizal; Kurakazu, T.; Sanchez, P.L.; Doi, K.; Yamagata, Y.; Yasui, H. Introgression lines of rice (*Oryza sativa* L.) carrying a donor genome from the wild species, *O. glumaepatula* Steud. and *O. meridionalis* Ng. *Breed. Sci.* **2010**, *60*, 597–603. [CrossRef]
13. Arbelaez, J.D.; Moreno, L.T.; Singh, N.; Tung, C.-W.; Maron, L.G.; Ospina, Y.; Martinez, C.P.; Grenier, C.; Lorieux, M.; McCouch, S. Development and GBS-genotyping of introgression lines (ILs) using two wild species of rice, *O. meridionalis* and *O. rufipogon*, in a common recurrent parent, *O. sativa* cv. Curinga. *Mol. Breed.* **2015**, *35*, 81. [CrossRef] [PubMed]
14. Yano, M.; Katayose, Y.; Ashikari, M.; Yamanouchi, U.; Monna, L.; Fuse, T.; Baba, T.; Yamamoto, K.; Umehara, Y.; Nagamura, Y.; et al. *Hd1*, a major photoperiod sensitivity quantitative trait locus in rice, is closely related to the arabidopsis flowering time gene *CONSTANS*. *Plant Cell* **2000**, *12*, 2473–2483. [CrossRef]
15. Doi, K.; Izawa, T.; Fuse, T.; Yamanouchi, U.; Kubo, T.; Shimatani, Z.; Yano, M.; Yoshimura, A. *Ehd1*, a B-type response regulator in rice, confers short-day promotion of flowering and controls *FT-like* gene expression independently of *Hd1*. *Genes Dev.* **2004**, *18*, 926–936. [CrossRef] [PubMed]
16. Ichitani, K.; Okumoto, Y.; Tanisaka, T. photoperiod sensitivity gene of *Se-1* locus found in photoperiod insensitive rice cultivars of the northern limit region of rice cultivation. *Jpn. J. Breed.* **1997**, *47*, 145–152. [CrossRef]
17. Ichitani, K.; Okumoto, Y.; Tanisaka, T. Genetic analysis of the rice cultivar Kasalath with special reference to two photoperiod sensitivity loci, E_1 and *Se-1*. *Jpn. J. Breed.* **1998**, *48*, 51–57. [CrossRef]
18. Inoue, H.; Nishida, H.; Okumoto, Y.; Tanisaka, T. Identification of an early heading time gene found in the taiwanese rice cultivar Taichung 65. *Jpn. J. Breed.* **1998**, *48*, 103–108. [CrossRef]
19. Temnykh, S.; Park, W.D.; Ayres, N.; Cartinhour, S.; Hauck, N.; Lipovich, L.; Cho, Y.G.; Ishii, T.; McCouch, S.R. Mapping and genome organization of microsatellite sequences in rice (*Oryza sativa* L.). *Theor. Appl. Genet.* **2000**, *100*, 697–712. [CrossRef]
20. McCouch, S.R.; Teytelman, L.; Xu, Y.; Lobos, K.B.; Clare, K.; Walton, M.; Fu, B.; Maghirang, R.; Li, Z.; Xing, Y.; et al. Development and Mapping of 2240 New SSR Markers for Rice (*Oryza sativa* L.). *DNA Res.* **2002**, *9*, 199–207. [CrossRef]

21. Sano, Y. Genetic comparisons of chromosome 6 between wild and cultivated rice. *Jpn. J. Breed.* **1992**, *42*, 561–572. [CrossRef]
22. Matsubara, K.; Khin-Thidar; Sano, Y. A gene block causing cross-incompatibility hidden in wild and cultivated rice. *Genetics* **2003**, *165*, 343–352. [PubMed]
23. Koide, Y.; Ogino, A.; Yoshikawa, T.; Kitashima, Y.; Saito, N.; Kanaoka, Y.; Onishi, K.; Yoshitake, Y.; Tsukiyama, T.; Saito, H.; et al. Lineage-specific gene acquisition or loss is involved in interspecific hybrid sterility in rice. *Proc. Natl. Acad. Sci. USA* **2018**, *115*, E1955–E1962. [CrossRef] [PubMed]
24. Yanagihara, S.; McCouch, S.R.; Ishikawa, K.; Ogi, Y.; Maruyama, K.; Ikehashi, H. Molecular analysis of the inheritance of the *S-5* locus, conferring wide compatibility in Indica/Japonica hybrids of rice (*O. sativa* L.). *Theoret. Appl. Genet.* **1995**, *90*, 182–188. [CrossRef]
25. Liu, Y.S.; Zhu, L.H.; Sun, J.S.; Chen, Y. Mapping QTLs for defective female gametophyte development in an inter-subspecific cross in *Oryza Sativa* L. *Theor. Appl. Genet.* **2001**, *102*, 1243–1251. [CrossRef]
26. Yang, J.; Zhao, X.; Cheng, K.; Du, H.; Ouyang, Y.; Chen, J.; Qiu, S.; Huang, J.; Jiang, Y.; Jiang, L.; et al. A killer-protector system regulates both hybrid sterility and segregation distortion in rice. *Science* **2012**, *337*, 1336–1340. [CrossRef]
27. Long, Y.; Zhao, L.; Niu, B.; Su, J.; Wu, H.; Chen, Y.; Zhang, Q.; Guo, J.; Zhuang, C.; Mei, M.; et al. Hybrid male sterility in rice controlled by interaction between divergent alleles of two adjacent genes. *Proc. Natl. Acad. Sci. USA* **2008**, *105*, 18871–18876. [CrossRef]
28. McCouch, S.R.; CGSNL (Committee on Gene Symbolization, N. and L., Rice Genetics Cooperative). Gene Nomenclature System for Rice. *Rice* **2008**, *1*, 72–84. [CrossRef]
29. Juliano, A.B.; Naredo, M.E.B.; Lu, B.-R.; Jackson, M.T. Genetic differentiation in *Oryza meridionalis* Ng based on molecular and crossability analyses. *Genet. Resour. Crop Evol.* **2005**, *52*, 435–445. [CrossRef]
30. Yamagata, Y.; Yamamoto, E.; Aya, K.; Win, K.T.; Doi, K.; Sobrizal; Ito, T.; Kanamori, H.; Wu, J.; Matsumoto, T.; et al. Mitochondrial gene in the nuclear genome induces reproductive barrier in rice. *Proc. Natl. Acad. Sci. USA* **2010**, *107*, 1494–1499. [CrossRef]
31. Nguyen, G.N.; Yamagata, Y.; Shigematsu, Y.; Watanabe, M.; Miyazaki, Y.; Doi, K.; Tashiro, K.; Kuhara, S.; Kanamori, H.; Wu, J.; et al. Duplication and loss of function of genes encoding RNA polymerase III subunit C4 causes hybrid incompatibility in rice. *G3 Genes Genom. Genet.* **2017**, *7*, 2565–2575. [CrossRef] [PubMed]
32. Ichitani, K.; Takemoto, Y.; Iiyama, K.; Taura, S.; Sato, M. Chromosomal location of *HCA1* and *HCA2*, hybrid chlorosis genes in rice. *Int. J. Plant Genom.* **2012**, *2012*. [CrossRef] [PubMed]
33. Bomblies, K.; Weigel, D. Hybrid necrosis: Autoimmunity as a potential gene-flow barrier in plant species. *Nat. Rev. Genet.* **2007**, *8*, 382–393. [CrossRef] [PubMed]
34. Oka, H.I. Phylogenetic differentiation of cultivated rice. XV. Complementary lethal genes in rice. *Jpn. J. Genet.* **1957**, *32*, 83–87. [CrossRef]
35. Harushima, Y.; Yano, M.; Shomura, A.; Sato, M.; Shimano, T.; Kuboki, Y.; Yamamoto, T.; Lin, S.Y.; Antonio, B.A.; Parco, A.; et al. A high-density rice genetic linkage map with 2275 markers using a single F_2 population. *Genetics* **1998**, *148*, 479–494. [PubMed]
36. Huang, X.; Peng, X.; Sun, M.-X. *OsGCD1* is essential for rice fertility and required for embryo dorsal-ventral pattern formation and endosperm development. *New Phytol.* **2017**, *215*, 1039–1058. [CrossRef] [PubMed]
37. Huang, X.; Lu, Z.; Wang, X.; Ouyang, Y.; Chen, W.; Xie, K.; Wang, D.; Luo, M.; Luo, J.; Yao, J. Imprinted gene *OsFIE1* modulates rice seed development by influencing nutrient metabolism and modifying genome H3K27me3. *Plant J.* **2016**, *87*, 305–317. [CrossRef] [PubMed]
38. Hara, T.; Katoh, H.; Ogawa, D.; Kagaya, Y.; Sato, Y.; Kitano, H.; Nagato, Y.; Ishikawa, R.; Ono, A.; Kinoshita, T.; et al. Rice SNF2 family helicase ENL1 is essential for syncytial endosperm development. *Plant J.* **2015**, *81*, 1–12. [CrossRef] [PubMed]
39. Hong, S.K.; Kitano, H.; Satoh, H.; Nagato, Y. How is embryo size genetically regulated in rice? *Development* **1996**, *122*, 2051–2058.
40. Wambugu, P.W.; Brozynska, M.; Furtado, A.; Waters, D.L.; Henry, R.J. Relationships of wild and domesticated rices (*Oryza* AA genome species) based upon whole chloroplast genome sequences. *Sci. Rep.* **2015**, *5*, 13957. [CrossRef]
41. Yin, H.; Akimoto, M.; Kaewcheenchai, R.; Sotowa, M.; Ishii, T.; Ishikawa, R. Inconsistent diversities between nuclear and plastid genomes of aa genome species in the genus *Oryza*. *Genes Genet. Syst.* **2015**, *90*, 269–281. [CrossRef] [PubMed]

42. Sotowa, M.; Ootsuka, K.; Kobayashi, Y.; Hao, Y.; Tanaka, K.; Ichitani, K.; Flowers, J.M.; Purugganan, M.D.; Nakamura, I.; Sato, Y.-I.; et al. Molecular relationships between Australian annual wild rice, *Oryza meridionalis*, and two related perennial forms. *Rice* **2013**, *6*, 26. [CrossRef] [PubMed]
43. Moner, A.M.; Henry, R.J. *Oryza meridionalis* Ng. In *The Wild Oryza Genomes*; Mondal, T.K., Henry, R., Eds.; Springer-Verlag: Heidelberg, Germany, 2018; pp. 177–182.
44. Brozynska, M.; Copetti, D.; Furtado, A.; Wing, R.A.; Crayn, D.; Fox, G.; Ishikawa, R.; Henry, R.J. Sequencing of australian wild rice genomes reveals ancestral relationships with domesticated rice. *Plant Biotechnol. J.* **2017**, *15*, 765–774. [CrossRef] [PubMed]
45. Moner, A.M.; Furtado, A.; Chivers, I.; Fox, G.; Crayn, D.; Henry, R.J. Diversity and evolution of rice progenitors in australia. *Ecol. Evol.* **2018**, *8*, 4360–4366. [CrossRef] [PubMed]
46. Ichitani, K.; Taura, S.; Sato, M.; Kuboyama, T. Distribution of *Hwc2-1*, a causal gene of a hybrid weakness, in the World Rice Core collection and the Japanese Rice mini Core collection: Its implications for varietal differentiation and artificial selection. *Breed. Sci.* **2016**, *66*, 776–789. [CrossRef] [PubMed]
47. Muto, C.; Ishikawa, R.; Olsen, K.M.; Kawano, K.; Bounphanousay, C.; Matoh, T.; Sato, Y.-I. Genetic diversity of the *wx* flanking region in rice landraces in northern Laos. *Breed. Sci.* **2016**, *66*, 580–590. [CrossRef]
48. Ichitani, K.; Yamaguchi, D.; Taura, S.; Fukutoku, Y.; Onoue, M.; Shimizu, K.; Hashimoto, F.; Sakata, Y.; Sato, M. Genetic analysis of ion-beam induced extremely late heading mutants in rice. *Breed. Sci.* **2014**, *64*, 222–230. [CrossRef]
49. Wan, J.; Ikehashi, H. Identification of two types of differentiation in cultivated rice (*Oryza sativa* L.) detected by polymorphism of isozymes and hybrid sterility. *Euphytica* **1997**, *94*, 151–161. [CrossRef]
50. Kawahara, Y.; de la Bastide, M.; Hamilton, J.P.; Kanamori, H.; McCombie, W.R.; Ouyang, S.; Schwartz, D.C.; Tanaka, T.; Wu, J.; Zhou, S.; et al. Improvement of the *Oryza sativa* Nipponbare reference genome using next generation sequence and optical map data. *Rice* **2013**, *6*, 4. [CrossRef]
51. Yu, J.; Hu, S.; Wang, J.; Wong, G.K.-S.; Li, S.; Liu, B.; Deng, Y.; Dai, L.; Zhou, Y.; Zhang, X.; et al. A draft sequence of the rice genome (*Oryza sativa* L. ssp. indica). *Science* **2002**, *296*, 79–92. [CrossRef]
52. Ichitani, K.; Taura, S.; Tezuka, T.; Okiyama, Y.; Kuboyama, T. Chromosomal location of *HWA1*and *HWA2*, complementary hybrid weakness genes in rice. *Rice* **2011**, *4*, 29–38. [CrossRef]
53. Busungu, C.; Taura, S.; Sakagami, J.-I.; Ichitani, K. Identification and linkage analysis of a new rice bacterial blight resistance gene from XM14, a mutant line from IR24. *Breed. Sci.* **2016**, *66*, 636–645. [CrossRef] [PubMed]

4

Durum Wheat (*Triticum durum* Desf.): Origin, Cultivation and Potential Expansion

Amadou Tidiane Sall [1,2], Tiberio Chiari [3], Wasihun Legesse [4], Kemal Seid-Ahmed [2], Rodomiro Ortiz [5], Maarten van Ginkel [6] and Filippo Maria Bassi [2,*]

1 Institut Sénégalais de Recherches Agricoles (ISRA), Saint-Louis 46024, Senegal; tidianeassall11@gmail.com
2 International Center for Agricultural Research in the Dry Areas (ICARDA), Rabat 10000, Morocco; s.a.kemal@cgiar.org
3 Italian Agency for Development Cooperation (AICS), Addis Ababa 1000, Ethiopia; tiberio.chiari@aics.gov.it
4 Ethiopian Institute Agricultural Research (EIAR), Agricultural Research Center, Debre Zeit 1000, Ethiopia; wasihunl@yahoo.com
5 Sveriges lantbruksuniversitet (SLU), Institutionen för Växtförädling (VF), 23053 Alnarp, Sweden; rodomiro.ortiz@slu.se
6 International Center for Agricultural Research in the Dry Areas (ICARDA), Amman 1118, Jordan; maartenvanginkel1951@gmail.com
* Correspondence: F.Bassi@cgiar.org.

Abstract: Durum wheat is an important food crop in the world and an endemic species of sub-Saharan Africa (SSA). In the highlands of Ethiopia and the oases of the Sahara this crop has been cultivated for thousands of years. Today, smallholder farmers still grow it on marginal lands to assure production for their own consumption. However, durum wheat is no longer just a staple crop for food security but has become a major cash crop. In fact, the pasta, burghul and couscous industry currently purchase durum grain at prices 10 to 20% higher than that of bread wheat. Africa as a whole imports over €4 billion per year of durum grain to provide the raw material for its food industry. Hence, African farmers could obtain a substantial share of this large market by turning their production to this crop. Here, the achievements of the durum breeding program of Ethiopia are revised to reveal a steep acceleration in variety release and adoption over the last decade. Furthermore, the variety release for Mauritania and Senegal is described to show how modern breeding methods could be used to deliver grain yields above 3 t ha^{-1} in seasons of just 92 days of length and in daytime temperatures always above 32 °C. This review describes the potential of releasing durum wheat varieties adapted to all growing conditions of SSA, from the oases of the Sahara to the highlands of Ethiopia. This indicates that the new breeding technologies offer great promise for expanding the area of durum wheat production in SSA but that this achievement remains primarily dependent on the market ability to purchase these grains at a higher price to stimulate farmer adoption. The critical importance of connecting all actors along the semolina value chain is presented in the example of Oromia, Ethiopia and that success story is then used to prompt a wider discussion on the potential of durum wheat as a crop for poverty reduction in Africa.

Keywords: Agro-industry; Ethiopia; oasis wheat; pasta wheat; Senegal River; value chain

1. Introduction

Durum wheat (*Triticum durum* Desf.) is an important food crop of the world, with an estimated 36 million t of annual global production [1]. The largest producing countries are Turkey and Canada with estimated 2 million ha each [2,3], followed by Algeria, Italy and India, each cultivating over

1.5 million ha [4–6]. Syria belonged to this group of large producers but the recent unrest has strongly reduced crop production. France, Greece, Morocco, Pakistan, Portugal, Kazakhstan, Russia, Spain and Tunisia cultivate durum wheat on between 0.5 and 0.8 million ha annually [3]. Azerbaijan, Iraq and Iran combined grow durum wheat on over 0.7 million ha [6]. In addition, Egypt, Jordan and Lebanon grow it on relatively large areas [7–9]. The Sonora desert and other small areas of Mexico also target the production of this crop for the export market on approximately 0.2 million ha [10]. Australia is similarly exploring the cultivation of this crop with 0.1 million ha allocated annually to its production [11]. In sub-Saharan Africa (SSA), Ethiopia is the largest producer of durum wheat, with approximately 0.6 million ha [12].

A very large amount of genetic diversity exists for this crop and that diversity also extends to the many traditional ways of consuming it, including several unique dishes that represent with pride the national identities: pasta, couscous, bourghul, *freekeh*, *gofio* and unleavened breads, just to name a few [13]. Regardless of its tight connection to the dishes of the tradition, durum wheat today is cultivated in developed countries mainly as a cash crop to feed the booming food industry. The annual production of pasta was estimated at 14.3 million t in 2013, with a global market approximated at €14.9 billion and average global price of 1045 € t^{-1} [14]. On a global scale, most of its consumption and production are in Europe, South America and the United States of America. Africa accounts only for 5.6% of total pasta production, mainly in Egypt, South Africa and Tunisia [14] and Asia consumption is also on the raise. Detailed data for SSA are hard to obtain, as most statistics combine durum wheat with bread wheat into single "wheat" data points but the estimations that could be gathered from several sources suggest an import market of €337 million and an export market mostly within the continent of €40 million (Table 1). Reliable data on the size of the internal market were not found. In this review, the developing couscous and bourghul industrial markets are not included, as data are not readily available. Italy, North African, South Africa and Turkey are the largest exporters of pasta to SSA [15]. However, the total area dedicated to durum wheat in SSA is limited to 630,000 ha, of which 90% is cultivated in Ethiopia. Therefore, this is the only country capable of producing pasta using locally grown grain, while for all other SSA countries the bulk of pasta production required the import of €483 million worth of durum grain from Canada, Turkey and the USA (Table 1). It must be mentioned that the pasta industry in SSA often utilizes bread wheat flour for its production and typically only products from North Africa and developed countries meet the international standard definition of 'pasta' by using 100% durum semolina [14]. Clearly, there is huge agricultural and commercial scope for expanding domestic production and marketing of durum wheat in SSA countries.

Durum wheat and rice are the most lucrative among the cereals, with prices usually 20 to 40% higher than common wheat, millet, maize and sorghum [16]. While durum wheat remains a critical staple food for smallholder farmers in marginal lands, thanks to its exceptional adaptation to climatic stresses, its large-scale production is tightly linked to its greater monetary return. In the absence of governmental subsidies that push toward the cultivation of other crops, farmers tend to prefer durum wheat as long as the market continues to guarantee additional profits. In this regard, the existence of a strong value chain for the pasta, couscous and bourghul industry is quintessential to the success of durum wheat cropping.

Table 1. Economy and production of durum wheat in Africa and sub-Saharan Africa.

Country	Crop Use per Land			Durum Exports [b]		Durum Imports [b]	
	Rice [a]	Wheat (all) [a]	Durum	Grains	Pasta	Grains	Pasta
	(ha)	(ha)	(ha)	(€)	(€)	(€)	(€)
East	**74,069**	**1,931,714**	**599,552**	**365,889**	**247,159**	**129,119,247**	**46,339,452**
Eritrea	-	25,000	12,500 [c]	-	-	-	8,210,000
Ethiopia	33,820	1,605,654	561,979 [d]	-	28,079	129,113,334 [n]	24,433,488 [n]
Kenya	31,349	162,900	16,290 [e]	365,889	219,080	2177	13,695,964
Somalia	1338	2500	na	-	-	3735	-
Sudan	7562	135,660	6783 [f]	-	-	na	na
Central	**206,592**	**44,687**			**1,636,218**	**11,864,150**	**21,908,766**
Burundi	21,670	8828	na	-	1324	3136	226,183
Cameroon	166,734	660	na	-	1,487,086	11,858,665 [n]	13,442,925 [n]
Gabon	620	-	-	-	69,532	2349	5,665,685
Rwanda	17,568	35,199	na	-	78,276	-	2,573,974
South	**1,437,257**	**608,622**	**26,521**	**38,590,262**	**11,794,879**	**148,416,094**	**86,270,994**
Angola	29,510	3420	720 [g]	-	-	1229	-
Madagascar	909,492	2087	na	-	2730	142,521	26,445,917
Malawi	65,275	1269	na	-	3862	34,292,325 [n]	808,085
Mozambique	300,000	18,081	-	1295	-	-	3,568,930
South Africa	1150	505,500	23,456 [h]	35,183,511	11,702,825	33,889,190	31,108,885
Uganda [1]	93,000	14,000	100 [i]	86,843	76,540	3,673,000	3,935,998
Zambia	38,520	41,810	0 [j]	3,318,613	2177	-	5,982,422
Zimbabwe	310	22,455	2246 [k]	-	6745	76,417,829	14,420,757
West	**7,394,599**	**102,033**	**4302**	**1,123,513**	**87,653,612**	**193,535,341**	**155,068,054**
Benin	68,259	-		474,234	222,295	8,365,599 [n]	50,781,356 [n]
Burkina Faso	138,852	-	-	-	65,747	272,228	22,915,826 [n]
Gambia	66,380	-	-	-	536,030	-	5,094,377
Ghana	215,905	-	-	2340	3,633,396	78,376	7,847,000 [o]
Guinea	1,642,687	-	-	-	111,862	17,673,754	3,780,845
Ivory Coast	791,691	-	-	646,939	32,545,382	28,513,879 [o]	3,024,576 [o]
Mali	604,745	9947	1100 [k]	-	211,989	-	16,050,025
Mauritania	43,900	1700	na [l]	-	19,077	51,161,839	14,377,271
Niger	21,572	1883	na	-	7,328,976	3,101,758	20,659,562 [n]
Nigeria	2,931,400	80,000	na	-	40,917,371	38,300,083 [n]	2,633,648
Senegal	108,547	3	2 [m]	-	2,061,487	46,058,643 [n]	7,903,568
Sierra Leone	671,422	-	-	-	-	9182	-
Togo	92,239	-	-	-	2,099,984	-	27,284,750 [n]
sub-Saharan Africa	**9,112,517**	**2,687,056**	**632,375**	**40,079,663**	**101,331,868**	**482,934,832**	**309,587,266**
Africa	**9,714,796**	**9,960,981**	**3,557,740**	**40,346,664**	**215,258,164**	**4,166,972,506**	**375,508,531**

[1] The former Sudan is not a true Sub-Saharan country, but it has agro-environmental conditions that differ from North Africa and therefore it is presented here. Data consider Sudan and South Sudan together. [2] Uganda is reported among Southern Africa countries instead of East Africa for its closer similarity in the use of durum wheat. [a] Data obtained from FAOSTAT for the 2013 season [17]. [b] Data obtained from The Economic Complexity Observatory of the year 2013 [15], except when otherwise indicated. [c] [18]. [d] [19]. [e] [20] [f] [21]. [g] [22]. [h] [23]. [i] [24]. [j] [25]. [k] [26]. [l] [17]. [m] Land surface utilized on-station. [n] Data confirmed on Fact-Fish [27]. [o] Data confirmed on Index Mundi [28]. The **bold** rows indicate sub-totals for different African regions.

In this review, the current status of durum wheat production in SSA is discussed in comparison to the needs of the local pasta industry to better understand the potential of its expansion through the deployment of novel adapted varieties. Because of its industrial nature, durum wheat has often been disregarded by SSA policy makers in favour of bread wheat as a more direct "food security" approach. However, among the sustainable development goals set by the United Nations, "poverty reduction" is considered as a strategic way to tackle famine, without causing nutritional deficits due to mono-food diets. In this sense, durum wheat is at least as well suited as bread wheat in improving livelihoods. Both aspects of durum wheat, as a "food security" staple food for smallholder farmers, as well as a "poverty reduction" industrial crop will be considered here.

2. An Endemic Crop of SSA: Durum Wheat Second Centre of Origin in Ethiopia

Durum wheat originated from the domesticated form of a wild species named emmer wheat (*Triticum dicoccum* Koern.) between 12,000 and 10,000 years ago, in the West Levantine [29]. Phoenicians have traded it along the Mediterranean shores since historical times and throughout the rise of civilizations this crop has encountered several waves of expansion until today's global importance [30]. However, durum wheat did not originate solely in West Asia. Archaeological evidence suggests that naked emmer reached Ethiopia approximately 5000 years ago [31], probably arriving from the Levantine, through Egypt, along the Silk Road [32]. Today emmer wheat occupies approximately 7% of the wheat production in Ethiopia under the local name of *aja*. Recent molecular data [33] indicated that Ethiopian farmers repeated what had been achieved already in West Asia before, by deriving durum wheat anew through the further domestication of emmer. This new origin of the same crop gave rise to a subspecies known as *T. turgidum* ssp. *aethiopicum* or *abyssinicum*. Until relatively recently, landraces belonging to this subspecies were widely cultivated by smallholder farmers in Ethiopia, with up to 80% of the total durum land farmed with these unique biotypes [34]. The highlands of Ethiopia are known areas of rich biodiversity and durum wheat is no exception [35,36]. For instance, one of the unique characteristics identified among *T. aethiopicum* landraces is the purple colour of the grains, particularly rich in anthocyanins [37]. Anthocyanins act as anti-oxidants and provide other health benefits, hence these could be potentially exploited by the pasta industry to develop extra nutritious food products. Morphological and molecular characterization of these landraces has only just begun and already several traits such as resistance to diseases (e.g., stem rust, powdery mildew), drought tolerance, long coleoptile, high tillering and resistance sources to Hessian fly have been identified [38,39]. This biodiversity has already started attracting strong interest by the international community for utilization, pushing the Ethiopian Government to protect it under strict germplasm exchange policy [18]. In order to conserve these resources, the Ethiopian Biodiversity Institute (EBI) has established a holding of over 7000 accessions collected from different parts of Ethiopia [34]. These collections have been extensively investigated for their morphological and molecular diversity by many researchers and useful traits were identified and are now utilized by breeders and plant genetic conservationists in Ethiopia and beyond [40–48]. In the past two decades, the acreage of traditional tetraploid wheat has drastically diminished due to displacement by improved bread wheat varieties, extensive cultivation of *Tef* and Kabuli chickpea, farmland fragmentation, policies favouring bread wheat and the absence of a strong seed supply system [49]. To reduce this genetic erosion, EBI has established in situ conservation sites to conserve the agro-biodiversity at the farm level in different parts of Ethiopia. Community biodiversity practices were established in East Shoa and South Wollo zones with the aim of establishing community seed banks, participatory variety selection and the re-introduction of local durum wheat biotypes, food legumes and sorghum into the cropping system [50,51]. Regardless of their specific uses, these landraces represent a treasure chest of potentially new and useful traits that breeders could be able to exploit to deliver superior varieties with added market values.

3. Durum Wheat in East Africa as a Staple and Cash Crop

East African countries cultivate almost 2 million ha of wheat, of which only 630,000 ha are farmed with durum wheat (Table 2). Eritrea, Kenya, Somalia and Sudan combined harvested as little as 37,000 ha of durum wheat in 2014. Yet, these countries have maintained in their culinary taste the influence of the past Italian presence in the region, with pasta imports reaching 40 million USD in 2017 in Ethiopia only. In the case of Kenya, national production is sufficient to support the export of €0.5 million worth of pasta and durum grains.

Table 2. Durum wheat varieties currently cultivated in Sub-Saharan Africa.

Country	Variety Name	Adoption	Pedigree	Origin
Ethiopia	'Cocorit71'	Old variety, still cultivated	Enano/4*Tehuacan60//Stewart63/3/Anhinga	CIMMYT
Ethiopia	'Langdon(LD)357'	Old variety, still cultivated	LD308/Nugget	USA
Ethiopia	'Gerardo'	Old variety, still cultivated	GerardoVZ466/3/ND61130/Leeds//Grulla	CIMMYT
Ethiopia	'Ejersa'	Variety utilized by farmers in Oromia	Labud/Nigris3// Gan	CIMMYT
Ethiopia	'Bakalcha'	Widely cultivated variety, now replaced due to susceptibility to stem rust	Gedirfa/Gwerou15	CIMMYT
Ethiopia	'Ude'	Variety that replaced Bakalcha in most zones	Chen/Altar// Jori69	CIMMYT
Ethiopia	'Mangudo'	Covers several districts in Oromia	Omruf1/Stojocri2/3/1718/BeadWheat24//Karim	ICARDA
Ethiopia	'Asasa'	Low moisture area in Rift Valley	Cho/Taurus//Yav/3/Fg/4/Cra/5/Fg/Dom/6/Hui	national
Ethiopia	'Utuba'	New favorite by farmers, cultivated already on 10,000 ha	Omruf1/Stojocri2/3/1718/BeadWheat24//Karim	ICARDA
Ethiopia	'Sinana1'	18,000 ha	Emmer selection from landraces	national
Ethiopia	'Lemesso'	18,000 ha	Emmer selection from landraces	national
Mauritania	'Karim'	Cultivated by farmers along the Senegal river and in oasis	Jori/Anhinga//Flamingo	CIMMYT
Mauritania, Senegal	'Haby'	New release under fats-track multiplication	Mrb5/T.dico Aleppo Col//Cham1	ICARDA
Mauritania, Senegal	'Elwaha'	New release under fats-track multiplication	Oslks/5/Azn/4/BezHF/3/SD19539//Cham1/Gdr2	ICARDA
Mauritania, Senegal	'Bani Suef 5'	New release under fats-track multiplication	Dipperz/Bushen3	CIMMYT
Senegal	'Amina'	New release under fats-track multiplication	Korifla/AegSpeltoidesSyr//Loukos	ICARDA
Mali	'Biskri-Bouteille'	Old variety, still cultivated. Only available recorded release	Biskri/Bouteille	national
South Africa	'Kronos'	Most cultivated variety	APB MSFRS pop selection	USA
Kenya	'Mwewe'	Old variety, still cultivated	Flamingo/Leads	CIMMYT
Sudan	'Sham1'	Old variety, still cultivated	Plc/Ruff// Gta/Rtte	CIMMYT ICARDA
Eritrea	'Mindum XA10'	Old variety, still cultivated	Mindum/Asmara10	USA
Nigeria	'Anser8'	Holds potential for adoption	Altar84/Alondra//Sula	CIMMYT

The durum varieties used for production are old bred-lines from Centro Internacional de Mejoramiento de Maíz y Trigo (CIMMYT) and International Centre for Agricultural Research in the Dry Areas (ICARDA) such as 'Mwewe' (Flamingo/Leads), Mindum XA10 (Mindum/Asmara 10) and Sham 1 (Plc/Ruff//Gta/Rtte), in Eritrea, Kenya and Sudan, respectively (Table 2). The most critical traits of these varieties are earliness and tolerance to heat in irrigated Sudan and resistance to rust diseases under rainfed cultivation in Eritrea and Kenya. Information from Somalia is scarce and hard to obtain. Considering that the most cultivated durum varieties listed above are more than 30 years old, there is a significant genetic yield gap that could be filled through the release and commercialization of more modern varieties.

The Ethiopian case is presented in some detail, including critical historical steps, as it provides valuable lessons for other SSA countries planning to grow their durum wheat sector. In Ethiopia, durum wheat is produced predominantly in the Gojam, Gonder, Shewa, Tigrai and Wollo regions [52]. The main growers are smallholder farmers in the highlands, where the environmental characteristics are relatively low temperatures and high rainfall on black swelling/shrinking vertisol soils, with water logging as a common problem. The crop is planted late in the growing season to avoid early water logging and it continues to grow during the dry period on residual moisture at altitudes between 1800 and 2800 m.a.s.l. [44]. Due to late planting, it forfeits some of its additional potential yield in favour of higher protein content. The crop is consumed in several different forms such as unleavened breads, pancakes, macaroni, spaghetti, biscuits and pastries. The most common of the Ethiopian and Eritrean recipes include *dabo* (Ethiopian home-made bread), *hambasha* (bread from northern Ethiopia), *kitta* (unleavened bread), *injera* (thin bread normally made with *Tef*), *nifro* (boiled whole grains), *kolo* (roasted whole grains), *dabo kolo* (round and seasoned dough) and *kinche* (crushed kernels, cooked with milk or water and mixed with spiced butter). Besides the role of grain in traditional food and processed products, durum wheat straw is also greatly appreciated for its high palatability for livestock in the mixed farming systems of the highlands of Ethiopia [53]. Ethiopia today cultivates 562,000 ha of durum wheat [12], accounting for the vast majority of the cultivation of this crop in SSA (Table 2). Still, today's value represents just half of the land that was dedicated to durum wheat in 1967 [54] and this reduction continues in favour of more extensive farming of bread wheat [23]. This is the combined result of political will, the introduction of modern bread wheat cultivars that have replaced the traditional durum wheat landraces and the absence until now of vocal local industry demand of high quality pasta made from durum semolina. Ethiopia's push toward bread self-sufficiency has resulted in a monoculture of bread wheat (as well as maize), often cultivated in both the long (*meher*) and short (*belg*) rainy seasons, which in turn created a favourable environment of continuous host presence for the spread of damaging rust diseases and for the surge of tenacious weeds [55,56]. Tef, the largest cultivated crop in Ethiopia, also contributes to an expansion of monoculture in Ethiopian agriculture.

Durum wheat research in Ethiopia started back in 1949 at the Paradiso Experimental Station near Asmara [57]. Among several local durum landrace collections tested for productivity and stem and leaf rust resistance, four selections (A10, H23, P20 and R18) were developed and released to farmers in Eritrea in 1952. In 1956 and 1957, several crosses were made between local and exotic varieties mainly for the purpose of transferring the stem and leaf rust resistance of A10 and R18 to cv. 'Mindum' from the USA (Table 2). This resulted in two new varieties, which unfortunately had to be rapidly retracted due to susceptibility to new leaf rust races [58]. In the 1980s, the wheat research activities at the Paradiso station were discontinued and durum wheat breeding was transferred to the Debre Zeit Agricultural Research Centre [42]. At the Centre, many cultivars were developed and released, derived from landrace selections, local crosses and introductions from the international durum wheat breeding programs at CIMMYT and ICARDA. For clarity, in this review the word 'cultivar' has been used to define germplasm cultivated on large amounts of land, while the word 'variety' is reserved to define germplasm officially registered in the variety catalogue of one country. The first durum cultivars released from local breeding selections were 'Arendeto' (DZ04-118) and 'Marou' (DZ04-688), obtained by mass selection [59]. These were followed by the varieties 'Cocorit-71', 'LD-357' and 'Gerardo' obtained from the international agricultural research centres. Since 1982, a formal variety release system has been put in place, which also rationalized the previous work into a variety catalogue, which accounts today for 40 durum wheat cultivars (Figure 1). In the last two decades, many federal and regional agricultural research centres have become involved in durum wheat improvement to respond to the demand by 300 local flour and pasta manufacturers as well as the local consumers. This push by the national food industry, combined with a stronger presence in the region of international development agencies involved in breeding against the emerging Ug99 stem rust race threat [60], has resulted in an increase in the release of durum cultivars, with 20 varieties inscribed in the last 10 years [61]. These new varieties are more responsive to chemical inputs, resistant to diseases and can reach average yields of 4–5 t ha^{-1} under rainfed conditions [62]. 'Utuba' was released in 2015 as an alternative variety to 'Mangudo' and 'Mukuye' because of its amber seeds, high protein content and high yield potential. The grain yield performance on research station ranged from 3.4 to 6.5 t ha^{-1} and from 2.5 to 4.5 t ha^{-1} in farmers' fields [63]. 'Utuba' (Omruf1/Stojocri2/3/1718/BeadWheat24//Karim) takes on average 62 days to flower, 108 days to mature and it is also appreciated for its height (80–90 cm), which ensures good amount of straw for the livestock. A survey conducted by ICARDA has indicated that farmers that abandon the widely-grown durum cultivar 'Ude' (Chen/Altar//Jori69) to grow the recent release 'Utuba' (Table 2), obtain an average yield gain of 32% and an equivalent monetary return. Regardless of this clear advantage, adoption by farmers remains very low [64], mainly because of the high cost of purchasing quality seeds, scarce access to agriculture micro-credits and a national seed system incapable of reaching the more remote areas [65]. To solve some of these issues, international agricultural research centres and development agencies together with the national agricultural research institutes have launched a project to develop informal "Community Based Seed Enterprises" [66–69]. This informal system promotes farmers' aggregation around the possibility to gain access to improved seeds from their neighbours. Lead farmers are designated and provided free-of-charge with certified seeds of improved varieties. These leaders are then responsible for multiplying the seeds and providing them to their neighbours for a reasonable price agreed among each other, often involving exchange of livestock, land rental or payments after harvest. A significant effort has been made to expand the production of improved durum wheat cultivars to supply raw materials to the food industries. For example, in the 2018–2019 cropping season, the Bale Zone Bureau of Agriculture up-scaled the cultivation area of two durum wheat cultivars (cvs. Utuba and Mangudo) in nine districts covering over 6244 ha. In the north Shoa-Amhara region, the Africa RISING project, in partnership with the North Shoa Zone Bureau of Agriculture, expanded the area under these two cultivars to over 700 ha. This fast adoption pace is due to the national and international effort of promoting the new varieties but also the great farmers appreciation. Further, the recent contractual agreement between Minjar farmers and the ALVIMA pasta processing factory is predicted to provide an additional push to its adoption.

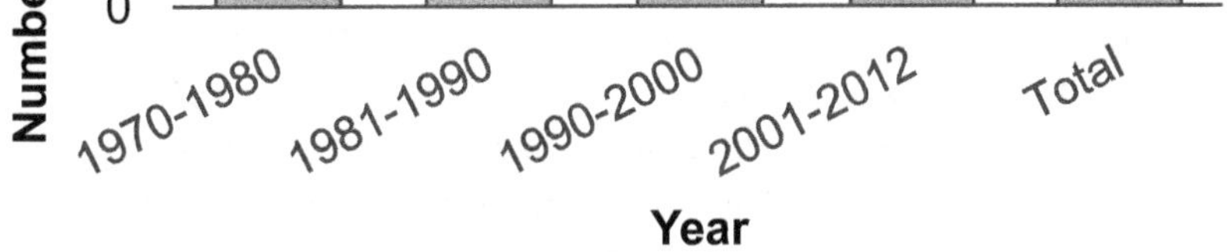

Figure 1. Durum variety releases in Ethiopia since 1970–2012.

Until today, Ethiopia still cultivates emmer wheat, the ancestor of durum wheat. Its cultivation is mainly restricted to marginal areas by about 300,000 households, covering 36,000 ha with an average productivity of 1.7 t ha^{-1} as recorded during the 2013–2014 season [70,71]. This area also continues to be drastically reduced due to expansion of modern bread wheat cultivars. Improvement of emmer wheat is given little attention and only two cultivars ('Sinana-1' and 'Lemesso') have been released through selection from landraces [61]. This crop is mainly used for the preparation of local food products such as *defo* or *dabo* (bread), *injera* (flat pancake bread), porridge, *kita* (flat steamed bread),

Kinche (boiled coarse grain) and local drinks [72]. Emmer wheat is recommended for mothers as a special diet to maintaining their health and strength after childbirth because of its high protein content and digestibility [73]. In fact, its grain protein content ranges from 8.5 to 21.5%, which is 5–35% higher than in grain from oats or barley and it has a very low glycaemic index [74]. Emmer wheat is also a good source of resistance to leaf and stem rusts, powdery mildew, *Septoria* glume blotch, *Fusarium* head blight, Russian Wheat Aphid, in order of importance and tolerance to drought and heat [75–79].

4. Durum Wheat Value Chain in Oromia Region, Ethiopia

Recent investments in the pasta industry are proving extremely promising in Ethiopia thanks to new food habits of the growing urban populations, which are looking for fast and tasty foods, while still cheap and nutritious. Pasta has represented a ready-to-use option since its first introduction in Ethiopia in 1938 by the pioneering Italian enterprise Colonalpi (currently called Kaliti Food Share Company), later followed by the establishment of state-owned industries. Today the state industries have been privatized and grouped, together with numerous others, as members of the Ethiopian Millers Association. These pasta producers used to rely on massive importation of durum wheat grains, which was not a sustainable long-term business strategy due to high and volatile costs. Further, the purchase of foreign grains competed with other national priorities for the use of governmental hard currency stocks. Indeed, the revamping of national durum wheat production has caused the reduction of imports to negligible amounts in 2015 [80], after having equalled €129 million in 2013 (Table 1). However, at the same time, pasta import increased two-fold between 2011 and 2015, when it reached 50,000 t at a cost of about €40 million [80]. To reverse this trend, the Ethiopian Millers Association has eagerly explored the possibility to procure the needed raw material directly from local farmers to reduce production costs and increase competitiveness against foreign pasta imports. Unfortunately, the local production did not guarantee sufficient rheological grain quality to satisfy the industrial needs. In fact, grain of tetraploid landraces does not meet industrial standards in terms of colour or protein quality, while the high-yielding modern varieties tend to produce bleached and 'chalky' grains when grown on waterlogged vertisols in the absence of abundant nitrogen fertilization [81]. Hence, specific incentives needed to be provided to farmers to obtain industrial-grade harvests. The scope of the Ethiopian-Italian cooperation project for the Agricultural Value Chain in Oromia (AVCPO) was to re-direct some of the already existing bread wheat production system of the Bale zone toward the more lucrative farming of durum wheat for the industry. The process acted on the key elements required by the pasta industry to stabilize and self-sustain the value chain: competitive price, high rheological quality for conversion into pasta, easy and timely delivery, consistent stock of grains and predictable increases over years (Figure 2). Launched in April 2011, the initial steps relied on just two durum varieties (Table 2), identified as highly productive, resistant to prevailing diseases in the Bale zone and with good gluten strength: 'Ejersa' (Labud/Nigris3//Gan) and 'Bakalcha' (980SN Gedirfa/Gwerou15). A total of 40 t of certified seed were purchased from the Sinana Agricultural Research Centre (SARC). The dialogue with the pasta industries resulted in the signing of an innovative supply contract that set the purchase value to the prevailing bread wheat price, with the addition of a 'premium' strictly proportional to kernel protein content. This contract provided the needed incentive to farmers for the application of adequate fertilization strategies and has ensured high grain quality since. Furthermore, to supply the industry with large and uniform stocks of grains, AVCPO promoted farmer aggregation into 15 cooperatives and four unions and provided each with warehouses for temporary storage of grain. To measure the required quality, AVCPO equipped the SARC durum quality laboratory and trained researchers and technicians. Small-holder farmers cultivating around 0.5 to 2.0 ha of land were able to deliver their small sales to the warehouses and from there the industry could purchase large bulked stocks, as needed. Technical assistance to farmers and needed continuous research efforts were delivered by regional research and development institutions both from central and district-commune branches.

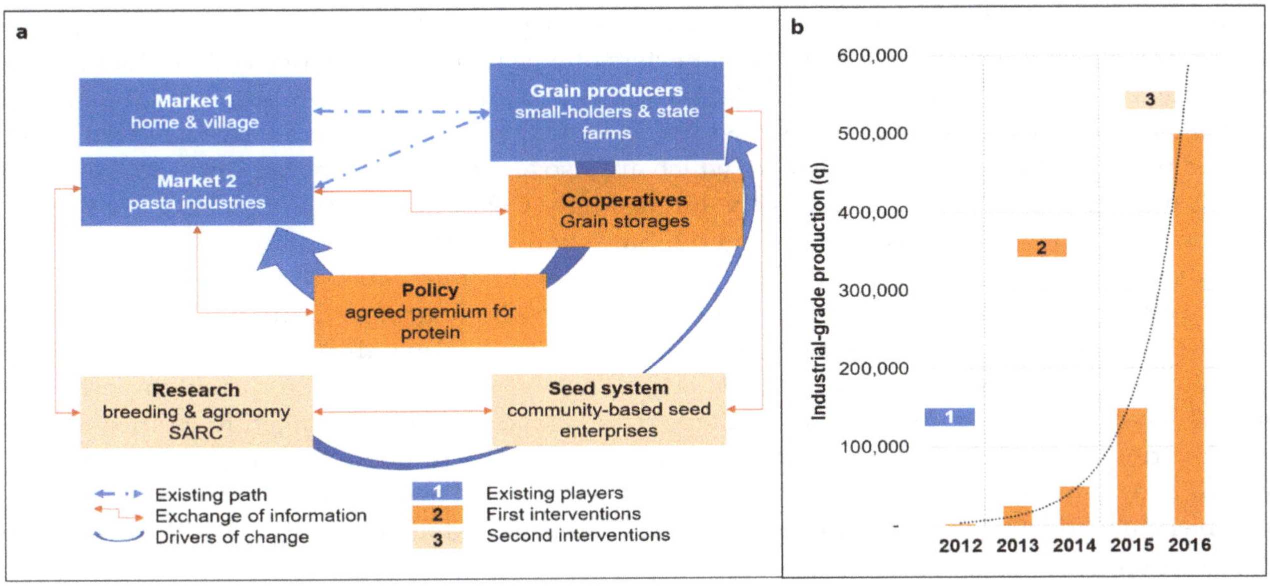

Figure 2. Durum wheat value chain in Oromia, Ethiopia. (**a**) Schematic of the intervention and value chain key actor relationships: to promote increased income in rural areas, the industrial requirements (market 2) were recorded and used to first promote contractual agreements for the sale of industrial-grade seeds and to assemble the farmers into cooperatives around grain storages and second to drive the research agenda with the release of superior cultivars and their multiplication via community-based seed enterprises; (**b**) Success indicator measured as the amount of durum grain sold to the food industry from the Oromia region since the inception of the project. SARC refers to the Sinana Agricultural Research Centre.

Highly innovative contractual relationships were created among farmer cooperatives and industries, pushing the surrounding authorities and public institutions to provide support and surveillance on proper accomplishment of duties. Among these, SARC formally acted as neutral third party for measuring the protein content and determining the final price. The emphasis on the highest level of participation and ownership by all involved stakeholders was considered as the key element for the success and sustainability of the development process [82]. Since the first harvest, durum wheat provided to farmers a significant monetary gain per ha of 25 to 30% over concurrent bread wheat and the industries were greatly satisfied with good rheological quality and reduced prices over imports. The availability of seed stocks of the two selected varieties enabled for prompt expansion of area planted through newly adopting farmers and cooperatives. Over time, the self-sufficient nature of the AVCPO's complex of cooperatives and institutions has created the premises for a vibrant market-oriented community eager to absorb and valorise new varieties and technologies developed by their research partners. Especially in the current situation of evolving rust races dramatically affecting bread wheat in the Bale and other wheat belts, farmers attribute to durum wheat the role of a rescue crop. By the convergence of all these factors, durum wheat production has exponentially increased from 500 t in 2011–2012, to a record harvest of 4.6 million t in 2017 due mainly to the 'Utuba' recent release and cultivation on large scale. 'Utuba' was christened and released as Ethiopian durum wheat variety in 2015 [83]. In the meantime, the value chain is already expanding to nearby Arsi and Shewa zones. The example of Oromia can be considered a successful approach on integration of the whole durum wheat value chain [84], with a proven rapid and sustainable impact. Hence, it provides a good example to follow for other SSA countries that rely today on durum wheat and pasta imports.

5. Durum Wheat in West Africa as a Future Cash Crop

West African countries cultivate over 7 million ha of irrigated rice but only 100,000 ha of wheat (mainly bread wheat) and mostly in Nigeria. A recent steep increase of wheat area has been reported for Nigeria but these data are not yet available from FAO statistics, the main source used for compiling Table 1. Still, all West African countries are importers of wheat grain and its derived products. A total of €155 million worth of pasta and €193 million worth of durum grains were imported in 2013 (Table 1). Benin is the largest importer of pasta in West Africa with almost €51 million worth imported in 2013, followed by Niger, Burkina Faso and Togo, which are also among the largest importers in Africa with €20, €22 and €27 million worth, respectively. Interestingly, €87 million worth of pasta are re-exported each year, mostly by Côte d' Ivoire and Nigeria. Since national durum production is close to zero, it means that large quantities of durum wheat grain are imported internationally, converted by the local industry into pasta products and then sold locally and to neighbouring countries. Hence, as was the case for the Oromia region in Ethiopia, there is potential for national durum cultivation to support this strong local industry, while sharing the €180 million worth per year of the current import market with the local growers. In Nigeria, initial steps have already been undertaken to identify suitable durum varieties at the Kadawa Kano field station. Here, 12 candidate varieties from CIMMYT's breeding program were assessed over two seasons. Trials revealed that grain yields exceeding 6.2 t ha^{-1} could be achieved in 100 days by the top performer 'Anser8' (Altar84/Alondra//Sula) under gravity irrigation [85].

Mauritania is the largest importer of durum grain in West Africa with over €51 million spent every year. This country has one of the most challenging agro-environments in West Africa, with farming substantially restricted to the narrow band along the Senegal River, where rainfall of up to 600 mm per year and irrigation water from the river sustain crop production (Agriculture in Mauritania, 2009). The Senegal River basin has a potential of irrigating 135,000 ha [86], of which less than 20% are currently utilized. The main crops are rice, pearl millet and cowpea. Wheat cultivation along the river is estimated at 8200 ha, of which approximately 5000 ha are grown with durum wheat. The only cultivated durum variety is 'Karim' (syn: 'Yavaros79', Jori/Anhinga//Flamingo), a widely adapted +35 years old CIMMYT-derived variety. Wheat is cultivated during the winter season in rotation with rice and cowpea under gravity irrigation. The window for growing wheat is rather narrow to avoid interfering with the cultivation of the two seasons of rice. Sowing has to occur between the end of November and the middle of December. The harvest is just 80 to 100 days later in early March. Regardless of this short season, two recent projects carried on at the experimental stations of Daara and Kaedi (U-Forsk2013 and SARD-SC) have revealed that yields of 3 t ha^{-1} could be reached along the Senegal River Valley. In response to these results, three new durum wheat varieties ('Haby' [Mrb5/T.dico Aleppo Col//Cham1], 'Elwaha' [Oslks/5/Azn/4/BezHF/3/SD19539//Cham1/Gdr2] and 'Bezater' [Ossl1/Stj5/5/Bicrederaa1/4/BEZAIZSHF//SD19539/Waha/3/Stj/Mrb3/6/Stj3//Bcr/Lks4/3/Ter3]) were released in 2016 (Table 2) and their seed multiplication has begun [87]. On the opposite shore of the river, the field station of Fanaye in Senegal obtained yields as high as 6 t ha^{-1}, when early planting towards the end of November was achieved. The irrigable agricultural land of Senegal is divided along three rivers, (in order of importance): Senegal, Faleme and Casamance, thus providing a total estimated irrigable land of 350,000 ha [88]. The Senegal River valley alone accounts for 240,000 ha of potential arable land [89], of which 110,000 ha are currently used for rice cultivation. Since 2017 small-holder farmers started growing improved heat tolerant durum wheat varieties after completion of the rice harvest, during the winter season, which is typically left at fallow. This research achievement, if sustained by polices and market demand, could help replace the €46 million worth of annual durum import by the national pasta industry. Furthermore, if the total rice area was to be converted to durum wheat instead of the fallow period, then this would be sufficient to generate an overproduction of durum grains to be exported to neighbouring countries for an interesting price. Just as the wheat-rice rotation system has been the cornerstone of India's food self-sufficiency with over 10 million ha still cultivated today [90], it can also become a new boost for the West African agriculture. In addition, the integration of a legume crop in the rotation with durum wheat and rice would be desirable to also increase long-term soil health and agro-ecosystem stability. In this regard, a suggestion is made to replace one rice season with cowpea, an excellent source of food and feed, with very high market value. The cropping model suggested would then become rice-durum wheat-cowpea. This expansion into considering a pulse such as cowpea as part of the durum wheat production system is, however, beyond the scope of this review and shall not be discussed further.

A third country relying on the Senegal River for irrigation is Mali, whose production is concentrated along this and the Niger River. The total irrigable land is estimated at 340,000 ha [91] with a potential to further expand. The vast majority of the land is utilized for the production of rice and maize during the warm months. Wheat is cultivated during winter on just 10,000 ha, of which a very small portion is durum wheat (Table 1). The old variety 'Biskri-Bouteille' (Biskri/Bouteille) is the only reported release for Mali [92]. It is likely that the breeding activities and import from neighbouring countries have resulted in more modern releases but no document could be located. Similar to its neighbours, Mali imports large quantities of pasta (€14 million) and part of it is further exported (€0.2 million). Hence, local production of durum wheat is a viable option for all three countries along the Senegal River. Their total area currently cultivated with rice reaches 754,000 ha. Assuming the same conditions apply to the whole surface, cultivation of durum wheat during the short winter fallow season has the potential to generate additional food, without reducing the current production of their main staple food. The newly identified super-early and heat tolerant durum varieties released in Mauritania and Senegal ('Haby', 'Elwaha', 'Bezater' and 'Amina': Korifla/AegSpeltoidesSyr//Loukos) can provide good industrial grain for the national industry and hold the potential to generate more than 1 million t of additional food in Sub-Saharan Africa [87].

The situation in Nigeria is no different than that observed for the Senegal River countries, even though, with over 80,000 ha farmed to wheat in 2013, it is already the largest bread wheat producer in West Africa (Table 1). A recent push by the Nigerian government, such as the removal of subsidies for the imported grains, has incentivized farmers to increase their wheat production and 2018/19 area harvested and production are estimated at 60,000 hectares and 60,000 tons, respectively [93]. Wheat is typically planted in November or December and harvested around April. The land used for wheat production is then rotated for other rainfed crops during the rainy season, which lasts in northern Nigeria from April to September. Rice is sometimes grown after wheat. The amount of land occupied by durum wheat is not declared in any of the available documents. Certainly, Nigeria imports €38 million per year of durum wheat grain to be converted into pasta for the national and export market (€41 million worth). Hence, the local industry could certainly benefit from an increase in national production. Considering that the area cultivated with rice exceeds 2.9 million ha and that irrigation water is readily available in many parts of the countries, it certainly suggests great potential for expansion.

Similarly, Guinea is a large importer of durum grain (for €17 million EUR) but none is currently produced on the 1.6 million ha of rice cultivation. Côte d'Ivoire is the largest exporter of pasta (€32 million worth per year) but also one of the largest importers of durum grain (€28 million worth), with no production of wheat recorded on the 790,000 ha of rice cultivation (Table 1).

In summary, West African countries have the potential to convert their off-season of their 7.2 million ha of rice fields into durum wheat cultivation, instead of having an unproductive winter fallow. New, super-early and heat tolerant varieties have been developed, tested and confirmed along the Senegal River [87] and their seed is readily available through the CGIAR WHEAT program. Their cultivation could turn an annual import market of €185 million worth of grain and almost €200 million worth of pasta into a national income to improve industrialization, create jobs and reduce poverty in rural areas.

6. Southern and Central Africa Durum Wheat Use in the Industry with Limited Cultivation

Southern and Central African countries cultivate 1.6 million ha of rice and 0.65 million ha of wheat. Unfortunately, data on wheat cultivation in Central Africa are few and unsubstantial. Among Southern African countries, durum wheat is cultivated on just 26,500 ha, mostly in South Africa and Zimbabwe. The most widely cultivated varieties are the 'Desert' durum developed in Arizona and California, with "Kronos" (Arizona Plant Breeders male sterile-facilitated recurrent selection population selection) as the preferred one (Table 2). All countries obtain yields above 4 t ha^{-1}, which only partially meets the national industry demand. Still, part of the grains is exported for generating an income of €38 million.

All countries combined imported €160 million worth of durum grain in 2013 (Table 1). The largest importers of grain were Malawi, South Africa and Zimbabwe, which use it to sustain their national pasta industry. In fact, South Africa utilizes the grain to generate pasta for re-export with a value addition of over €11 million, while Cameroon reaches €1.4 million of pasta exports annually. Interestingly, some SSA countries do not apply import taxes on durum wheat, which in turn has promoted cases of illegal false labelling of bread wheat grain as durum wheat to avoid custom costs [25]. The import of pasta products in 2013 was €108 million worth and the biggest importers were South Africa and Madagascar, with €31 and 26 million worth, respectively. Therefore, a business opportunity exists for the local pasta industry, while creating the chance for growers to improve their livelihoods. Considering an average price per ton of durum wheat grain of €300 on local markets and attainable yields of 3 ton ha^{-1}, approximately 160,000 ha of the currently cultivated 650,000 ha of bread wheat would need to be converted to fill the production gap. Obviously, the reduction of bread wheat would in turn open a gap in the availability of national bread flour, pushing the country to further imports. However, import prices of bread wheat flour is significantly cheaper than durum wheat imports, especially when considering that durum wheat production is a trade that does not require government subsidies to be profitable. Hence, the national economy would overall benefit from a production shift toward durum wheat, as long as this does not upset the higher price paid for semolina. Furthermore, durum-bread wheat flour blends are commonly used in North Africa for the baking of affordable and protein-rich breads.

A second consideration is in regard to the spread of diseases. In fact, South Africa has been monitoring a growing threat of Karnal Bunt disease [94], while Uganda is the first country where the devastating stem rust race Ug99 was observed, before it spread to the neighbouring countries [95]. Both of these diseases affect prevalently bread wheat, while durum wheat has thus far remained resistant [96,97]. Hence, replacement of bread wheat by durum wheat would not only have a potential valuable effect on the economy but also reduce the incidence of damaging diseases on the wheat crop. Alternatively, durum wheat could be cultivated on part of the 1.8 million ha dedicated to rice during the fallow off-season period, assuming that adequate rainfall or irrigation water is available. This could be the case for Madagascar, where durum wheat could be cultivated during the off-season in the same terrace fields grown with paddy rice [98]. In fact, a recent study on wheat suitability in SSA [23] using geospatial analysis revealed that Angola, Mozambique, Zambia and Zimbabwe are the countries with the largest potential extension of suitable land for establishing wheat production. The suitable mega environments identified were highlands with high rainfall and frequent diseases (ME2A [99]) and drought prone rainfall with cold winter months (ME4A).

7. Durum Wheat Cultivation in the Saharan Oases: A Staple Food of Tradition

The Sahara oases are unique environments that remained impervious to modernization. In this review, both types of oases are considered; those areas of desert where water surfaces from the soil or where it can be collected by human activities through dams (*barrage*) or other methods as defined by Zaharieva et al. [100]. Semi-nomadic tribes live in these locations and developed self-sustaining agricultural systems based on the sporadic rainfalls and underground or aboveground water accumulations. Several major oases can be found in SSA in Chad, Mali, Mauritania, Niger and Sudan but also in Algeria, Egypt, Libya, Morocco and Tunisia. There are no extensive records of the total area cultivated. The Saharan oases are estimated at a total surface of 900,000 ha, of which approximately half is used for intensive agriculture [100]. Further, average sizes for oases are between 5 and 200 ha of cultivated land, depending on the abundance of yearly rainfall or available percolated water and can sustain the life of up to 1000 people per oasis [101]. In Mauritania, 350 oases account for a total surface of cultivation by wheat (bread and durum) of over 2000 ha [102]. Roughly the same area is cultivated in the oases of Mali [103], while the five largest oases in Algeria (Ghardaia region) account for 2200 ha of cereal culture [104] and in Morocco the major oasis region of Errachidia cultivates an approximate 5000 ha of cereals [105]. Cultivated crops include sorghum and millet as rainfed crops, both types of wheat and cowpea as irrigated crops. Larger oases have access to a constant water supply, allowing irrigation by pivot or drip irrigation, such as in the regions of east of Morocco and Algeria [106]. In these cases, water is pumped as needed and wheat is often cultivated among date palms with the moisture used by both cultures. In most other cases, large quantities of water are available only during specific times of the year and to collect it in sufficient amounts for cultivation, it is necessary to build temporary dams with clay, sand and stones. The dam is then opened at the beginning of the winter and as the water recedes, holes are dug into the mud and cereal grains are placed inside (Figure 3). Growing on residual moisture and with high temperatures, the yields rarely exceed 0.5 t ha^{-1}, while under pumped irrigation yields of 4 to 5 t ha^{-1} are common [107].

Figure 3. Wheat cultivation in oasis in Mauritania. (**a**) Holes in the mud for the planting of durum wheat as the water retreats; (**b**) Gradient on plant maturity caused by the difference in planting time following the retreat of the water.

The farmers of the desert cultivate mostly wheat biotypes of unique morphology defined as *oasiensis* types, which represent mixtures of several tetraploid and hexaploid wheat species (for review see Reference [108]). Durum wheat cultivation in the oases dates to the initial trade routes between the Nile Valley and West Africa [100]. Several traditional dishes are made from this crop and its straw is very important as feed for the small ruminants and camels. The 'Alkama Binka' is one of the most frequently found landraces in the Saharan oases of Algeria and Morocco [109]. Modern cultivars have also been introduced, such as 'Waha' (syn. 'Cham1', Plc/Ruff//Gta/Rtte) in Algeria and 'Karim' in Mauritania and their superior yields are causing a contraction in the use of landraces (Table 2). The wealth of genetic diversity of germplasm from the Saharan oases has been recognized by several authors and several calls for better collection and conservation have been made but with limited success [108]. In consideration of the harsh environment where these landraces thrive and the fact that durum production will be increasingly stressed due to climate issues [110], they certainly represent a valuable resource of useful alleles for heat, drought and salinity tolerance, which can be deployed in breeding for stress adaptation. Furthermore, the oases represent fragile ecosystems, where land availability is dependent on rainfall and maximum yields per unit of land are more critical than anywhere else. In that sense, the introduction of modern agronomy and irrigation practices, in integration with targeted breeding efforts could deliver true game changers. Alternatively, the reduced available land surface could be used as an advantage to generate very exclusive durum products. In fact, the 'rarity' could be exploited through well integrated value chains to deliver products at elevated prices on the occidental markets, as is already the case for the oases dates. Considering that oases produce less than 5% of their needs in cereals [104] and the rest is purchased from neighbouring towns, the possibility of generating larger incomes would be a suitable strategy to tackle famine. In that sense, the already high value of durum grains could be further exploited via smart-marketing to increase the revenues.

8. Future Prospects: A South-South Collaboration to Expand Durum Wheat Cultivation in Africa

All of Africa accounts for an annual import of €4.1 billion worth of durum grain to supply the national pasta and couscous market. These are mostly imported to North Africa (NA) from Canada, USA and Turkey (Table 1). North Africa already cultivates durum wheat on 2.9 million ha and the area for further expansion is limited. This opens an opportunity for SSA to gain access to an €3.7 billion annual market by filling part of the grain needs of NA. The current area dedicated to wheat cultivation in SSA is limited to 2.6 million ha, mostly in Ethiopia, South Africa and Sudan. In Ethiopia, new interest has sprung toward the promotion of industrial crops such as durum wheat to provide the local manufacturers with prime raw material without the need of relying on expensive imports. In addition, urbanization has shifted the food habits of many countries and pasta has gained steadily in appreciation by African consumers. Furthermore, the case presented for cultivation of durum wheat in rotation with rice along the Senegal River, matches what is already customary on over 10 million ha of wheat-rice or wheat-rice-rice rotations in India [90]. In that sense, there is large potential for wheat expansion on the 9.1 million ha of rice land in SSA. Since further expansion of the wheat areas will require additional investments and will face the risk of reduced yields, it appears logical to seek the wheat type that would provide the maximum monetary return for unit of land converted. Durum wheat in this case would represent an ideal cash crop to help reduce poverty in SSA. For comparison, the average import prices of major cereals to South Africa [111] for the year 2015 were at: US$ 502 t^{-1} aromatic rice, US$ 330 t^{-1} durum wheat, US$ 278 t^{-1} malt barley, US$ 209 t^{-1} hard red bread wheat, US$ 171 t^{-1} sorghum and US$ 150 t^{-1} feed maize. While it is true that import prices change for each country based on access to trade, existence of infrastructure and specific import policies, South Africa provides a good example of a reactive trading nation in SSA. On this basis, it is evident that durum wheat remains one of the most income advantageous winter cereals, significantly more expensive than bread wheat and malt barley. However, to succeed in the utilization of the financial return of this crop, it is necessary to have a well-integrated value chain capable of delivering profitable economic returns to farmers. The example of the value chain in the Oromia region of Ethiopia could be repeated in several other regions and should provide a good guideline to follow for out-scaling to other countries. Still, the industrial machinery and the strategy for production need to be harmonized among African countries to generate a fair and vibrant market. The desire for semolina-based food is expected to increase in the years to come [112] but the national industry will be successful in targeting the demand only if their products can compete not just in price but also in quality with the imported ones. In that sense, great traditional and modern knowledge for cultivation and production of this crop exist already in North Africa and Ethiopia. Breeding programs for this crop have been successful in targeting the harsh drought conditions of North Africa and the disease pressure in Ethiopia. In order to expand the production of this crop to non-traditional territories, the expertise gathered there could be transferred to SSA in the form of novel and adapted varieties. It is therefore desirable that Ethiopian breeders could produce varieties well adapted to the SSA mega-environment of type 2A, with high rainfall and high disease pressure. Instead, Egyptian breeders could help in delivering varieties targeted to the hot and irrigated areas of mega-environment type ME1, such as West Africa and Sudan. The other North African countries could target ME4A, with low rainfall and cold winters, as well as help in the further development of the Saharan oases. Altogether, this envisioned South-South collaboration could ensure that varieties developed in traditional durum growing areas such as North Africa and Ethiopia, would adapt to the conditions of the southern partners. Harvests could then be sold to those African countries with strong pasta industries and the finished semolina products would be sold all over Africa. This integrated value chain would ensure a steep increase in monetary circulation and an overall reduction in the poverty of Africa. Recent publicly funded projects like Africa Rising [113], SARD-SC [114], TAAT [115] and U-Forsk2018 have targeted the increase in production of wheat in SSA and created the basis to hope for a comprehensive "durum wheat revolution" in SSA.

Author Contributions: Writing—original draft preparation, A.T.S., F.M.B., T.C., W.L.; Writing—review and editing, F.M.B., A.T.S., M.v.G., and K.S.; supervision, F.M.B., M.v.G., R.O.; project administration, R.O.; funding acquisition, F.M.B., R.O.

References

1. Chris, G. World Durum Outlook. Available online: http://www.internationalpasta.org/resources/IPO%20BOARD%202013/2%20Chris%20Gillen.pdf (accessed on 7 April 2017).
2. Statistic Canada. Canada: Outlook for Principal Field Crops. Available online: http://www.agr.gc.ca/eng/industry-markets-and-trade/statistics-and-market-information/by-product-sector/crops-industry/outlook-for-principal-field-crops-in-canada/canada-outlook-for-principal-field-crops-february-16-2016/?id=1455720699951 (accessed on 7 April 2017).
3. USDA Foreign Agricultural Service. *Grain and Feed Annual*; GAIN Report No: TR5016; USDA Foreign Agricultural Service: Ankara, Turkey, 2015.
4. Nagarajan, S. Quality characteristics of Indian wheat. In *Future of Flour*; Popper, L., Schäfer, W., Freund, W., Eds.; AgriMedia GmbH: Clenze, Germany, 2006; pp. 79–86.
5. Le Lamer, O.; Rousselin, X. The durum wheat market. In *Studies of FranceAgriMer*; Bova, F., Ed.; FranceAgriMer: Montreuil-sous-Bois CEDEX, France, 2011; pp. 1–46.
6. Bonjean, A.P.; Angus, W.J.; van Ginkel, M. *The World Wheat Book: A History of Wheat Breeding*; Lavoisier: Paris, France, 2016; Volume 3.
7. Al-Issa, T.A.; Samarah, N.H. Tillage practices in wheat production under rainfed conditions in Jordan: An economic comparison. *World J. Agric. Sci.* **2006**, *2*, 322–325.
8. Karam, F.; Kabalan, R.; Breidi, J.; Rouphael, Y.; Oweis, T. Yield and water-production functions of two durum wheat cultivars grown under different irrigation and nitrogen regimes. *Agric. Water Manag.* **2009**, *96*, 603–615. [CrossRef]
9. El-Areed, S.; Nachit, M.M.; Hagaras, A.; El-Sherif, S.; Hamouda, M. Durum wheat breeding for high yield potential in Egypt. In *Proceedings of the International Symposium on Genetics and Breeding of Durum Wheat*; Porceddu, E., Damania, A.B., Qualset, C.O., Eds.; CIHEAM: Bari, Italy, 2014; pp. 291–294.
10. Juarez, B.; Wolf, D. Grain and feed annual Mexico. In *Global Agricultural Information*; USDA Foreign Agricultural Service: Washington, DC, USA, 2015; pp. 1–17.
11. John, K. Durum Wheat Production. Available online: http://www.nvtonline.com.au/wp-content/uploads/2013/03/Crop-Guide-NSW-Durum-Wheat-Production.pdf (accessed on 24 December 2016).
12. Evan School Policy Analysis and Research (EPAR). Wheat Value Chain: Ethiopia. Available online: https://evans.uw.edu/sites/default/files/EPAR_UW_204_Wheat_Ethiopia_07272012.pdf (accessed on 24 December 2016).
13. Elias, E.M. Durum wheat products. In *Durum Wheat Quality in the Mediterranean Region*; Di Fonzo, N., Kaan, F., Nachit, M., Eds.; CIHEAM: Zaragoza, Spain, 1995; Volume 22, pp. 23–31.
14. International Pasta Organisation (IPO). The Truth about Pasta Toolkit. Available online: http://www.internationalpasta.org/index.aspx?id=47 (accessed on 27 December 2016).
15. Simoes, A.J.G.; Hidalgo, C.A. The Economic Complexity Observatory: An analytical tool for understanding the dynamics of economic development. In *Scalable Integration of Analytics and Visualization*; San Francisco, CA, USA, 2011.
16. Food Price Index (FPI). Available online: http://www.fao.org/worldfoodsituation/foodpricesindex/en/ (accessed on 22 December 2016).
17. Food and Agriculture Organization (FAO). *FAO Statistical Yearbook 2014: Africa*; Tijani, B., Gennari, P., Eds.; FAO: Rome, Italy, 2014.

18. Food and Agriculture Organization (FAO). Eritrea: Country Report. In Proceedings of the FAO International Technical Conference on Plant Genetics, Leipzig, Germany, 17–23 June 1996.
19. Bill and Melinda Gates Foundation (BMGF). Multi Crop Value Chain Phase II: Wheat Ethiopia. Available online: https://agriknowledge.org/files/8k71nh141 (accessed on 10 January 2017).
20. What Farming in Kenya. Available online: https://informationcradle.com/kenya/wheat-in-kenya/ (accessed on 15 May 2019).
21. Saunders, D.A.; Hettel, G.P. *Wheat in Heat-Stressed Environments: Irrigated, Dry Areas and Rice-Wheat Farming Systems*; UNDP; ARC; BARI; CIMMYT: Mexico, 1993; ISBN 968-6127-87-9.
22. Mistretta, G. Angola: Alcune caratteristiche del mercato della pasta. Ricadute sul Prodotto Italiano. Ambassador letter on 06 November 2013 to InfoMercatiEsteri.it. Available online: http://www.infomercatiesteri.it/public/images/paesi/4/files/PASTA%20di%20grano.doc (accessed on 15 May 2019).
23. Negassa, A.; Shiferaw, B.; Koo, J.; Sonder, K.; Smale, M.; Braun, H.J.; Gbegbelegbe, S.; Guo, Z.; Hodson, D.; Wood, S.; et al. *The Potential for Wheat Production in Africa: Analysis of Biophysical Suitability and Economic Profitability*; CIMMYT: DF, Mexico, 2013.
24. Ssekitoleko, G.W.; Vogel, W.O. Wheat marketing and pricing policy in Uganda: A critical assessment. In *Proceedings of Theninth Regional Wheat Workshop for Eastern, Central and Southern Africa, Addis Abeba, Ethiopia, 2–6 October 1995*; Tanner, D.G., Payne, T.S., Abdalla, O.S., Eds.; CIMMYT: Mexico City, Mexico, 1995.
25. United States Agency for International Development (USAID). *Staple Foods Value Chain Analysis: Country Report Zambia*; Chemonics International Inc.: Washington, DC, USA, 2009.
26. Kayoed-Anglade, S.; Aaron, R. Projet du Moulin Modern du Mali. Available online: https://www.afdb.org/fileadmin/uploads/afdb/Documents/Environmental-and-Social-Assessments/Mali_-_Projet_du_moulin_moderne_du_Mali_-_R%c3%a9sum%c3%a9_PGES.pdf (accessed on 4 January 2017).
27. Factfish: Research Made Simple. Available online: www.factfish.com (accessed on 15 May 2019).
28. Index Mundi. Available online: www.indexmundi.com (accessed on 15 May 2019).
29. Hakan, O.; Willcox, G.; Graner, A.; Salamini, F.; Kilian, B. Geographic distribution and domestication of wild emmer wheat (*Triticum dicoccoides*). *Genet. Resour. Crop Evol.* **2010**, *58*, 11–53.
30. Benbelkacem, A.; Khaldounl, A.; Djenadi, C. The history of wheat breeding in Algeria. In *World Wheat Book: A History of Wheat Breeding*; Bonjean, A.P., Angus, W.J., van Ginkel, M., Eds.; Lavoisier: Paris, France, 2016; Volume 3, pp. 477–495.
31. National Research Council. *Lost Crops of Africa: Volume I: Grains*; The National Academies Press: Washington, DC, USA, 1996.
32. Luo, M.C.; Yang, Z.L.; You, F.M. The structure of wild and domesticated emmer wheat populations, gene flow between them and the site of emmer domestication. *Theor. Appl. Genet.* **2007**, *114*, 947–959. [CrossRef]
33. Kabbaj, H.; Sall, A.T.; Al-Abdallat, A.; Geleta, M.; Amri, A.; Filali-Maltouf, A.; Belkadi, B.; Ortiz, R.; Bassi, F.M. Genetic Diversity within a Global Panel of Durum Wheat (*Triticum durum*) Landraces and Modern Germplasm Reveals the History of Alleles Exchange. *Front. Plant Sci.* **2017**, *8*, 1277. [CrossRef] [PubMed]
34. Mengistu, D.K.; Kirosa, Y.A.Y.; Pè, M.E. Phenotypic diversity in Ethiopian durum wheat (*Triticum turgidum* var. durum) landraces. *Crop J.* **2015**, *3*, 190–199. [CrossRef]
35. Vavilov, N.I. The origin, variation, immunity and breeding of cultivated plants, trans from the Russian by Starr Chester, K., Waltham, Mass.: Chronica Botanica. *Science* **1951**, *115*, 364.
36. Pecetti, L.; Annicchiarico, P.; Damania, A.B. Biodiversity in a germplasm collection of durum wheat. *Euphytica* **1992**, *60*, 229–238.
37. Ficco, D.B.M.; Mastrangelo, A.M.; Trono, D.; Borrelli, G.M.; De Vita, P.; Fares, C.; Beleggia, R.; Platani, C.; Papa, R. The colors of durum wheat: A review. *Crop Pasture Sci.* **2014**, *65*, 1–15. [CrossRef]
38. Mass, F.B.; Paterson, F.L.; Foster, J.E.; Ohm, H.W. Expression and inheritance of resistance of ELS 6404-160 durum wheat to Hessian Fly. *Crop Sci.* **1989**, *29*, 23–28. [CrossRef]
39. Belay, G.; Merker, A.; Tesfaye, T. Cytogenetic studies in Ethiopian landraces of tetraploid wheat (Tritium turgidum L.), spike morphology and ploidy level and karyomorphology. *Hereditas* **1994**, *121*, 45–52. [CrossRef]
40. Bekele, E. Analysis of regional patterns of phenotypic diversity in the Ethiopian tetraploid and hexaploid wheats. *Hereditas* **1984**, *100*, 131–154. [CrossRef]
41. Negassa, M. Estimates of phenotypic diversity and breeding potential of Ethiopian wheats. *Hereditas* **1986**, *104*, 41–48. [CrossRef]

42. Tesemma, T.; Belay, G. Aspects of Ethiopian tetraploid wheats with emphasis on durum wheat genetics and breeding research. In *Wheat Research in Ethiopia: A Historical Perspective*; Hailu, G., Tanner, D.G., Mengistu, H., Eds.; CIMMYT: El Batan, Mexico, 1991; pp. 47–72.
43. Belay, G.; Tesfaye, T.; Becker, H.C.; Merker, A. Variation and interrelationships of agronomic traits in Ethiopian tetraploid wheat landraces. *Euphytica* **1993**, *71*, 181–188. [CrossRef]
44. Bechere, E.; Belay, G.; Mitiku, D.; Merker, A. Phenotypic diversity of tetraploid wheat landraces from northern and north-central regions of Ethiopia. *Hereditas* **1996**, *124*, 165–172. [CrossRef]
45. Tsegaye, S.; Tesemma, T.; Belay, G. Relationships among tetraploid wheat (*Triticum turgidum* L.) landrace populations revealed by isozyme markers and agronomic traits. *Theor. Appl. Genet.* **1996**, *93*, 600–605. [CrossRef]
46. Eticha, F.; Bekele, E.; Belay, G.; Börner, A. Phenotypic diversity in tetraploid wheats collected from Bale and Wello regions of Ethiopia. *Plant Genet. Resour.* **2005**, *3*, 35–43. [CrossRef]
47. Mengistu, D.K.; Kidane, Y.G.; Catellani, M.; Frascaroli, E.; Fadda, C.; Pe, M.E.; Dell'Acqua, M. High-density molecular characterization and association mapping in Ethiopian durum wheat landraces reveals high diversity and potential for wheat breeding. *Plant Biotechnol. J.* **2016**, *14*, 1800–1812. [CrossRef] [PubMed]
48. Mengistu, D.K.; Kidane, Y.G.; Fadda, C.; Pè, M.E. Genetic diversity in Ethiopian Durum Wheat (*Triticum turgidum* var durum) inferred from phenotypic variations. *Plant Genet. Resour.* **2016**, *16*, 39–49. [CrossRef]
49. Tsegaye, B.; Berg, T. Utilization of durum wheat landraces in East Shewa, central Ethiopia: Are home uses an incentive for on-farm conservation? *Agric. Hum. Values* **2007**, *24*, 219–230. [CrossRef]
50. Fassil, K.; Sehaye, Y.T.; McNeilly, T. Diversity of durum wheat (Triticum durum Desf.) at in situ conservation sites in North Shewa and Bale, Ethiopia. *J. Agric. Sci.* **2001**, *136*, 383–392.
51. Israel, O.K.; Bhandari, B.; Guilherme de Azevedo, S.; Natarajan, K.; Gezu, G.; Gashu, S.; Desalegn, T. *Global Study on CBM and Empowerment, Ethiopia Exchange Report*; Wageningen University: Wageningen, The Netherlands, 2010; p. 42.
52. Tesemma, T.; Belay, G.; Worede, M. Morphological diversity in tetraploid wheat landrace populations from the central highlands of Ethiopia. *Hereditas* **1991**, *114*, 171–176.
53. Tolera, A.; Tsegaye, B.; Berg, T. Effects of variety, cropping year, location and fertilizer application on nutritive value of durum wheat straw. *J. Anim. Physiol. Anim. Nutr.* **2007**, *92*, 121–130. [CrossRef]
54. Bezabeh, E.; Haregewoin, T.; Giorgis, D.H.; Daniel, F.; Belay, B. Change and growth rate analysis in area, yield and production of wheat in Ethiopia. *Res. J. Agric. Environ. Manag.* **2015**, *4*, 189–191.
55. Bassi, F.M.; Chiari, T.; Verma, R.P.S.; Kumar, S. Crop Diversification: The 'Old' Weapon in the Hands of Ethiopian Farmers, Borlaug Global Rust Initiative. Available online: http://www.globalrust.org/blog/crop-diversification-%E2%80%98old%E2%80%99-weapon-hands-ethiopian-farmers (accessed on 10 January 2017).
56. Global Milling. Ethiopia Battles Wheat Rust Disease Outbreak in Critical Wheat-Growing Regions. Available online: http://globalmilling.com/ethiopia-battles-wheat-rust-disease-outbreak-in-critical-wheat-growing-regions/ (accessed on 27 December 2016).
57. Tesemma, T.; Mohammed, J. Review of wheat breeding in Ethiopia. *Ethiopian J. Agric. Sci.* **1982**, *1*, 11–24.
58. Nastasi, V. *Wheat Production in Ethiopia*; Information Bulletin on the Near East Wheat and Barley Improvement and Production Project; FAO: Rome, Italy, 1964; Volume 1, pp. 13–24.
59. Tesemma, T. Durum wheat breeding in Ethiopia. In *Fifth Regional Wheat Workshop for Eastern, Central and Southern Africa and the Indian Ocean*; van Ginkel, M., Tanner, D.G., Eds.; CIMMYT: El Batan, Mexico, 1988; pp. 18–22.
60. Consultative Group on International Agricultural Research (CGIAR). Virulent new strains of Ug99 stem rust, a deadly wheat pathogen. *Science Daily*, 28 May 2010.
61. Ministry of Agriculture (MoA). Plant Variety Release, Protection and Seed Quality Control Directorate. *Crop Var. Regist.* **2013**, *16*, 301.
62. Leta, G.; Belay, G.; Walelign, W. Nitrogen fertilization effects on grain quality of durum wheat (*Triticum turgidum L.* var. durum) varieties in central Ethiopia. *Agric. Sci.* **2013**, *4*, 123–130.
63. Mekuria, T.; Wasihun, L.; Shitaye, H.; Asenafi, G. Durum wheat (Triticum durum Desf) Variety "Utuba" Performance in Ethiopia. *Agric. Res. Technol. Open Access J.* **2018**, *18*, 556063. [CrossRef]

64. De Groote, H.; Gitonga, Z.; Mugo, S.; Walker, T.S. Assessing the effectiveness of maize and wheat improvement from the perspectives of varietal output adoption in east and southern Africa. In *Crop Improvement, Adoption and Impact of Improved Varieties in Food Crops in Sub Saharan Africa*; Walker, T.S., Alwang, J., Eds.; CABI: Egham, UK, 2015; pp. 207–227.
65. Dawit, A. *Farmer-Based Seed Multiplication in the Ethiopian Seed System: Approaches, Priorities and Performance*; FAC Working Paper 36; Future Agricultures Consortium: Brighton, UK, 2011.
66. Bishaw, Z. Wheat and Barley Seed Systems in Ethiopia and Syria. Ph.D. Thesis, Wageningen University, Wageningen, The Netherlands, 2004; pp. 1–401.
67. Joshi, A.K.; Azab, M.; Mosaad, M.; Braun, H.J. Delivering rust resistant wheat to farmers: A step towards increased food security. *Euphytica* **2011**, *179*, 187–196. [CrossRef]
68. Ojiewo, C.O.; Kugbei, S.; Bishaw, Z.; Rubyogo, J.C. Community seed production. FAO & ICRISAT: 2015. In *Workshop Proceedings, 9–11 December 2013*; FAO: Rome, Italy; ICRISAT: Addis Ababa, Ethiopia, 2013; 176p.
69. Bishaw, Z.; Niane, A.A. Are farmer-based seed enterprises profitable and sustainable? Experiences of VBSEs from Afghanistan. In Proceedings of the Community Seed Production, Rome, Italy, 9–11 December 2013; FAO: Rome, Italy; ICRISAT: Addis Ababa, Ethiopia, 2013; pp. 55–64.
70. Board on Science and Technology for International Development (BOSTID). *Lost Crops of Africa, Vol. I Grains*; National Academic Press: Washington, DC, USA, 1996.
71. Central Statistical Agency (CSA). *Agricultural Sample Survey 2013/2014, Statistical Bulletin No 536*; Federal Democratic Republic of Ethiopia Central Statistical Agency: Addis Ababa, Ethiopia, 2014; pp. 1–121.
72. D'Andrea, C.A.; Haile, M. Traditional emmer processing in highland Ethiopia. *J. Ethnobiol.* **2002**, *22*, 179–217.
73. Cooper, R. Re-discovering ancient wheat varieties as functional foods. *J. Tradit. Complement. Med.* **2015**, *5*, 138–143. [CrossRef] [PubMed]
74. Stallknecht, G.F.; Gilbertson, K.M.; Romey, J.E. Alternative wheat cereals as good grains: Einkorn, emmer, spelt, kamut and triticale. In *Progress in New Crops*; Janick, J., Ed.; ASHS Press: Alexandria, VA, USA, 1997; pp. 156–170.
75. Negassa, M. Possible new genes for resistance to powdery mildew, septoria glume blotch and leaf rust of wheat. *Plant Breed.* **1987**, *98*, 37–46. [CrossRef]
76. Damania, A.B.; Hakim, S.; Moualla, M.Y. Evaluation of variation in *T. dicoccum* for wheat improvement in stress environment. *Hereditas* **1992**, *116*, 163–166. [CrossRef]
77. Robinson, J.; Skovmand, B. Evaluation of emmer wheat and other Triticeae for resistance to Russian wheat aphid. *Genet. Resour. Crop Evol.* **1992**, *39*, 159–163.
78. Naod, B.; Fininsa, C.; Badebo, A. Sources of stem rust resistance in Ethiopian tetraploid wheat accessions. *Afr. Crop Sci. J.* **2007**, *15*, 51–57.
79. Oliver, R.E.; Cai, X.; Friesen, T.L.; Halley, S.; Stack, R.W.; Xu, S.S. Evaluation of Fusarium head blight resistance in tetraploid wheat (*Triticum turgidum* L.). *Crop Sci.* **2008**, *48*, 213–222. [CrossRef]
80. Ethiopian Revenues and Customs Authority (ERCA). 2014. Available online: http://www.erca.gov.et/index.php (accessed on 5 January 2017).
81. Geleto, T.; Tanner, D.G.; Mamo, T.; Gebeyehu, G. Response of rainfed bread and durum wheat to source, level and timing of nitrogen fertilizer on two Ethiopian vertisols: N uptake, recovery and efficiency. *Fertil. Res.* **1996**, *44*, 195–204. [CrossRef]
82. Biggeri, M.; Ciani, F.; Ferrannini, A. Aid effectiveness and multilevel governance: The case of a value chain development project in rural Ethiopia. *Eur. J. Dev. Res.* **2016**, *29*, 843–865. [CrossRef]
83. Ministry of Agriculture (MOA). Plant Variety Release, Protection and Seed Quality Control Directorate. *Crop Var. Regist.* **2015**, *17*, 330.
84. Biggeri, M.; Burchi, F.; Ciani, F.; Herrmann, R. Linking small-scale farmers to the durum value chain in Ethiopia: Assessing the effects on production and wellbeing. *Food Policy* **2018**, *79*, 77–91. [CrossRef]
85. Falaki, A.M.; Mohammed, I.B. Performance of some durum wheat varieties at Kadawa, Kano state of Nigeria. *Bayero J. Pure Appl. Sci.* **2011**, *4*, 48–51. [CrossRef]
86. Agriculture in Mauritania. Encyclopedia of the Nations—Information about Countries of the World, United Nations and World Leaders. Available online: http://www.nationsencyclopedia.com/Africa/Mauritania-AGRICULTURE.html (accessed on 15 December 2016).

87. Sall, A.T.; Kabbaj, H.; Cisse, M.; Gueye, H.; Ndoye, I.; Maltouf, A.F.; El-Mourid, M.; Ortiz, R.; Bassi, F.M. Heat Tolerance of Durum Wheat (*Tritcum durum* Desf.) Elite Germplasm Tested along the Senegal River. *J. Agric. Sci.* **2018**, *10*, 217–233. [CrossRef]
88. Food and Agriculture Organization (FAO). *Senegal: Irrigation Market Brief*; Report No. 26; FAO: Rome, Italy, January 2016.
89. Manikowski, S.; Strapasson, A. Sustainability Assessment of Large Irrigation Dams in Senegal: A Cost-Benefit Analysis for the Senegal River Valley. *Front. Environ. Sci.* **2016**, *4*, 18. [CrossRef]
90. Mahajan, A.; Gupta, R.D. *Integrated Nutrient Management (INM) in a Sustainable Rice-Wheat Cropping System*; Springer: Heidelberg, Germany, 2009; p. 268.
91. Club du Sahel (CILSS). *Development of Irrigated Agriculture in Upper Volta: Proposals for a Second Programme 1980-85*; Permanent Interstate Committee for Drought Control in the Sahel: Niamey, Republic of Niger, 1979.
92. Chabrolin, R. Programme pour les Recherches Concernant le blé au Mali; Fonds IRD: B22431. 1967.
93. USDA Foreign Agricultural Service. 2018 Grain and Feed Annual Nigeria-USDA GAIN Reports. Available online: https://gain.fas.usda.gov/Recent%20GAIN%20Publications/Grain%20and%20Feed%20Annual_Lagos_Nigeria_4-12-2018.pdf (accessed on 5 March 2019).
94. Stansbury, C.D.; Pretorius, Z.A. Modelling the potential distribution of Karnal bunt of wheat in South Africa. *S. Afr. J. Plant Soil* **2001**, *18*, 159–168. [CrossRef]
95. Singh, R.P.; Hodson, D.P.; Huerta-Espino, J.; Jin, Y.; Bhavani, S.; Njau, P.; Herrera-Foessel, S.; Singh, P.K.; Singh, S.; Govindan, V. The emergence of Ug99 races of the stem rust fungus is a threat to World wheat production. *Annu. Rev. Phytopathol.* **2011**, *49*, 465–481. [CrossRef]
96. Duveiller, E.; Mezzalama, M. *Karnal Bunt: Screening for Resistance and Distributing KB Free Seed*; CIMMYT: El Batan, Mexico, 2009.
97. Letta, T.; Olivera, P.; Maccaferri, M.; Jin, Y.; Ammar, K.; Badebo, A.; Noli, E.; Crossa, J.; Tuberosa, R. Genome-wide search of stem rust resistance loci at the seedling stage in durum wheat. *Plant Genome* **2014**, *7*, 1–13. [CrossRef]
98. Ravelomamtsoa, S.H.; Randrianaivoarivony, J.M.; Ramalanjaona, V.L. Overview of wheat in Madagascar. In Proceedings of the Wheat for Food Security in Africa, Addis-Ababa, Ethiopia, 8–12 October 2012.
99. Rajaram, S.; van Ginkel, M.; Fischer, R.A. CIMMYT's wheat breeding mega-environments (ME). In Proceedings of the 8th International Wheat Genetics Symposium, Beijing, China, 19–24 July 1994.
100. Zaharieva, M.; Bonjean, A.; Monneveux, P. Alert: Saharan Oases wheat genetic resources in danger. In *World Wheat Book: A History of Wheat Breeding*; Bonjean, A.P., Angus, W.J., van Ginkel, M., Eds.; Lavoisier: Paris, France, 2016; Volume 3, pp. 543–588.
101. Saharan Development. Agriculture and Farming. Available online: http://www.sahara-developpement.com/Western-Sahara/AgricultureEtElevage--117.aspx (accessed on 26 December 2016).
102. Food and Agriculture Organization (FAO). La Mauritanie et la FAO: Partenariat pour Améliorer la Résilience et Renforcer la Sécurité Alimentaire et Nutritionnelle. Available online: http://www.fao.org/3/a-au199f.pdf (accessed on 25 December 2016).
103. Food and Agriculture Organization (FAO)-ISESCO. In Proceedings of the International Workshop on Globally Important Agriculture Heritage Systems (GIAHS) for the Islamic Countries, Rome, Italy, 4–5 November 2014.
104. Houichiti, R.; Bissati, S.; Bouammar, B. Oasis agriculture and food insecurity in Algeria: The case of Ghardaia region. *Pensee J.* **2014**, *76*, 1–7.
105. Ait, H. Systemes de production et stratégies des agriculteurs dans les oasis de la region d'Errachidia au Maroc. *New Medit.* **2003**, *2*, 37–43.
106. El Abbass, S. Systèmes d'irrigation dans les oasis de Mauritanie: Problèmes de pompage et tentatives de réalimentation des nappes phréatiques. In Proceedings of the Journées Internationales sur l'Agriculture et la Gastronomie des Oasis, Elche, Spain, 14–15 October 2009.
107. Merouche, A.; Debaeke, P.; Messahel, M.; Kelkouli, M. Response of Durum wheat varieties to water in semi-arid Algeria. *Af. J. Agric. Res.* **2014**, *9*, 2880–2893.
108. Zaharieva, M.; Bonjean, A.; Monneveux, P. Saharan wheats: Before they disappear. Genet. *Resour. Crop Evol.* **2014**, *61*, 1065–1084. [CrossRef]
109. Benlaghlid, M.; Bouattoura, N.; Monneveux, P.; Borries, C. Etude de la diversité génétique et de la physiologie de l'adaptation au milieu. *Options Méditerranéennes* **1990**, *11*, 171–194.

110. Constantinidou, K.; Zittis, G.; Hadjinicolaou, P. Variations in the Simulation of Climate Change Impact Indices due to Different Land Surface Schemes over the Mediterranean, Middle East and Northern Africa. *Atmosphere* **2019**, *10*, 26. [CrossRef]
111. South African Revenue Service. Available online: www.sagis.org.za/sars.html (accessed on 7 February 2017).
112. William, A.R. Market Trends Category Analysis: A Look into the Pasta. Available online: www.preparedfoods.com/articles/103693-market-trends-category-analysis-a-look-into-the-pasta (accessed on 5 January 2017).
113. Karaimu, P. Africa Research in Sustainable Intensification for the Next Generation. Available online: https://africa-rising.net/ (accessed on 15 May 2019).
114. Hauser, J.F. Support to Agricultural Research for Development of Strategic Crops in Africa. Available online: http://sard-sc.org/ (accessed on 15 May 2019).
115. IITA. TAAT Launch Signifies New Day for African Agriculture. Available online: https://www.iita.org/news-item/taat-launch-signifies-new-day-african-agriculture/ (accessed on 15 May 2019).

5

HWA1- and *HWA2*-Mediated Hybrid Weakness in Rice Involves Cell Death, Reactive Oxygen Species Accumulation and Disease Resistance-Related Gene Upregulation

Kumpei Shiragaki [1], Takahiro Iizuka [1], Katsuyuki Ichitani [2], Tsutomu Kuboyama [3], Toshinobu Morikawa [1], Masayuki Oda [1] and Takahiro Tezuka [1,4,*]

1 Graduate School of Life and Environmental Sciences, Osaka Prefecture University, 1-1 Gakuen-cho, Nakaku, Sakai, Osaka 599-8531, Japan; ma201027@edu.osakafu-u.ac.jp (K.S.); ddlfte@gmail.com (T.I.); d-morikawa@hannan-u.ac.jp (T.M.); masa-oda@d.email.ne.jp (M.O.)

2 Faculty of Agriculture, Kagoshima University, 1-21-24 Korimoto, Kagoshima, Kagoshima 890-0065, Japan; ichitani@agri.kagoshima-u.ac.jp

3 College of Agriculture, Ibaraki University, 3-21-1 Chuo, Ami, Ibaraki 300-0393, Japan; tsutomu.kuboyama.a@vc.ibaraki.ac.jp

4 Education and Research Field, College of Life, Environment, and Advanced Sciences, Osaka Prefecture University, 1-1 Gakuen-cho, Nakaku, Sakai, Osaka 599-8531, Japan

* Correspondence: tezuka@plant.osakafu-u.ac.jp.

Abstract: Hybrid weakness is a type of reproductive isolation in which F_1 hybrids of normal parents exhibit weaker growth characteristics than their parents. F_1 hybrid of the *Oryza sativa* Indian cultivars 'P.T.B.7' and 'A.D.T.14' exhibits hybrid weakness that is associated with the *HWA1* and *HWA2* loci. Accordingly, the aim of the present study was to analyze the hybrid weakness phenotype of the 'P.T.B.7' × 'A.D.T.14' hybrids. The height and tiller number of the F_1 hybrid were lower than those of either parent, and F_1 hybrid also exhibited leaf yellowing that was not observed in either parent. In addition, the present study demonstrates that SPAD values, an index correlated with chlorophyll content, are effective for evaluating the progression of hybrid weakness that is associated with the *HWA1* and *HWA2* loci because it accurately reflects degree of leaf yellowing. Both cell death and H_2O_2, a reactive oxygen species, were detected in the yellowing leaves of the F_1 hybrid. Furthermore, disease resistance-related genes were upregulated in the yellowing leaves of the F_1 hybrids, whereas photosynthesis-related genes tended to be downregulated. These results suggest that the hybrid weakness associated with the *HWA1* and *HWA2* loci involves hypersensitive response-like mechanisms.

Keywords: *Oryza sativa*; hybrid weakness; cell death; reactive oxygen species; leaf yellowing; SPAD; hypersensitive response

---

1. Introduction

The traits of existing crop cultivars can be improved by crossing cultivars or lines to introduce beneficial traits, such as resistance or tolerance to disease or stress, to susceptible cultivars. However, because reproductive isolation mechanisms can hinder the production of hybrids, methods must be developed to overcome the underlying mechanisms of such reproductive isolation.

One type of post-zygotic reproductive isolation, namely hybrid weakness (i.e., hybrid lethality or hybrid necrosis). F_1 hybrids that exhibit this phenomenon are characterized by weaker growth than their parents, and the phenomenon has been reported to occur in the offspring of crosses involved a number of species, including *Oryza sativa* [1–4], *Nicotiana* spp. [5], *Capsicum* spp. [6], *Arabidopsis thaliana* [7], *Triticum* spp. [8,9], *Gossypium* spp. [10], and *Phaseolus vulgaris* [11]. The genetic mechanisms of hybrid weakness are explained by the Bateson–Dobzhanzky–Muller model [12–14], which posits that the reduced hybrid vigor is driven by deleterious interactions between genes at different loci. In many cases, one of the causal genes is related to disease resistance (*R*), and interactions between the *R* gene and other causal gene cause autoimmune responses in the hybrid offspring [2,7]. The autoimmune responses include the accumulation of reactive oxygen species such as H_2O_2, cell death, upregulation of disease resistance-related genes, and downregulation of photosynthesis-related genes [2,7,15,16].

In rice, hybrid weakness has been reported to result from interactions between the *HWI1* locus, which encodes the LRR-RLK gene (*R* gene), and the *HWI2* locus, which encodes a subtilisin-like protease, and hybrids have been reported to exhibit localized programmed cell death (PCD), the high accumulation of salicylic and jasmonic acids, and amplified heat-related weakness symptoms [2]. These results demonstrate that the interaction of causal genes can activate downstream immune responses, such as hypersensitive response-like mechanisms [2,7].

The hybrid weakness that results from the interaction of *Hwa1-1*, a dominant allele of the *HWA1* locus, and *Hwa2-1*, a dominant allele of the *HWA2* locus, was firstly reported in rice by Oka [4]. In that study, F_1 hybrid seedlings that exhibited hybrid weakness were reported to exhibit normal germination and seedling growth until developing three to four leaves, after which plant growth halted and the leaves yellowed. Then, unless the environment was particularly favorable, the plants died before reaching anthesis. The distributions of the *Hwa1-1* and *Hwa2-1* alleles were limited to Indian cultivars [4], and both the *HWA1* and *HWA2* loci were located in a 1637-kb region of the long arm of chromosome 11 [17]. However, the causal genes have not been identified, and the molecular mechanism underlying the hybrid weakness associated with the *HWA1* and *HWA2* loci remain unclear.

Accordingly, the aim of the present study was to characterize the phenotypes of the hybrid weakness that is associated with the *HWA1* and *HWA2* loci, in order to understand the system's underlying mechanisms. The effectiveness of SPAD values, an index correlated with chlorophyll content [18], for determining the progression of the hybrid weakness was also evaluated. The occurrence of cell death and H_2O_2 accumulation was also evaluated, and the expression of disease resistance and photosynthesis-related genes in the leaves of F_1 hybrids exhibiting hybrid weakness were analyzed.

2. Results

2.1. Hybrid Weakness Phenotypes

The Oryza sativa Indian cultivars 'A.D.T.14' and 'P.T.B.7' carry homozygous Hwa1-1 and Hwa2-1 alleles, respectively. All the F_1 hybrids of a cross between 'A.D.T.14' and 'P.T.B.7' exhibited dwarfism, reduced tiller number, and leaf yellowing (Figure 1). Increases in the height of the F_1 hybrids nearly halted at 50 days after sowing (DAS), whereas that of the parents continued increasing (Figure 2A). The progression of plant age in leaf number in the F_1 hybrids was the same as that in both parents (Figure 2B). The tiller number of both parents continued increasing and reached >35 tillers at 70 DAS, whereas that of the F_1 hybrids increased little and only reached five tillers by 70 DAS (Figure 2C). In addition, both parents headed by 80 DAS, whereas none of the F_1 hybrids had started heading after 140 DAS (Figure 1).

Leaf yellowing was first observed in the F_1 hybrids that had developed seventh or eighth leaves at 30 DAS. Afterward, the leaves turned yellow sequentially, from the lower to the upper leaves. At 60 DAS, the fourth, fifth, and sixth leaves of the F_1 hybrids turned yellow, starting from the leaf tip, and progressing toward the leaf base, whereas those of both parents remained green (Figure 3A–C). Meanwhile, the SPAD values of the fourth, fifth, and sixth leaves of the F_1 hybrids were lower than those of the parents (Figure 3D–F). Furthermore, in the leaves of the F_1 hybrids, the SPAD values of the lower leaves were lower than those of the upper leaves (fourth vs. fifth and sixth leaves and fifth vs. sixth leaves), and within each leaf, the SPAD values of the leaf tips were lower than those of the leaf bases (Figure 3D–F).

Figure 1. Parental and F_1 hybrid phenotypes at 80 days after sowing. (**A**) *Oryza sativa* 'A.D.T.14'; (**B**) F_1 hybrid; and (**C**) *O. sativa* 'P.T.B.7'. Arrows indicate emerging panicles. Scale bars indicate 50 cm.

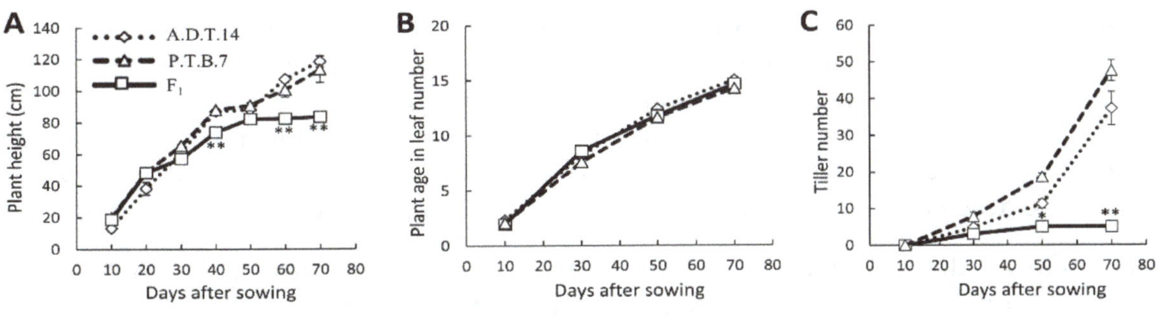

Figure 2. Phenotypic traits of parental and F_1 hybrid rice. (**A**) plant height; (**B**) plant age in leaf number; and (**C**) tiller number. Values and error bars indicate mean ± SE values (n = 5), although some error bars are hidden by the symbols. Mid-parental and hybrids values were compared using two-tailed Student's t-test. Significance: ** $P < 0.01$, * $P < 0.05$.

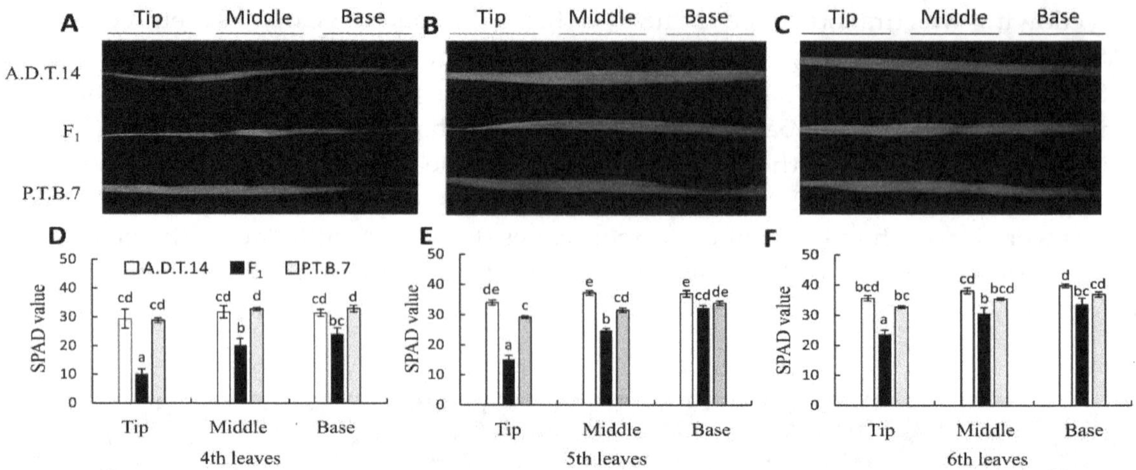

Figure 3. Phenotypes and SPAD values of leaves (fourth to sixth) from parental and F_1 hybrid rice. The phenotypes (**A–C**) and SPAD values (**D–F**) of fourth (**A, D**), fifth (**B, E**), and sixth (**C, F**) leaves were assessed at 60 days after sowing. SPAD value was measured at the tip, middle, and base of each leaf. Values and error bars indicate mean ± SE values (n = 3). Different lowercase letters in each plot (**D–F**) indicate significant differences (Tukey HSD test, $P < 0.05$).

2.2. *Cell Death and* H_2O_2 *Accumulation*

The physiological changes that accompanied leaf yellowing were surveyed by analyzing F_1 hybrid leaves that had been classified into four stages based on degree of yellowing (Figure 4A). Chlorophyll content was assessed by SPAD analysis and spectrophotometry. The SPAD values of Stage-1, -2, and -3 leaf tips, Stage-3 leaf middles, and Stage-3 leaf bases were lower than those of Stage-0 leaves (Figure 4B). The SPAD values of Stage-3 leaf tips, Stage-3 leaf middles, and Stage-3 leaf bases were lower than those of Stage-1 leaves (Figure 4B). The SPAD values of Stage-3 leaf middles and Stage-3 leaf bases were lower than those of Stage-2 leaves (Figure 4B). Total chlorophyll content also decreased in all leaf parts (tip, middle, and base) as yellowing progressed (Figure 4C). Because of the usefulness of SPAD value as discussed later, the progression of yellowing of leaves used in the subsequent experiments were evaluated based on SPAD value.

To determine whether cell death occurred in F_1 hybrid leaves, cellular ion leakage, owing to ion permeability by cell death, was measured. Ion leakage increased slightly and significantly in the tips of Stage-2 and Stage-3 leaves, respectively (Figure 4D). Cell death in F_1 hybrid leaves was also evaluated using trypan blue staining, which is used to identify the highly permeable membranes of dead cells. Only Stage-3 leaves contained dead cells (Figure 5A), and the analysis of transverse sections of Stage 3 revealed that the dead cells were located around vascular and epidermal cells (Figure 5B).

Meanwhile, 3,3-diaminobenzidine (DAB) staining revealed the presence of H_2O_2, which, as a reactive oxygen species, is an important regulator of cell death. Plant tissue is stained brown when DAB is oxidized by H_2O_2 into an insoluble polymer. Hydrogen peroxide (H_2O_2) was detected in the leaves of all stages, except Stage 0 (Figure 6).

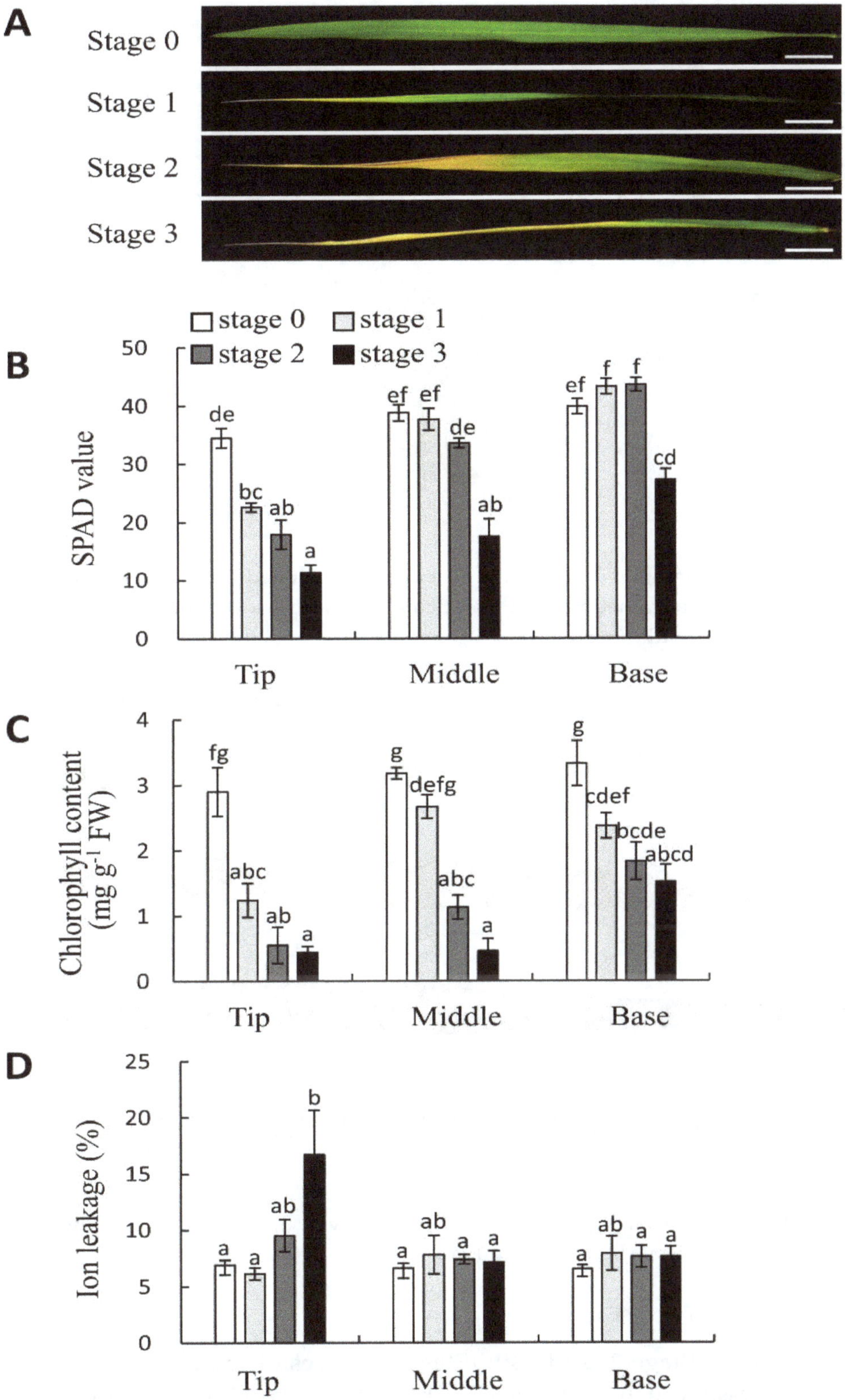

Figure 4. Physiological changes of yellowing hybrid leaves. (**A**) Stages of yellowing: Stage 0, no yellowing; Stage 1, 1/4 of leaf yellow; Stage 2, 1/2 of leaf yellow; Stage 3, 3/4 of leaf yellow. Scale bars indicate 2 cm. (**B**) Changes in the SPAD values at the tip, middle, and base of leaves during the progression of yellowing. (**C**) Changes in chlorophyll content during the progression of yellowing. (**D**) Changes in ion leakage during the progression of yellowing. Values and error bars (**B**–**D**) indicate mean ± SE values (n = 3), and different lowercase letters in each plot (**B**–**D**) indicate significant differences (Tukey HSD test, $P < 0.05$).

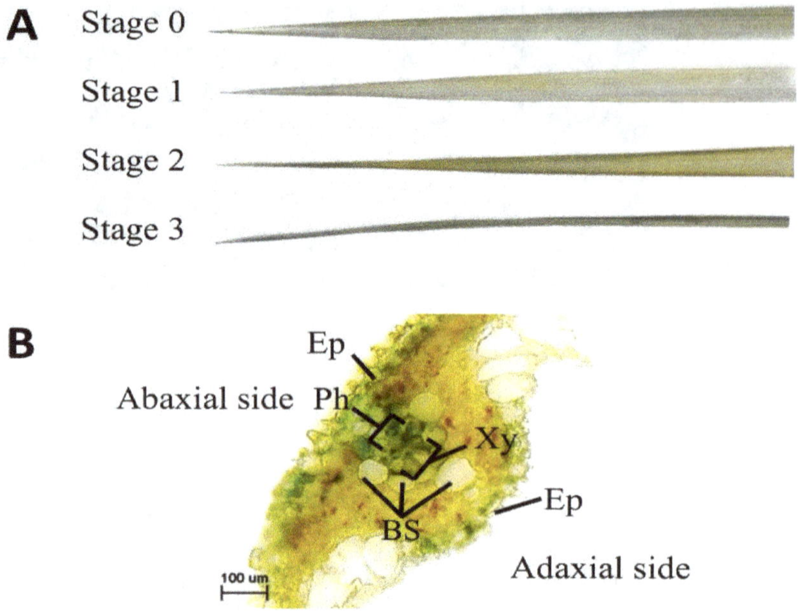

Figure 5. Trypan blue staining of dead cells in F_1 leaves. (**A**) Stained leaves from each yellowing stage. (**B**) Transverse section of a stained Stage-3 leaf. Scale bar indicates 100 μm. Xy: Xylem; Ph: Phloem; Ep: Epidermal cell; BS: Bundle sheath cell.

Figure 6. Presence of reactive oxygen species in hybrid leaves. 3,3-diaminobenzidine (DAB) staining was used to detect H_2O_2. Scale bars indicate 1 cm.

2.3. Hybrid Weakness-Related Gene Expression

At 70 DAS, the 11th (Stage 3) and 13th (Stage 0) leaves of the parents and F_1 offspring were collected for gene expression analysis (Figure 7). The 13th leaves of both the parents and hybrids were entirely green, as indicated by high SPAD values, even though the SPAD values of the F_1 hybrids were somewhat lower than those of either parent (Figure 7A,C). Meanwhile, the 11th leaves of the F_1 hybrids exhibited significant yellowing, as indicated by low SPAD values, whereas those of both parents were entirely green, as indicated by high SPAD values (Figure 7B,D).

The expression of 11 disease resistance-related genes and four photosynthesis-related genes were surveyed (Table 1 and Table S1). The PR1 genes (PR1A and PR1B), the expression of which is induced by salicylic acid [19,20], were upregulated in the 11th leaves of the F_1 hybrids (Figure 8), as were several PR2 genes (Gns5, Gns2, and OsEGL2), which encode glucanase-related proteins that degrade fungal cell walls [21] (Figure 8). Meanwhile, of several genes that encode chitinase-related proteins (PR4, CHT9, CHT11, and RIXI), which also degrade fungal cell walls [22], PR4, CHT9, and RIXI were all upregulated in the 11th leaves of the F_1 hybrids; however, only PR4 was upregulated significantly (Figure 8). The expression of ACO2, which encodes an enzyme related to ethylene production [23], was similar in the 11th and 13th leaves of the F_1 hybrids (Figure 8). Finally, PDC1, the expression of which is induced by jasmonic acid [24], was upregulated in the 11th leaves of the F_1 hybrids, although not significantly (Figure 8).

Of the four photosynthesis-related genes, PSAF, LHCB, and OsRbcL were somewhat downregulated in the 11th leaves of the F_1 hybrids; however, none of these trends were significant (Figure 8).

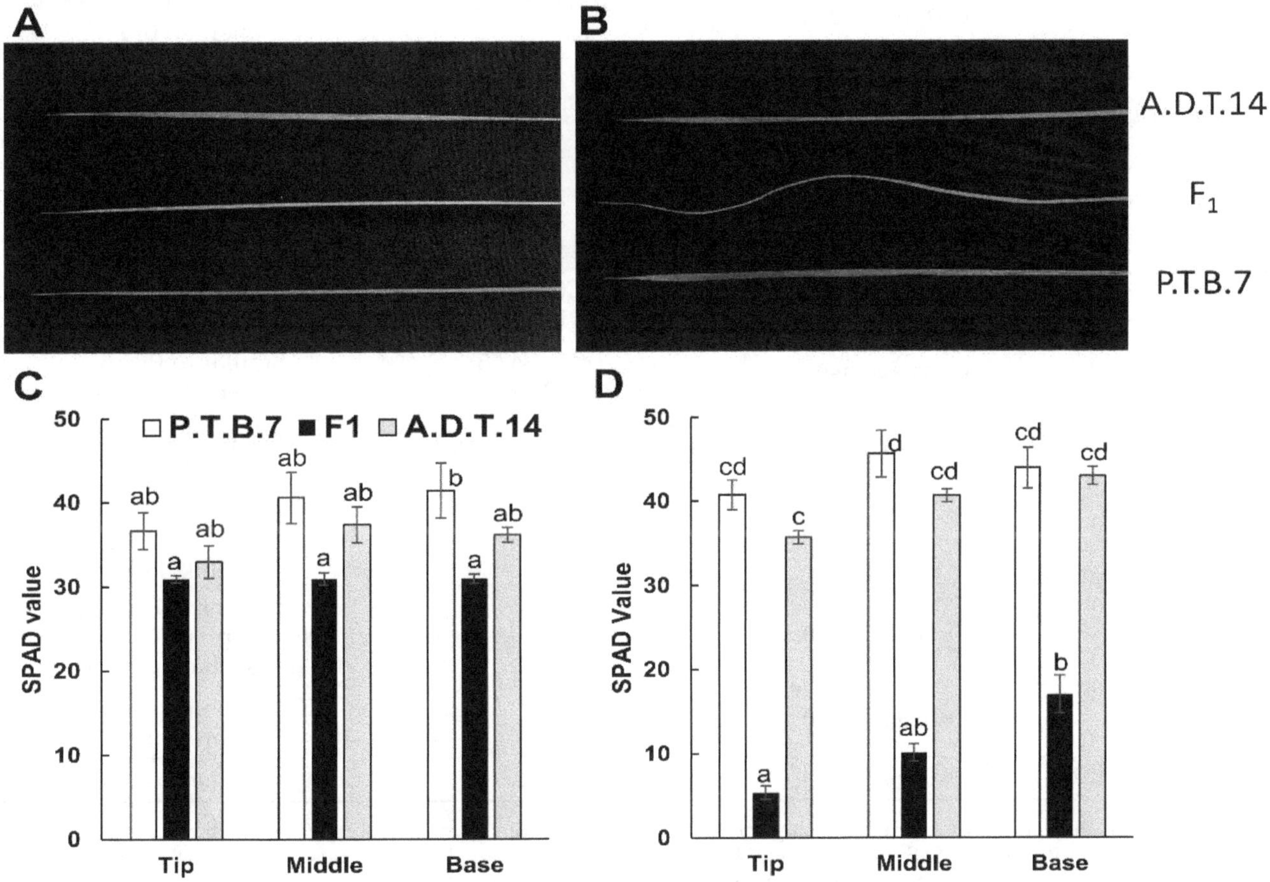

Figure 7. Phenotypes and SPAD values of the parental and F_1 leaves (11th and 13th) used for gene expression analysis. The phenotypes (**A, B**) and SPAD values (**C, D**) of 11th (**A, C**) and 13th (**B, D**) leaves were assessed at 70 d after sowing. SPAD value was measured at the tip, middle, and base of each leaf. Values and error bars indicate mean ± SE values (n = 3), and different lowercase letters in each plot (**C, D**) indicate significant differences (Tukey HSD test, $P < 0.05$).

Table 1. Genes analyzed for RT-PCR.

Gene Symbol	RAP-DB locus ID [a]	CGSNL Gene Name [b]	Description
PR1A	Os07g0129200	*PATHOGENESIS-RELATED GENE 1A*	Similar to Pathogenesis-related protein PR1a
PR1B	Os01g0382000	*PATHOGENESIS-RELATED GENE 1B*	Similar to Pathogenesis-related protein PRB1-2 precursor
Gns5	Os01g0940700		Similar to Glucan endo-1,3-beta-glucosidase GII precursor
Gns2	Os01g0944900		Similar to Glucan endo-1,3-beta-D-glucosidase
OsEGL2	Os01g0942300		Similar to Beta glucanase precursor
PR4	Os11g0592200	*PATHOGENESIS-RELATED GENE 4*	Similar to Chitin-binding allergen Bra r 2
CHT9	Os05g0399400	*CHITINASE 9*	Chitinase 9
CHT11	Os03g0132900	*CHITINASE 11*	Similar to Chitinase 11
RIXI	Os11g0701800		Chitinase III C10701-rice (Class III chitinase homologue)
ACO2	Os09g0451000	*AMINOCYCLOPROPANE-1-CARBOXYLIC ACID OXIDASE 2*	Similar to 1-aminocyclopropane-1-carboxylase 1
PDC1	Os05g0469600	*PYRUVATE DECARBOXYLASE 1*	Similar to Pyruvate decarboxylase
PSAF	Os03g0778100	*PHOTOSYSTEM I SUBUNIT*	Similar to Photosystem-1 F subunit
LHCB	Os03g0592500		Similar to Photosystem II type II chlorophyll a/b binding protein
OsRbcL	Os06g0598500		Similar to Ribulose bisphosphate carboxylase large chain precursor
Fd1	Os08g0104600		Ferredoxin I, chloroplast precursor
Actin	Os05g0438800		Similar to Actin1

[a] Identity of each gene was referenced using the Rice Annotation Project database (https://rapdb.dna.affrc.go.jp/); [b] CGSNL (Committee on Gene Symbolization, Nomenclature and Linkage, Rice Genetics Cooperative) gene names were referenced using Oryzabase (https://shigen.nig.ac.jp/rice/oryzabase/) [25].

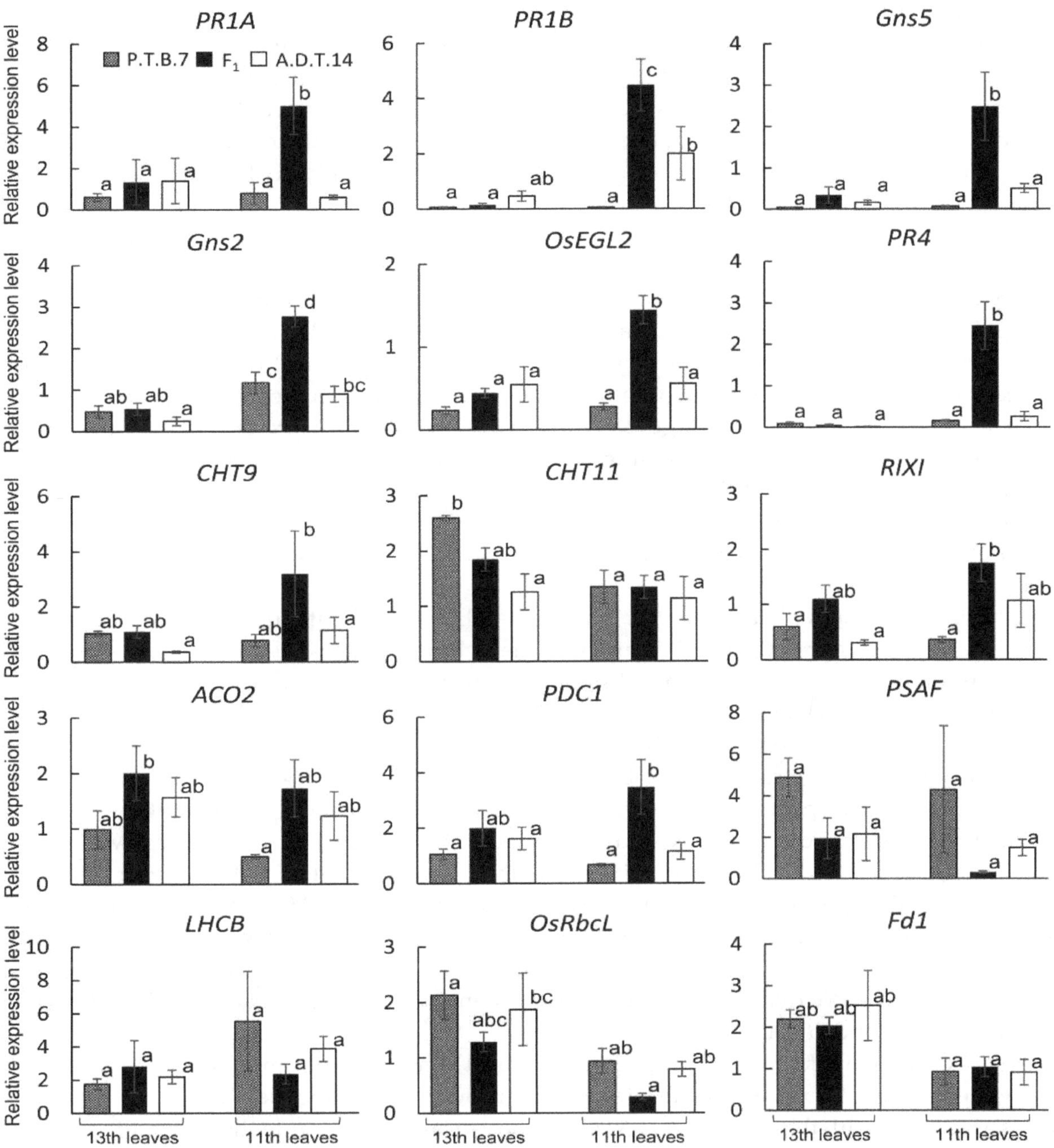

Figure 8. Relative gene expression levels of parental and F_1 hybrid leaves between 'P.T.B.7' and 'A.D.T.14'. Values and error bars indicate mean ± SE values ($n = 3$), and different lowercase letters in each plot indicate significant differences (Tukey HSD test, $P < 0.05$).

3. Discussion

Oka [4] reported that F_1 hybrids that exhibit hybrid weakness associated with the *HWA1* and *HWA2* loci exhibit growth termination and leaf yellowing after the seedlings developed three or four leaves. However, the plant growth phenotypes were not described in detail. In contrast, the present study determined that F_1 hybrids from the cross of 'A.D.T.14' and 'P.T.B.7' rice exhibited limited growth and tiller number, as well as and leaf yellowing (Figures 1–3). Even though leaf yellowing was also reported by Oka [4], the timing of the yellowing process was different [4]. In the present study, leaf yellowing was observed in F_1 hybrids that had developed seven or eight leaves at 30 DAS and, furthermore, was associated with the downregulation of photosynthesis-related genes (Figure 8).

In *O. sativa*, three other gene sets have been reported to cause hybrid weakness, and the phenotypes associated with each system are different. More specifically, the hybrid weakness associated with the *HWC1* and *HWC2* loci is characterized by short stature, short roots, and rolled leaves [26], whereas that associated with the *HWI1* and *HWI2* loci is characterized by short stature and impaired root formation [2], and that associated with the *HW3* and *HW4* loci is characterized by short culms, fewer panicles, pale green leaves, and chlorotic leaf spots [3]. Remarkably, leaf yellowing has only been reported for the hybrid weakness associated with the *HWA1* and *HWA2* loci. Together, these reports suggest that either the causal genes of each system have different functions or the processes downstream of the causal gene interactions are different.

In the present study, the usefulness of SPAD value for determining the progression of leaf yellowing during hybrid weakness associated with the *HWA1* and *HWA2* loci were evaluated. SPAD values generally corresponded with leaf yellowing (Figure 4A,B) but failed to identify significant differences between the chlorophyll content of Stage-0 leaves and that of either the bases of Stage-1 leaves or the middles or bases of Stage-2 leaves. These results indicate that spectrophotometry is more sensitive than SPAD values to changes in chlorophyll content (Figure 4B,C). However, it is important to note that, because SPAD value accurately reflected degree of leaf yellowing and because spectrophotometry requires leaf destruction (Figure 4), SPAD value measurement is an effective and nondestructive method that can be used to quickly and easily evaluate hybrid weakness associated with the *HWA1* and *HWA2* loci.

The hybrid weakness phenotype that was studied by the present study also exhibited cell death in the yellow leaves (Figures 4D and 5). Similarly, the hybrid weakness associated with the *HWI1* and *HWI2* loci involved cell death at the basal nodes [2], and the hybrid weakness associated with the *HW3* and *HW4* loci involved cell death in leaves [3]. Cell death has been also detected in the leaves of intraspecific *Arabidopsis* hybrids that exhibit hybrid necrosis [7,27] and in the leaves, stems, and roots of interspecific *Nicotiana* hybrids that exhibit hybrid lethality [28,29]. Therefore, despite differences in localization, cell death appears to be a common feature of hybrid weakness in plants.

In the present study, cell death was only detected in Stage-2 and Stage-3 leaves, which indicates that the timing of cell death does not coincide with that of either leaf yellowing or reductions in SPAD value or chlorophyll content (Figure 4). On the other hand, H_2O_2 was detected in Stage-1, -2, and -3 leaves that exhibited yellow leaf tips and low SPAD values (Figure 6). H_2O_2 commonly triggers plant cell death during hypersensitive reactions, senescence, abiotic stress responses, and development [30–32]. We detected that reactive oxygen species would lead to cell death on hybrid weakness by *HWA1* and *HWA2*.

In many cases of hybrid weakness, one of the causal genes encodes an *R* gene, and the interaction of the *R* gene with another causal gene triggers an autoimmune response [7,27,33,34]. In the hybrid weakness associated with the *HWI1* and *HWI2* loci, the causal genes include the LRR-RLK gene (*R* gene) and a subtilisin-like protease gene, respectively, and the interaction of the causal genes results in an autoimmune response [2]. Meanwhile, in the hybrid weakness associated with the *HW3* and *HW4* loci, *HW3* encodes a calmodulin-binding protein, and even though the gene is a defense-response gene, not an *R* gene, the interaction of *HW3* with *HW4* results in an autoimmune response [3]. Furthermore, in the hybrid weakness associated with the *HWA1* and *HWA2* loci, a candidate region, which harbored both loci, also contained 12 *R* genes, along with many other genes [17]. During hypersensitivity reactions, reactive oxygen species are produced, thereby mediating cell death, chloroplast disruption, and the upregulation of defense-related genes [35,36]. In the present study, cell death was detected in the F_1 leaves after H_2O_2 generation and leaf yellowing (Figure 3, Figure 5, and Figure 6), the yellow leaves of the F_1 hybrids exhibited upregulated defense-related genes (Figure 8). These results suggest that the hybrid weakness associated with the *HWA1* and *HWA2* loci involves hypersensitive reaction-like responses. However, because leaf senescence is also associated with H_2O_2 production, cell death, the upregulation of certain defense genes, and leaf yellowing [37], it is possible that the hybrid weakness associated with the *HWA1* and *HWA2* loci involves premature senescence. Additional molecular studies will reveal the exact mechanism underlying the hybrid weakness associated with the *HWA1* and *HWA2* loci.

4. Materials and Methods

4.1. Plant Materials and Growth Conditions

The *Oryza sativa* Indian cultivars 'A.D.T.14' (*indica* [17]), which is homozygous for the *Hwa1-1* allele, and 'P.T.B.7' (*aus* [17]), which is homozygous for the *Hwa2-1* allele, were crossed to generate F_1 hybrid offspring. The genotypes of the two cultivars were previously reported by Oka [4]. F_1 seeds were obtained by crossing 'P.T.B.7' (♀) and 'A.D.T.14' (♂) parents. Seeds of 'A.D.T.14', 'P.T.B.7', and the F_1 offspring were sown on 17 July 2011. After the seeds were germinated on moistened filter paper in Petri dishes, the seedlings were transplanted to soil (Sukoyaka-Jinko-Baido; Yanmar Co., Ltd., Osaka, Japan) in Wagner pots of 1/5000 a. The seedlings were grown under natural light conditions in a greenhouse at Osaka Prefecture University, Sakai, Japan. The temperature and humidity of the greenhouse were recorded using a data logger (Ondotori; T&D Co., Ltd., Matsumoto, Japan), and the plants were fertilized weekly using Otsuka-A prescription (OAT Agrio Co., Ltd., Tokyo, Japan), which contained 18.6 mM N, 5.1 mM P, 8.6 mM K, 8.2 mM Ca, and 0.4 mM Mg. The plants were cultivated for 140 DAS to survey plant height, plant age in leaf number, tiller number, days to heading, and SPAD value. The plant height was measured from the surface of the soil to the tip of the tallest leaves. To evaluate the relationship between leaf yellowing and physiology, the leaves were classified according to degree of leaf yellowing. Leaves in which 0%, 25%, 50%, or > 75% of the blade had turned yellow were assigned to Stages 0, 1, 2, and 3, respectively (Figure 4A). These leaves classified according to degree of leaf yellowing were used to measured SPAD value and chlorophyll content, as well as to detect dead cells and H_2O_2. Parts of seedlings were cultivated in Wagner pots of 1/10,000 a in an incubator (14 h natural light and 10 h dark, 28 °C, light intensity: 512 μmol m^{-2} s^{-1}), and at 70 DAS, these plants were used as material for gene expression analysis.

4.2. SPAD and Chlorophyll Measurement

A SPAD meter (SPAD-502; Konica Minolta, Inc., Tokyo, Japan) was used to measure the SPAD values of the leaves without causing damage. SPAD values were obtained from the tip, middle, and base of each leaf. Meanwhile, total chlorophyll content was measured using a previously described spectrophotometric method [38]. Briefly, the leaves were cut into small pieces, weighed, treated with 20 mL 80% acetone, and ground using a pestle until bleached. The resulting solutions were transferred to 1.5 mL tubes and centrifuged at 10,000 g for 5 min. Each supernatant was transferred to a cuvette, and the absorbance of each supernatant was measured at 663.6 and 646.6 nm, after the spectrophotometer (V-530; JASCO Corp., Hachioji, Japan) was zeroed at 750 nm. Total chlorophyll concentration (mg g^{-1} FW) was calculated using the following equation: $[(17.76 \times OD_{646.6} + 7.34 \times OD_{663.6}) \times$ extraction volume in a cuvette]/fresh weight (g).

4.3. Ion leakage Measurement

Ion leakage was measured, as described previously [39]. Leaf disks (3 cm^2) were taken from the tips, middles, and bases of the leaves, floated for 5 min in water that contained 0.2% (v/v) Tween 20 for removing ion generating on making leaf disks, transferred to Petri dishes that contained fresh water with Tween 20 (0.2%), and incubated for 3 h for leaking out ions by cell death. Their conductivity (value A) of the solutions was measured using a conductivity meter (Twin Cond B-173; Horiba, Ltd., Kyoto, Japan). The leaf disks were then incubated at 95 °C for 25 min, for leaking out ions of whole leaf disks by destroying whole organization, and cooled to room temperature, and their conductivity of the solutions was also measured (value B). Finally, ion leakage (%) was calculated using the following equation: (value A/value B) × 100%.

4.4. Trypan Blue Staining

Trypan blue staining was performed as described previously [40]. Detached leaves were stained by boiling for 8 min in a 1:1 (v:v) mixture of ethanol and lactophenol (i.e., alcoholic lactophenol) that contained 0.1 mg ml^{-1} trypan blue, cleared in 70% chloral hydrate solution overnight, and then preserved in 70% glycerol. Trypan blue stains dead cells. Transverse slices were prepared using a hand-section method and visualized using a light microscope (Olympus BX50; Olympus, Co. Ltd., Tokyo, Japan).

4.5. Detection of Hydrogen Peroxide Accumulation

Hydrogen peroxide was detected visually, using previously described methods [41]. Briefly, leaves were soaked in a 3,3-diaminobenzidine (DAB) solution for 24 h, transferred to boiling 96% ethanol until bleaching, and then visualized. The presence of H_2O_2 was indicated by brown staining.

4.6. Real-Time qRT-PCR

Total RNA was isolated from leaves using an RNAiso PLUS kit (Takara Bio, Inc., Shiga, Japan), according to the manufacturer's protocol and then treated with RNase-free DNase (Promega Co., Madison, USA), and first-strand cDNA was synthesized from total RNA (2 μg) using oligo $(dT)_{18}$ primers and ReverTra Ace (Toyobo Co., Ltd., Osaka, Japan). Real-time RT-PCR was carried out to analyze the expression of 11 defense-related genes and four photosynthesis-related genes (Table 1), using *Actin* as an internal control. The primers used to amplify *PR1A, PR1B, Gns5, PR4*, and *Actin* had been reported previously [42], and the other primers were designed based on RAP-DB locus ID using the Primer-BLAST design tool [43] (Table S1). Real-time RT-PCR was performed in 20 μL reaction mixtures that contained 10 μL KAPA SYBR FAST qPCR Master Mix (2×) ABI PRISM (Takara Bio), 10 μM of each forward and reverse primer (0.4 μL each), and 1 μL cDNA template, and the real-time PCR amplification was performed under the following conditions: Initial denaturation at 94 °C for 10 min, followed by 40 cycles of 15 s at 94 °C and 1 min at 60 °C, with a final 30 s extension at 72 °C using an Applied Biosystems 7300 Real-Time PCR System (Applied Biosystems, Foster, CA, USA). The results were analyzed using ABI Prism software (Applied Biosystems). Each gene expression level was divided by the expression level of *Actin* to calculate relative expression level.

4.7. Statistical Analysis

Data were analyzed using SPSS (version 22; IBM, Co., Armonk, USA). Tukey HSD tests were used to compare SPAD, chlorophyll content, and ion leakage values, and two-tailed Student's t-tests were used to compare mid-parental (mean of 'A.D.T.14' and 'P.T.B.7') and hybrid values of plant height, foliar age, and tiller number.

Author Contributions: Conceptualization, K.I., T.K., and T.T.; formal analysis, K.S.; funding acquisition, K.I., T.K., and T.T.; investigation, K.S. and T.I.; resources, K.I.; supervision, K.I., T.K., T.M., M.O., and T.T.; writing—original draft, K.S.; writing—review and editing, T.T.

Acknowledgments: We are grateful to the Genebank of the National Institute of Agrobiological Sciences (Tsukuba, Japan) for providing seeds of the parent lines ('A.D.T.14' and 'P.T.B.7'). We would like to thank Editage (www.editage.com) for English language editing.

References

1. Amemiya, A.; Akemine, H. Biochemical genetic studies on the root growth inhibiting complementary lethal in rice plant (Studies on the embryo culture in rice plant. 3). *Bull. Natl. Inst. Agric. Sci.* **1963**, *10*, 139–226.
2. Chen, C.; Chen, H.; Lin, Y.S.; Shen, J.B.; Shan, J.X.; Qi, P.; Shi, M.; Zhu, M.Z.; Huang, X.H.; Feng, Q.; et al. A two-locus interaction causes interspecific hybrid weakness in rice. *Nat. Commun.* **2014**, *5*, 3357. [CrossRef] [PubMed]
3. Fu, C.Y.; Wang, F.; Sun, B.R.; Liu, W.G.; Li, J.H.; Deng, R.F.; Liu, D.L.; Liu, Z.R.; Zhu, M.S.; Liao, Y.L.; et al. Genetic and cytological analysis of a novel type of low temperature-dependent intrasubspecific hybrid weakness in rice. *PLoS ONE* **2013**, *8*, e73886. [CrossRef] [PubMed]
4. Oka, H.I. Phylogenetic differentiation of cultivated rice. XV Complementary lethal genes in rice. *Jpn. J. Genet.* **1957**, *32*, 83–87. [CrossRef]
5. Tezuka, T.; Marubashi, W. Genomic factors lead to programmed cell death during hybrid lethality in interspecific hybrids between *Nicotiana tabacum* and *N. debneyi*. *SABRAO J. Breed Genet.* **2006**, *38*, 69–81.
6. Inai, S.; Ishikawa, K.; Nunomura, O.; Ikehashi, H. Genetic analysis of stunted growth by nuclear-cytoplasmic interaction in interspecific hybrids of *Capsicum* by using RAPD markers. *Theor. Appl. Genet.* **1993**, *87*, 416–422. [CrossRef]
7. Bomblies, K.; Lempe, J.; Epple, P.; Warthmann, N.; Lanz, C.; Dangl, J.L.; Weigel, D. Autoimmune response as a mechanism for a Dobzhansky-Muller-type incompatibility syndrome in plants. *PLoS Biol.* **2007**, *5*, e236. [CrossRef]
8. Tsunewaki, K. Monosomic and conventional analyses in common wheat. III Lethality. *Jpn. J. Genet.* **1960**, *35*, 71–75. [CrossRef]
9. Tsunewaki, K. Aneuploid analysis of hybrid necrosis and hybrid chlorosis in tetraploid wheats using the D genome chromosome substitution lines of durum wheat. *Genome* **1992**, *35*, 594–601. [CrossRef]
10. Silow, R.A. The comparative genetics of *Gossypium anomalum* and the cultivated Asiatic cottons. *J. Genet.* **1941**, 259–358. [CrossRef]
11. Reiber, J.M.; Neuman, D.S. Hybrid weakness in *Phaseolus vulgaris* L. II. Disruption of root-shoot integration. *J. Plant Growth Regul.* **1999**, *18*, 107–112. [CrossRef] [PubMed]
12. Bateson, W. Heredity and variation in modern science. In *Darwin and Modern Science*; Seward, A.C., Ed.; Cambridge University Press: Cambridge, UK, 1909; pp. 85–101.
13. Dobzhansky, T. *Genetics and the Origin of Species*; Columbia University Press: New York, NY, USA, 1937.
14. Muller, H.J. Isolating mechanisms, evolution, and temperature. *Biol. Symp.* **1942**, *6*, 71–124.
15. Mizuno, N.; Hosogi, N.; Park, P.; Takumi, S. Hypersensitive response-like reaction is associated with hybrid necrosis in interspecific crosses between tetraploid wheat and *Aegilops tauschii* Coss. *PLoS ONE* **2010**, *5*, e11326. [CrossRef] [PubMed]
16. Okada, M.; Yoshida, K.; Takumi, S. Hybrid incompatibilities in interspecific crosses between tetraploid wheat and its wild diploid relative *Aegilops umbellulata*. *Plant Mol. Biol.* **2017**, *95*, 625–645. [CrossRef]
17. Ichitani, K.; Taura, S.; Tezuka, T.; Okiyama, Y.; Kuboyama, T. Chromosomal location of *HWA1* and *HWA2*, complementary hybrid weakness genes in rice. *Rice* **2011**, *4*, 29–38. [CrossRef]
18. Markwell, J.; Osterman, J.C.; Mitchell, J.L. Calibration of the Minolta SPAD-502 leaf chlorophyll meter. *Photosyn. Res.* **1995**, *46*, 467–472. [CrossRef]
19. Agrawal, G.K.; Rakwal, R.; Jwa, N.S.; Agrawal, V.P. Signalling molecules and blast pathogen attack activates rice *OsPR1a* and *OsPR1b* genes: A model illustrating components participating during defence/stress response. *Plant Physiol. Biochem.* **2001**, *39*, 1095–1103. [CrossRef]
20. Van Loon, L.C.; Rep, M.; Pieterse, C.M.J. Significance of inducible defense-related proteins in infected plants. *Annu. Rev. Phytopathol.* **2006**, *44*, 135–162. [CrossRef]
21. Romero, G.O.; Simmons, C.; Yaneshita, M.; Doan, M.; Thomas, B.R.; Rodriguez, R.L. Characterization of rice endo-β-glucanase genes (Gns2–Gns14) defines a new subgroup within the gene family. *Gene* **1998**, *223*, 311–320. [CrossRef]
22. Klarzynski, O.; Plesse, B.; Joubert, J.-M.; Yvin, J.-C.; Kopp, M.; Kloareg, B.; Fritig, B. Linear β-1,3 glucans are elicitors of defense responses in tobacco. *Plant Physiol.* **2000**, *124*, 1027–1038. [CrossRef]
23. Kende, H. Ethylene biosynthesis. *Annu. Rev. Plant. Biol.* **1993**, *44*, 283–307. [CrossRef]

24. Sugimori, M.; Kiribuchi, K.; Akimoto, C.; Yamaguchi, T.; Minami, E.; Shibuya, N.; Sobajima, H.; Cho, E.M.; Kobashi, N.; Nojiri, H.; et al. Cloning and characterization of cDNAs for the jasmonic acid-responsive genes *RRJ1* and *RRJ2* in suspension-cultured rice cells. *Biosci. Biotechnol. Biochem.* **2002**, *66*, 1140–1142. [CrossRef]
25. McCouch, S.R. Gene nomenclature system for rice. *Rice* **2008**, *1*, 72–84. [CrossRef]
26. Saito, T.; Ichitani, K.; Suzuki, T.; Marubashi, W.; Kuboyama, T. Developmental observation and high temperature rescue from hybrid weakness in a cross between Japanese rice cultivars and Peruvian rice cultivar 'Jamaica'. *Breed. Sci.* **2007**, *57*, 281–288. [CrossRef]
27. Alcázar, R.; García, A.V.; Kronholm, I.; de Meaux, J.; Koornneef, M.; Parker, J.E.; Reymond, M. Natural variation at Strubbelig Receptor Kinase 3 drives immune-triggered incompatibilities between *Arabidopsis thaliana* accessions. *Nat. Genet.* **2010**, *42*, 1135–1139. [CrossRef]
28. Tezuka, T.; Marubashi, W. Apoptotic cell death observed during the expression of hybrid lethality in interspecific hybrids between *Nicotiana tabacum* and *N. suaveolens*. *Breed. Sci.* **2004**, *54*, 59–66. [CrossRef]
29. Yamada, T.; Marubashi, W.; Niwa, M. Apoptotic cell death induces temperature-sensitive lethality in hybrid seedlings and calli derived from the cross of *Nicotiana suaveolens* × *N. tabacum*. *Planta* **2000**, *211*, 614–622. [CrossRef]
30. Huysmans, M.; Lema, A.S.; Coll, N.S.; Nowack, M.K. Dying two deaths—Programmed cell death regulation in development and disease. *Curr. Opin. Plant Biol.* **2017**, *35*, 37–44. [CrossRef]
31. Petrov, V.; Hille, J.; Mueller-Roeber, B.; Gechev, T.S. ROS-mediated abiotic stress-induced programmed cell death in plants. *Front Plant Sci.* **2015**, *6*, 69. [CrossRef]
32. Woo, H.R.; Kim, H.J.; Nam, H.G.; Lim, P.O. Plant leaf senescence and death—Regulation by multiple layers of control and implications for aging in general. *J. Cell Sci.* **2013**, *126*, 4823. [CrossRef]
33. Huang, X.; Li, J.; Bao, F.; Zhang, X.; Yang, S. A gain-of-function mutation in the *Arabidopsis* disease resistance gene *RPP4* confers sensitivity to low temperature. *Plant Physiol.* **2010**, *154*, 796–809. [CrossRef] [PubMed]
34. Yang, S.; Hua, J. A haplotype-specific resistance gene regulated by *BONZAI1* mediates temperature-dependent growth control in *Arabidopsis*. *Plant Cell* **2004**, *16*, 1060–1071. [CrossRef] [PubMed]
35. Coll, N.S.; Epple, P.; Dangl, J.L. Programmed cell death in the plant immune system. *Cell Death Differ.* **2011**, *18*, 1247. [CrossRef] [PubMed]
36. Overmyer, K.; Brosché, M.; Kangasjärvi, J. Reactive oxygen species and hormonal control of cell death. *Trends Plant Sci.* **2003**, *8*, 335–342. [CrossRef]
37. Jajic, I.; Sarna, T.; Strzalka, K. Senescence, stress, and reactive oxygen species. *Plants* **2015**, *4*, 393–411. [CrossRef]
38. Porra, R.J.; Thompson, W.A.; Kriedemann, P.E. Determination of accurate extinction coefficients and simultaneous equations for assaying chlorophylls a and b extracted with four different solvents: Verification of the concentration of chlorophyll standards by atomic absorption spectroscopy. *Biochim. Biophys. Acta* **1989**, *975*, 384–394. [CrossRef]
39. Rizhsky, L.; Shulaev, V.; Mittler, R. Measuring programmed cell death in plants. In *Apoptosis Methods and Protocols*; Brady, H.J.M., Ed.; Humana Press: Totowa, NJ, USA, 2004; pp. 179–189.
40. Zhang, H.K.; Zhang, X.; Mao, B.Z.; Li, Q.; He, Z.H. Alpha-picolinic acid, a fungal toxin and mammal apoptosis-inducing agent, elicits hypersensitive-like response and enhances disease resistance in rice. *Cell Res.* **2004**, *14*, 27–33. [CrossRef]
41. Chandru, H.K.; Kim, E.; Kuk, Y.; Cho, K.; Han, O. Kinetics of wound-induced activation of antioxidative enzymes in *Oryza sativa*: Differential activation at different growth stages. *Plant Sci.* **2003**, *164*, 935–941. [CrossRef]
42. Yamamoto, E.; Takashi, T.; Morinaka, Y.; Lin, S.; Wu, J.; Matsumoto, T.; Kitano, H.; Matsuoka, M.; Ashikari, M. Gain of deleterious function causes an autoimmune response and Bateson–Dobzhansky–Muller incompatibility in rice. *Mol. Genet. Genomics* **2010**, *283*, 305–315. [CrossRef]
43. Ye, J.; Coulouris, G.; Zaretskaya, I.; Cutcutache, I.; Rozen, S.; Madden, T.L. Primer-BLAST: A tool to design target-specific primers for polymerase chain reaction. *BMC Bioinform.* **2012**, *13*, 134. [CrossRef]

6

Loci Controlling Adaptation to Heat Stress Occurring at the Reproductive Stage in Durum Wheat

Khaoula El Hassouni [1,2], Bouchra Belkadi [2], Abdelkarim Filali-Maltouf [2], Amadou Tidiane-Sall [1,3], Ayed Al-Abdallat [4], Miloudi Nachit [1,2] and Filippo M. Bassi [1,*]

1 International Center for the Agricultural Research in Dry Area (ICARDA), Rabat 10000, Morocco
2 Faculty of sciences, University of Mohammed V, Rabat 10000, Morocco
3 Institut Sénégalais de Recherches Agricoles (ISRA), Saint-Louis 46024, Senegal
4 Faculty of Agriculture, The University of Jordan, Amman 11942, Jordan
* Correspondence: f.bassi@cgiar.org.

Abstract: Heat stress occurring during the reproductive stage of wheat has a detrimental effect on productivity. A durum wheat core set was exposed to simulated terminal heat stress by applying plastic tunnels at the time of flowering over two seasons. Mean grain yield was reduced by 54% compared to control conditions, and grain number was the most critical trait for tolerance to this stress. The combined use of tolerance indices and grain yield identified five top performing elite lines: Kunmiki, Berghouata1, Margherita2, IDON37-141, and Ourgh. The core set was also subjected to genome wide association study using 7652 polymorphic single nucleotide polymorphism (SNPs) markers. The most significant genomic regions were identified in association with spike fertility and tolerance indices on chromosomes 1A, 5B, and 6B. Haplotype analysis on a set of 208 elite lines confirmed that lines that carried the positive allele at all three quantitative trait loci (QTLs) had a yield advantage of 8% when field tested under daily temperatures above 31° C. Three of the QTLs were successfully validated into Kompetitive Allele Specific PCR (KASP) markers and explained >10% of the phenotypic variation for an independent elite germplasm set. These genomic regions can now be readily deployed via breeding to improve resilience to climate change and increase productivity in heat-stressed areas.

Keywords: heat stress; durum wheat; yield; tolerance; fertility; climate change; resilience

1. Introduction

Heat stress is a major environmental constraint to crop production. Terminal heat stress is defined as a rise in temperature that occurs between heading and maturity. When this stress matches with the reproductive phase of the wheat plant, it affects anthesis and grain filling, resulting in a severe reduction in yield [1]. High temperatures at the time of flowering cause floret sterility via pollen dehiscence [2], decrease photosynthetic capacity by drying the green tissues, and reduce starch biosynthesis [1,3]. These in turn result in a negative effect on grain number and weight [4–7]. The optimum growing temperature for wheat during pollination and grain filling phases is 21 °C [8,9], and for each increase of 1 °C above it is estimated a decline of 4.1% to 6.4% in yield [10]. Environmental temperatures have been increasing over the last century and more frequent heat waves are predicted in the next decades [11–13]. Therefore, breeding for tolerance to chronic as well as short term heat stress is a major objective worldwide [14–19]. Breeding selection would benefit by a better understanding of traits associated with tolerance to high temperatures, as well as the identification of the genomic regions controlling these traits.

In wheat, a large number of quantitative trait loci (QTLs) has been identified under heat stress via linkage analysis and genome-wide association study (GWAS) for yield, yield related traits, and some physiological traits such as chlorophyll content, chlorophyll fluorescence, and canopy temperature [20–27]. Grain number per spike and chlorophyll content were found to be the most critical traits for adaptation to warm conditions [24,25,28]. Heat stress reduces leaf chlorophyll content [29] affecting the amount of carbohydrates transported to the grains and final grain weight and size. High temperatures around anthesis reduce the number of grains per spike due to a decrease in spike growth and development, and an increase in ovules abortion [2,25,29,30]. To the best of our knowledge, molecular markers associated with heat tolerance are not generally used in wheat breeding programs [31–33]. The limited understanding of genes underlying physiological mechanisms and the regulation of yield components in wheat, and the lack of cloned major QTL for traits associated with heat tolerance has restricted the improvement in breeding for tolerance to this stress.

In the current study, a set of durum wheat lines were heat stressed by imposing a > 10 °C raise in maximum daily temperatures via the deployment of plastic tunnels at the time of flowering. GWAS studies allowed the identification of major QTLs controlling the adaptation to this stress and these were validated for marker assisted selection (MAS) in an independent germplasm set for rapid deployment via breeding.

2. Materials and Methods

2.1. Plant Material

A subset of 42 durum wheat inbred lines were selected from a global collection of 384 genotypes based on their similarity in flowering time and identified genetic diversity [34]. Briefly, the complete collection is highly diverse and includes 96 durum wheat landraces from 24 countries, and 288 modern lines from nine countries and two International research centers CIMMYT and ICARDA. The subset selected for this study includes 34 ICARDA and CIMMYT lines, five cultivars and one landrace. The list of the 42 genotypes and their details are provided in Table S1.

A second subset of 208 modern entries was also obtained from the global collection and field tested under severe high temperatures during 2014–2015 and 2015–2016 seasons along the Senegal River in Kaedi, Mauritania. Full details on this field experiment have been published in Sall et al. [35].

The third and final set was used for Kompetitive Allele Specific PCR (KASP) markers validation and it was composed of 94 ICARDA's elite lines that constituted the 2017 international nurseries 40th International Durum Yield Trial (IDYT) and 40th International Durum Observation Nurseries (IDON). This set was also tested at the station of Kaedi along the Senegal River in season 2015–2016.

2.2. Field Experiment Conditions and Phenotyping

The first subset of 42 entries was grown at Marchouch station (33°34′3.1″ N, 6°38′0.1″ W) in Morocco during two successive crop seasons (2015–2016 and 2016–2017). Each entry was sown in mid-November on a plot surface of 1.5 m^2 per genotype at a sowing density of 300 plants per m^2. The experiment was an alpha lattice with two replications, block size of six, and two treatments arranged in split-plot. Each six genotypes were arranged in close proximity to maximize competition between the genotypes, and compose one block of 9 m^2. Each block was surrounded by a border of barley to avoid border effect. Each block was spaced 1 m apart to allow the application of the plastic tunnel. The two treatments were normal rainfed conditions and plastic tunnel-mediated heat stress. The normal treatment followed standard agronomic practices with a base pre-sowing application of 50 Kg ha^{-1} of N, P, and K. At stage 15 of Zadok's (Z) scale herbicide was applied in a tank mixture (Pallas + Mustang at 0.5 L ha^{-1}) to provide protection against both monocots and dicots. At Z17 ammonium nitrate was provided to add 36 kg ha^{-1} of N and a final application of urea was used to add 44 kg ha^{-1} of N before booting (Z39). Weeds were also controlled mechanically to ensure clean plots. The soil of the experimental station is clay-vertisol type. The available on season moisture was 234 and 280 mm for 2015–2016 and 2016–2017, respectively, during the growing season, whereas the average daily temperature was 14.1 °C for the first year and 13.5 °C for the second year. The heat-stress treatment followed the same agronomic practices, with the difference that at the time of booting (Z45) a 10 m^2 and 1.5 m high plastic tunnel was placed over each block (Figure 1) and left there until early dough stage (Z83). An electronic thermometer (temperature data logger) was placed in the middle of each block (normal and heat stressed) to reveal that the temperatures were up to 16° C higher inside the plastic tunnels, to reach a maximum of 49 °C (Figure 1). Marchouch is a drought prone site, and no rainfall occurred after Z45 in any of the two field seasons.

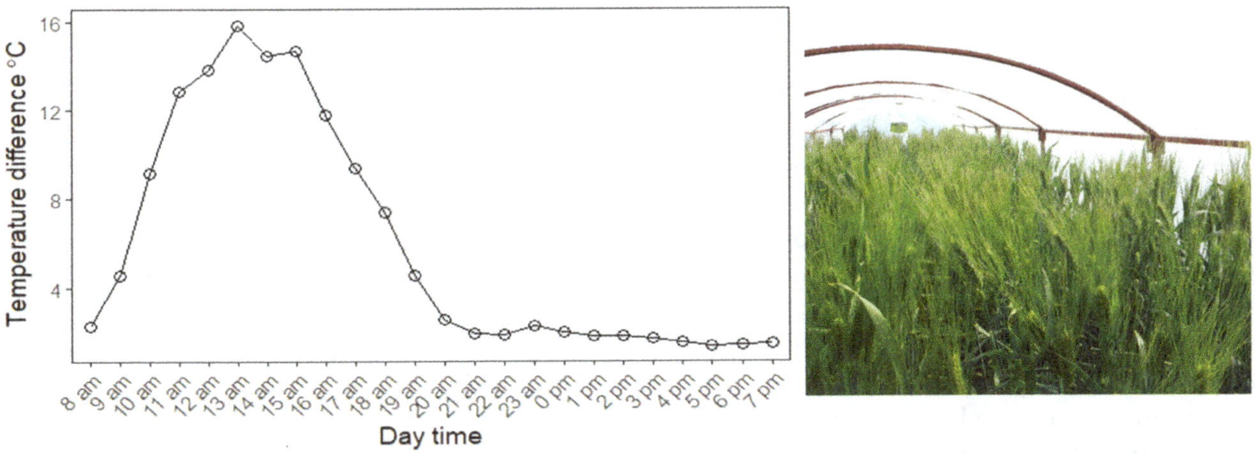

Figure 1. Mean temperature difference of 18 days over two seasons between the plastic tunnel-mediated heat stress and normal field conditions between 8 a.m. and 8 p.m., and a picture of the plastic tunnel at 9 a.m.

The following traits were recorded: days to heading (DTH) measured at the moment when the awns became visible, plant height (PH) measured from the ground to the top of the highest spike excluding the awns, and the number of fertile spikes per meter square ($Spkm^2$) was counted in a 0.25 m^2 area. The whole plot was harvested by hand and the dry biomass (Biom) was weighed before threshing. Grain yield (GY) was weighed for each plot and expressed as kg ha^{-1}. The weight of a thousand kernels (TKW) was expressed in grams. The harvest index (HI) was calculated as the ratio between GY and Biom. The grain number per spike (GNSpk) was derived from dividing grain number per meter square by $Spkm^2$ as follows:

$$\text{Grain number}/\text{m}^2 = \frac{\text{Grain weight of the plot}}{1.5\text{m}^2 \times \frac{\text{TKW}}{1000}} \quad (1)$$

$$\text{GNSpk} = \frac{\text{Grain number}/\text{m}^2}{\text{Spkm}^2} \quad (2)$$

The second and third sets were field tested in Kaedi, Mauritania (16°14″ N; 13°46″ W) during season 2014–2015 and 2015–2016 where the temperature reached a maximum of 41 °C and an average maximum daily temperature of 34 °C throughout the season. The trial was carried out under augmented design with a plot surface of 4.5 m^2. Standard agronomic management practices were adopted. Full details for this experiment are published elsewhere [35].

2.3. Data Analysis

A mixed linear model was run using the lme4 package [36] in R [37] to obtain best linear unbiased estimates (BLUEs) of the normally distributed traits. For count traits (DTH, $Spkm^2$, GNSpk), the generalized mixed linear model was used to get the BLUEs by Proc GLIMMIX in SAS. In both models, genotype, treatment, year, and replication were considered as fixed effects and block as random effect nested in treatment and year. Broad-sense heritability was calculated based on variance components from random model using the method suggested by DeLacy et al. [38]:

$$H^2 = \frac{\sigma^2 g}{\sigma^2 g + \frac{\sigma^2 GxT}{t} + \frac{\sigma^2 GxY}{y} + \frac{\sigma^2 GxTxY}{ty} + \frac{\sigma^2 e}{tyr}} \quad (3)$$

where: $\sigma^2_{G\times T}$ = genotype × treatment variance, $\sigma^2_{G\times Y}$ = genotype × year variance, $\sigma^2_{G\times Y\times T}$ = genotype × treatment × year variance, σ^2_e = residual variance, r is the number of replications per treatment, t is the number of treatments, and y is the number of years.

Box-and-whisker plots where constructed by ggplot2 package [39] using the BLUEs combined over year per each treatment. The relationship between the target trait GY and yield components (GNSpk, TKW, Biom, HI) was studied using the Pearson correlation coefficient and the additive regression model. The critical value of the correlation significance was determined at 0.30 for $p < 0.05$ and 0.39 for $p < 0.01$ for 40 df using the corrplot package [40]. The additive model incorporates flexible forms (i.e., splines) of the functions to account for non-linear relationship contrary to linear regression model estimated via ordinary least squares [41]. For the additive model, the effective degree of freedom term determines the nature of the relationship between the predictor and the response variables where EDF = 1 indicates linearity and EDF > 1 the non-linearity. The additive regression analysis was performed using the mgcv package [42].

Two stress tolerance indices were calculated to identify the heat tolerant genotypes. The stress susceptibility index (SSI) [43,44] was calculated as follows:

$$SSI = \frac{[1 - (Ys)/(Yp)]}{[1 - (\bar{Y}s)/(\bar{Y}p)]} \quad (4)$$

where Ys and Yp are yield values of the genotypes evaluated under heat stress and normal conditions, respectively, and $\bar{Y}s$ and $\bar{Y}p$ are the mean yields of the lines evaluated under heat stress and normal conditions, respectively.

The stress tolerance (TOL) [45] was calculated as follows:

$$TOL = Yp - Ys. \quad (5)$$

The classInt package [46] was used to identify the possible number of class intervals of the indices for the frequency distribution of the subset.

The cut-off value for tolerant vs. susceptible genotypes for SSI was equal to 1, with lines having SSI < 1 being stress tolerant. Regarding the TOL index, the smaller TOL values indicate the genotypes with low yield depression and hence more tolerant. The experiment-wide TOL mean (1608 kg ha^{-1}) was identified as the cut-off value for tolerant vs. susceptible. The emmeans package [47] based on ANOVA model was used to discriminate among the grain yield means of haplotypes.

2.4. Genotyping and Marker-Trait Associations

Details of the genotyping step of the core set and panel have been previously discussed in Kabbaj et al. [34] and Sall et al. [35]. Briefly, 7652 high-fidelity polymorphic single nucleotide polymorphism (SNPs) were obtained, showing less than 1% missing data, minor allele frequency (MAF) higher than 5%, and heterozygosity less than 5%. The sequences of these markers were aligned with a cut-off of 98% identity to the durum wheat reference genome [48] (available at: http://www.interomics.eu/durum-wheat-genome), to reveal their physical position. The average length of the Axiom probe is of 75 bp, hence the 2% allowed miss-match was set to account for the existence of 1 SNP within each sequence. A sub-set of 500 highly polymorphic SNPs were selected on the basis of even spread along the genome, and used to identify the existence of population sub-structure, which revealed the existence of 10 main sub-groups [34]. To avoid bias, these 500 markers were then removed from all downstream association analysis. Linkage disequilibrium was calculated as squared allele frequency correlations (r^2) in TASSEL V 5.0 software [49], using the Mb position of the markers along the bread wheat reference genome. Linkage disequilibrium (LD) decay was estimated and plotted using the "Neanderthal" method [50]. The LD decay was measured at 51.3 Mb for $r^2 < 0.2$ as presented in Bassi et al. [51].

The genome wide association study (GWAS) was based on BLUEs of all the traits that displayed a significant treatment effect and the two stress tolerance indices. Two models were fitted and compared using two covariate parameters, Q (population structure) and K (Kinship). Q model was performed using a general linear model (GLM), and Q + K model using a mixed linear model (MLM). The best model for each trait was selected based on the quantile-quantile (Q-Q) plots [52]. Flowering time (DTH) was used as covariate in all analyses to remove the strong effects of flowering genes from the study. The value calculated for the LD decay of 51.3 Mb indicated that this association panel interrogated the 12,000 Mb of the durum wheat genome via 248 "loci hypothesis," and hence the Bonferroni correction for this panel was set to 3.1 LOD for $p < 0.05$ as suggested by Duggal et al. [53]. Local LD decay for $r^2 < 0.2$ was calculated for a 100 Mbp window around the marker with highest LOD for all marker-trait associations (MTAs) identified at a distance inferior to 104 Mbp (twice the LD decay). The MTAs that occurred at a distance inferior to twice the local LD were considered to belong to the same QTL. QTL associated to flowering time were removed from all downstream analyses (Table S2). A regression analysis was performed between the haplotype of the peak marker of each QTL to determine possible duplicate or homeolog loci. In addition, all the MTAs analyses were performed using Tassel 5 software [49].

2.5. Markers Conversion to KASP (Kompetitive Allele Specific PCR)

The array sequences of 20 markers associated to traits (MTA) were submitted to LGC Genomics for in-silico design of KASP primers using their proprietary software. Those that passed the in-silico criteria were purchased and used to genotype the independent validation set. For each marker that amplified and showed polymorphism, the regression cut-off between phenotype and haplotype was imposed at $r = 0.105$ following Pearson's critical value [54]. KASP markers AX-95260810, AX-94432276, and AX-95182463 were tested for association with grain yield, while AX-94408589 for association with biomass. In addition, the top 20 and worst 20 lines were considered as the true positive and true negative for heat tolerance. Hence, the accuracy was calculated as the ratio of the correct allelic call among all, sensitivity as the ratio of the correct positive allelic among the top 20 yielding lines, and specificity as the ratio of the correct negative (wt) allelic calls among the 20 worst yielding lines. The sequence of the validated KASP markers is provided in Table S3, or the primers can be ordered directly at LGC Genomics indicating the Axiom code used in this article.

3. Results

3.1. Agronomic Performance of the Genotypes and Sensitivity of Traits to Heat Stress

The combined analysis of variance across four environments (two different temperature treatments over two crop seasons) revealed significant genotypic differences for all traits measured (Table 1). The yield performance of the genotypes across environments averaged 2171 kg ha^{-1} and ranged from 352 kg ha^{-1} obtained under heat stress conditions for the lowest yielding line DWAyT-0215, to 4658 kg ha^{-1} under normal conditions for the highest yielding line DWAyT-0217. The top yielding line under heat-stress was the ICARDA/Moroccan cultivar 'Faraj' with an average yield of 2249 kg ha^{-1} over the two seasons.

Table 1. Descriptive statistics, component of trait variation, and heritability (h^2) among a set of 42 durum genotypes (G) tested under two treatments (T): normal and plastic tunnel-mediated heat stress during seasons 2015–2016 and 2016–2017.

Trait	Acronym	Mean	Min	Max	Genetic Variance (%)	Treatment Variance (%)	G × T (%)	h^2
Days to heading	DTH	92	71	109	34 **	1^{ns}	1^{ns}	0.78
Plant height (cm)	PH	81	71	92	60 **	1^{ns}	16^{ns}	0.76
Biomass (kg ha^{-1})	Biom	8407	4792	13,108	49 **	7 **	7 **	0.79
Spikes number per m^2	Spkm^2	524	370	640	14 **	1^{ns}	2 **	0.50
Grain yield (kg ha^{-1})	GY	2171	352	4658	30 **	44 **	12 *	0.63
Harvest index (%)	HI	26	1	50	15 **	34 **	13^{ns}	0.20
Thousand kernel weight (g)	TKW	36	27	45	48 **	1^{ns}	18 **	0.72
Grain number per spike	GNSpk	13	3	24	19 *	29 **	16 **	0.46

*, ** Significant at the 0.05 and 0.01 probability levels, respectively.

The treatment effect was significant only for Biom, GY, HI, and GNSpk, whereas DTH, PH, Spkm^2, and TKW were not significantly affected by treatments (Figure 2). The yield components were all significantly reduced under heat stress except TKW that showed a slight increase for the genotypes exposed to heat. The genotypes tested under plastic-tunnels had 61%, 54%, 42%, and 17% lower average GNSpk, GY, HI and Biom, respectively, compared to control. Relatively high heritability was observed for all the phenological and agronomical traits except for HI that had the lowest heritability ($h^2 = 0.20$).

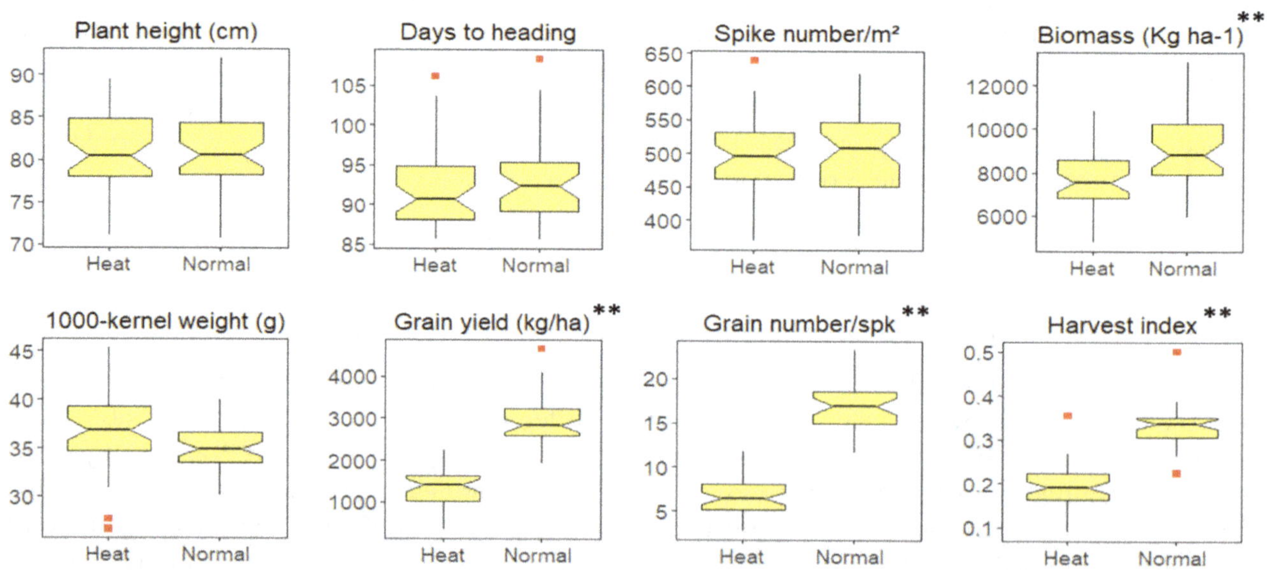

Figure 2. Boxplot of the best linear unbiased estimates (BLUEs) for various traits under two different environmental conditions (Heat: plastic tunnel-mediated heat stress and Normal) across two years. ** indicate significant difference between the means of control and heat-stressed plants at $p < 0.05$.

3.2. The Traits Interrelationship under Each Environmental Condition

Correlation analysis (Figure 3; Tables S4 and S5) was first conducted to investigate the interrelationship among all agronomic traits. Under both treatments, GNSpk had the highest association with GY ($r = 0.81$ under heat, $r = 0.67$ under normal), while Spkm^2 and TKW were the least correlated with GY. Biomass was also correlated with GY with $r = 0.61$ under heat and $r = 0.67$ under normal conditions. HI also showed a significant positive correlation with yield under both treatments, but its effect was stronger under heat stress ($r = 0.72$) than normal conditions ($r = 0.54$). DTH was not significantly correlated to any trait except HI ($r = -0.44$) under normal conditions.

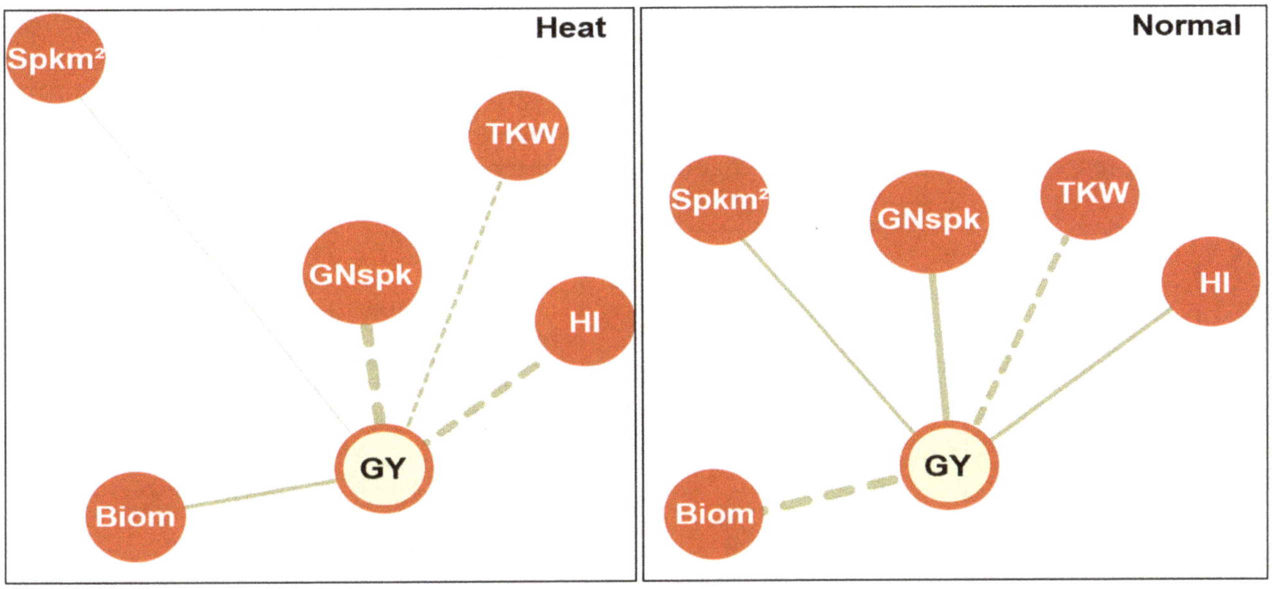

Figure 3. Relationships between grain yield (GY) and yield components (grain number per spike (GNspk), harvest index (HI), dry biomass (Biom), number of fertile spikes per meter square (Spkm^2), weight of a thousand kernels (TKW)) under plastic tunnel-mediated heat stress and normal conditions assessed by Pearson correlation and simple generalized additive model. The continuous grey line represents a linear relationship; the dashed grey line represents a non-linear relationship. The thickness of the line indicates the level of predictivity of the trait for GY. The length of the lines represents the correlation, the shorter the line the more the trait is correlated to GY.

Among yield components, the only significant and positive associations under the two environmental conditions were observed between Spkm^2, TKW, and Biom and between HI and GNSpk. Under heat conditions, a positive and significant correlation was noticed between GNSpk and Biom while under normal conditions HI was positively associated to TKW (Figure 3; Table S4).

The additive model was then used to further determine the nature of the relationship between GY and each predictor variable under normal and heat conditions (Figure 3; Table S5). The similarities observed between the two treatments in terms of the nature of relationship between GY and each of the predictors were the constantly linear and non-linear relationship between Spkm^2, TKW and the response variable GY, respectively.

GNSpk was considered the best predictor (deviance = 0.73%) with a complex relationship (EDF = 2.64) with GY under heat stress, whereas under normal conditions this trait was the second best predictor (deviance = 0.44%) with a linear relationship (EDF = 1). A similar trend was observed for HI in both treatments. Biom was found to be the best predictor (deviance = 0.52%) for GY with a non-linear relationship (EDF = 2.52) under normal conditions (Table S2; Figure S1).

3.3. Stress Tolerance Indices

Two different stress tolerance indices were calculated for GY: SSI and TOL (Figure 4). The genotypes showed wide variation for these indices. Seven SSI groups were identified with four having an SSI lower than 1 and the three remaining groups of genotypes having SSI > 1. The frequency distribution of the panel showed a wide variation and indicated the presence of susceptibility, with 45% of the genotypes falling in the very heat-susceptible class of SSI higher than 1, and only 7% of the lines showing high tolerance at SSI < 1. For TOL index, seven groups were also identified with 48% of the lines showing high yield depression and 5% of the genotypes presenting high stability. The smaller TOL values indicate the genotypes with low yield depression and hence more tolerant. However, good heat tolerance can also be reached by low yielding lines, but their value for breeding would be questionable. Hence, a scatterplot was devised to compare the GY under normal conditions and each of the heat indices (SSI and TOL). Five genotypes (four ICARDA lines, one Moroccan cultivar): Kunmiki, Berghouata1, Margherita2, IDON37-141, and Ourgh were found to have above average yield, low yield depression (low TOL values) and good heat tolerance (SSI < 1).

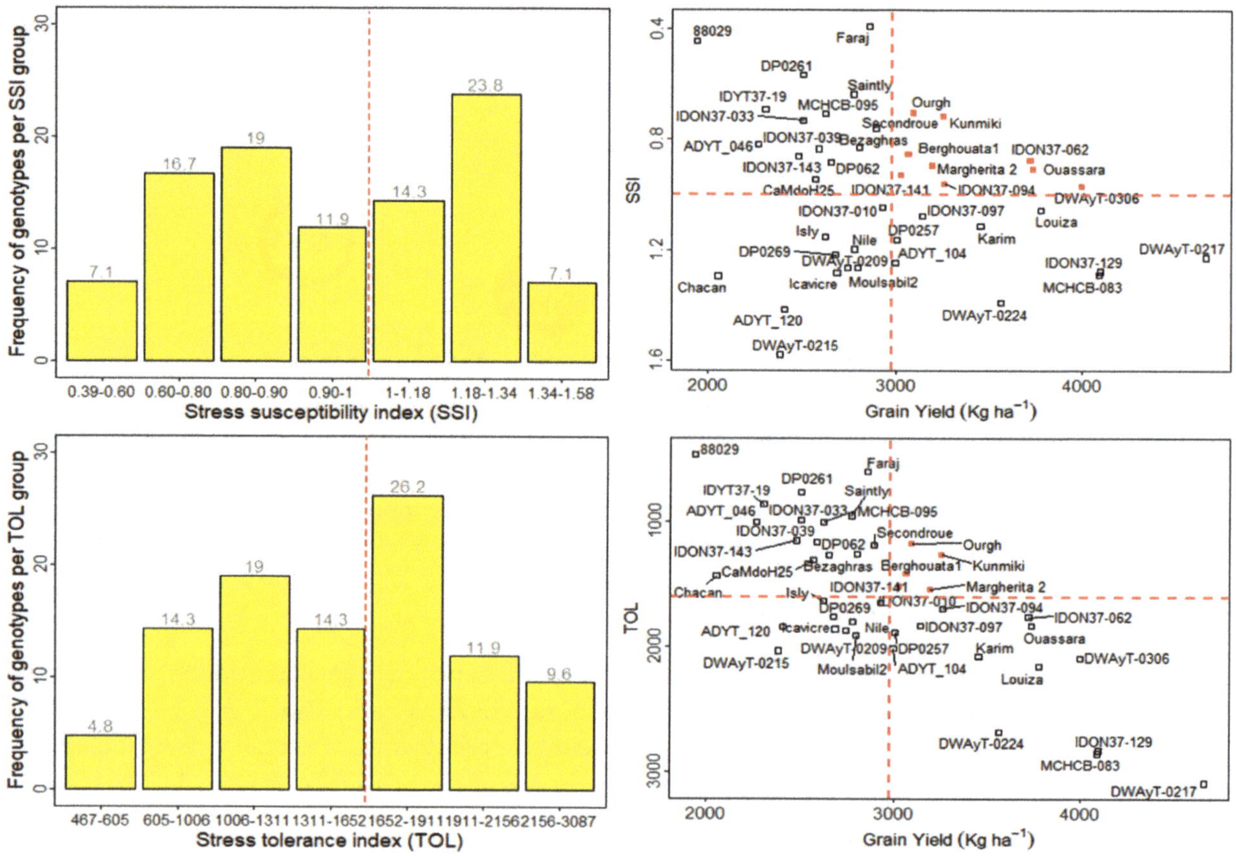

Figure 4. Two different stress tolerance indices SSI (stress susceptibility index) and TOL (tolerance index) of grain yield, comparing plastic tunnel-mediated heat stress with normal conditions for the 42 durum wheat genotypes. The bars plot shows the frequency distribution of SSI and TOL for the genotypes tested. The dashed red lines mark the separation between tolerant (left) and susceptible (right) genotypes. The scatter plot shows the yield performance of genotypes tested under normal conditions against each of SSI and TOL. The vertical dashed red lines indicate the average GY. The horizontal dashed red lines indicate the cut-off value for tolerant vs. susceptible genotypes for each index. Red dots indicate genotypes that were identified as superior by both bi-plots.

3.4. Markers Associated to Heat Stress Tolerance

A total of 204 MTAs were identified for four traits (GY, GNSpk, HI and Biom) under both stress and normal conditions and 49 MTAs were recorded for the two GY stress tolerance indices. Regression analysis and clustering based on local LD decay confirmed that these associations were distributed over 12 loci (Table 2 and Table S6). Chromosome 1A had the highest number of MTAs (27) while chromosome 4A had the lowest (6).

Under normal conditions, 56 MTAs were detected for three traits GY, GNSpk, and HI, with the third trait having the highest number of MTAs (48). No common region for these traits was identified under the non-stress environment. Under heat stress, a higher number of associations (148) were identified with trait variation (r^2) ranging from 0.25 to 0.36. The highest number of MTAs were detected for GNSpk distributed over 10 different loci, followed by HI on six loci. A common region for GY, GNSpk, HI, and Biom was identified under the heat condition on chromosome 6BS. Loci associated with both GNSpk and HI were detected on 1AL, 1BL, 2AL, 3AL, and 3BL. For heat tolerance indices (SSI-GY and TOL-GY), 49 MTAs were identified. The common loci associated with the two indices were on chromosomes 2AL, 5AL, and 5BL, while the loci on chromosomes 1AL and 6BS were identified only for TOL-GY and SSI-GY, respectively.

A comparison of the significant loci under each treatment and including the heat tolerance indices indicated a locus on chromosome 2AL, which was consistently identified for the indices, and both treatments for GNSpk and HI. Two loci on chromosomes 3AL and 3BL were associated with GNSpk and HI under both control and stress conditions, but were not associated with any of the indices. Three significant loci on chromosomes 1AL, 5BL, and 6BS were shared among heat stress treatment and stress tolerance indices, but not under normal conditions, making of these the most interesting genomic regions that specifically respond to heat stress. Overall, a total of 12 unique significant loci were identified (numbered QTL.ICD.Heat.01–QTL.ICD.Heat.12) and can be consulted in Table 2. Local LD decay was estimated for the 100 Mbp genomic region surrounding the peak marker. It varied between 31.7 and 108.7 Mbp, or a −38% to 112% variation compared to the average LD decay calculated for the whole panel (51.3 Mbp). This variation was accounted for to determine the correct physical size in each genomic region to assign multiple MTAs to the same QTL.

Table 2. Quantitative trait loci (QTLs) associated with multiple traits under plastic tunnel-mediated heat stress, normal conditions, and based on stress indices.

Locus	Trait	Chr. †	Main Marker	Position ‡ (bp)	Local LD (Mbp)	Max LOD	Max r^2	Heat Stress	Normal	Indices
QTL.ICD.Heat.01	GNspk, HI, TOL-GY	1AL	AX-94863732	570,040,339	31.7	3.38	0.27	*		*
QTL.ICD.Heat.02	GNspk, HI	1BL	AX-94447402	632,403,981	43.1	3.38	0.27	*		
QTL.ICD.Heat.03	GNspk, HI, SSI-GY, TOL-GY	2AL	AX-94538070	748,624,588	36.3	3.06	0.25	*	*	*
QTL.ICD.Heat.04	GY, HI	2BS	AX-95193898	6,012,904	36.0	3.67	0.36		*	
QTL.ICD.Heat.05	GNspk, HI	3AL	AX-95632723	562,421,267	75.4	3.39	0.27	*	*	
QTL.ICD.Heat.06	GNspk, HI	3BL	AX-95174625	788,551,042	85.4	3.38	0.27	*	*	
QTL.ICD.Heat.07	GNspk	5AS	AX-95247611	27,923,949	108.7	3.38	0.27	*		
QTL.ICD.Heat.08 §	SSI-GY, TOL-GY	5AS	AX-94631521	421,078,546	41.3	4.93	0.45			*
QTL.ICD.Heat.09 §	GNspk, SSI-GY, TOL-GY	5BS	AX-95182463	427,098,066	50.3	4.17	0.37	*		*
QTL.ICD.Heat.10 §	GNspk, HI, Biom, SSI-GY	6BS	AX-94408589	157,777,006	56.0	3.20	0.36	*		*
QTL.ICD.Heat.11	GNspk	7AL	AX-95074729	660,833,752	153.6	3.60	0.29	*		
QTL.ICD.Heat.12	GNspk, HI	7AS	AX-94381852	16,943,364	44.8	3.42	0.37		*	

† Chr.—Chromosome, based on alignment to durum wheat genome assembly [48].*—Significant QTL; ‡—Based on alignment to durum wheat genome assembly [48]; §—These QTLs have been converted into KASP markers and validated; GNspk—Grain number per spike; HI—Harvest index; TOL-GY—Tolerance index for grain yield; SSI-GY—Stress susceptibility index for grain yield; GY—Grain yield; Biom—Biomass.

3.5. *Effect of Different Allele Combination on Yield Performance*

The loci identified on chromosomes 1AL, 5BL, and 6BS appeared as the most critical for heat tolerance and were then tested further. These regions were associated with the control of multiple traits under heat stress: GY, GNspk, HI, Biom and the two indices SSI-GY and TOL-GY. A set of 208 modern lines were investigated for haplotype diversity at these three loci. Five groups with different allelic combinations were identified (Figure 5). Their allelic effect on GY was then assessed when field tested under high temperatures along the Senegal River [35]. The haplotype class with positive alleles at all three loci had the highest GY average reaching 2381 kg ha^{-1} with a maximum value of 3856 kg ha^{-1}. Genotypes of the haplotype classes with only two favorable alleles reached GY of 2199 and 2103 kg ha^{-1}, while lines that only carried one positive allele 2103 and 2023 kg ha^{-1} (Figure 5). ANOVA confirmed that the haplotype group with all three positive alleles was significantly superior to the others.

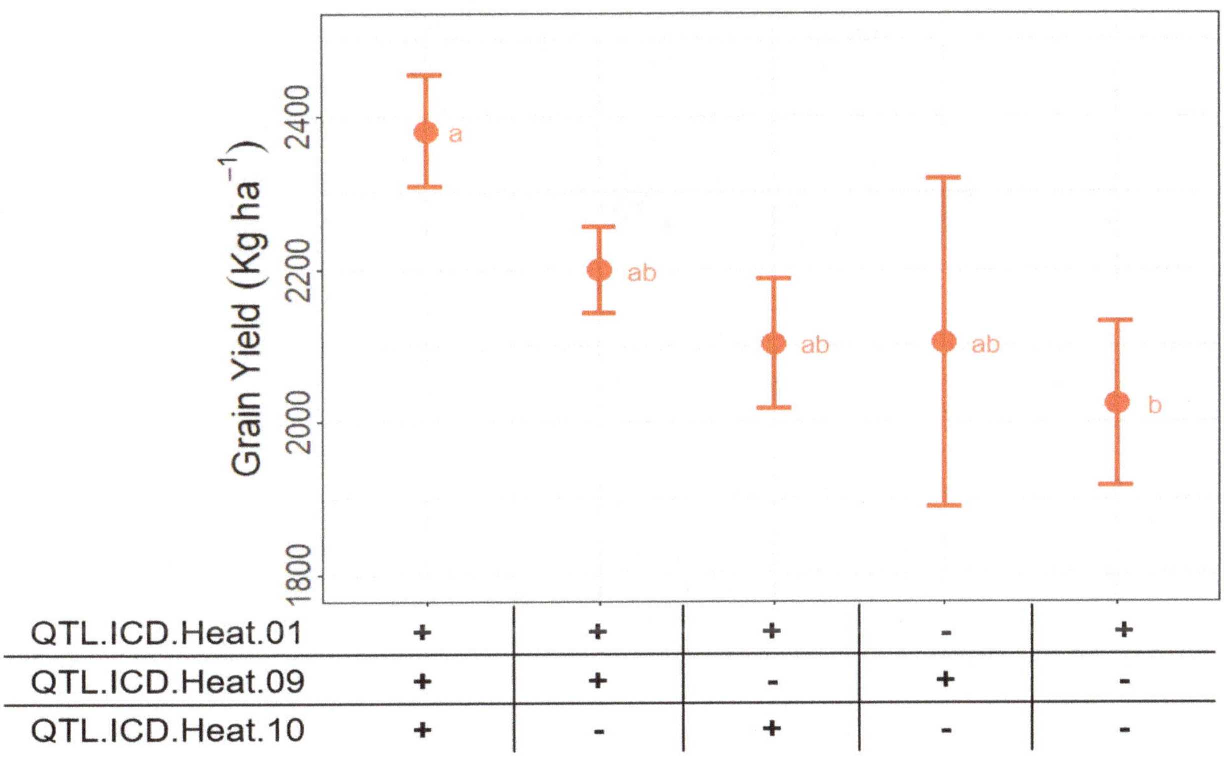

Figure 5. Effect of different allele combinations of the significant loci on yield performance of 208 accessions tested under heat stressed conditions along the Senegal River. The circle indicates the average of each class over 2 years, and the whiskers show the standard error of the mean. The accessions were divided into five clusters based on their haplotype for three major QTLs: "+" mark the positive and "-" the wild-type alleles. Letters (a, b, ab) indicate significant differences between the clusters.

3.6. *Validation of Markers for Marker Assisted Selection*

To effectively deploy in breeding the most interesting QTLs via MAS, it is first required a step of validation using more affordable marker methodologies and in different genetic backgrounds and environments. A total of 20 MTA sequences linked to important agronomical and spike fertility traits were submitted for KASP primers design. Among these, only 14 could be successfully designed, and 11 identified a polymorphism within the validation set. Four showed significant ($p < 0.05$) correlation to the test phenotype (Figure 6). Three QTLs were represented by these four markers, AX-95260810 and AX-94432276 tagged QTL.ICD.Heat.08 on chromosome 5AL, AX-95182463 underlines QTL.ICD.Heat.09 on chromosome 5BL, and AX-94408589 tags QTL.ICD.Heat.10 on chromosome 6BS. The latter two QTLs are among the three main effect regions identified in this study (Figure 5).

AX-95260810 reached 15% correlation to grain yield under heat, 74% accuracy, 43% sensitivity, and 100% specificity. Especially, its ability to identify 100% of non-heat tolerant entries is particularly remarkable. AX-95182463 and AX-94408589 also reached significant correlations of 14% and 32% for grain yield and biomass under severe heat, respectively, with sensitivities of 62% and 40%, accuracies of 30% and 65%, and specificities of 4% and 90%. Overall, AX-9526081 and AX-94408589 appeared as the most suitable for MAS application.

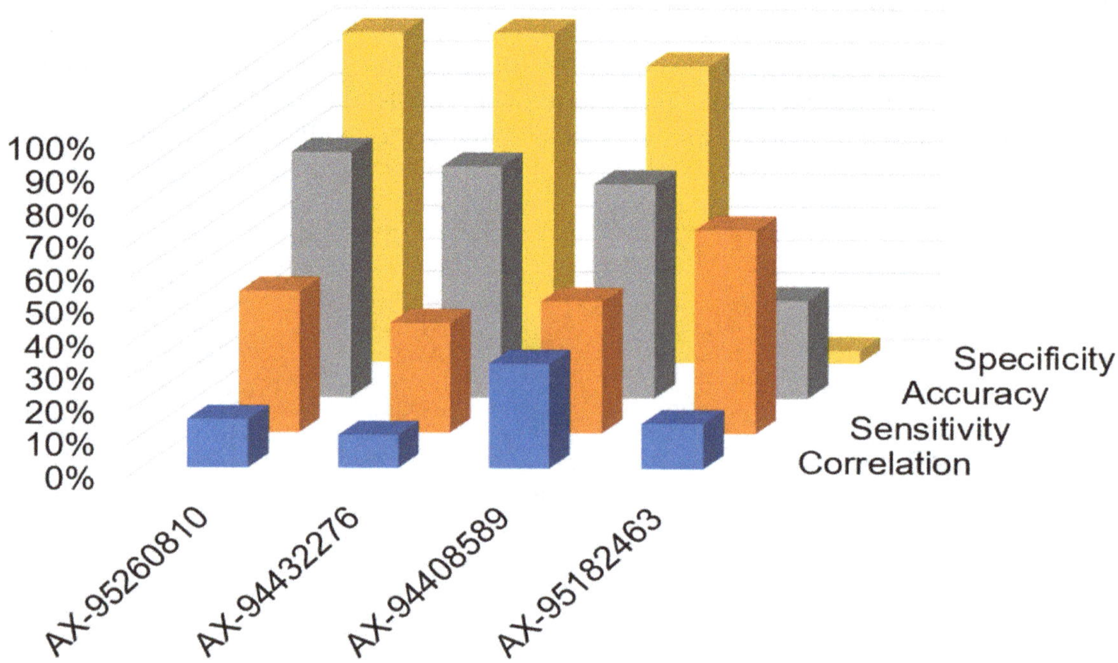

Figure 6. Kompetitive Allele Specific PCR (KASP) markers validation on an independent set of 94 elite lines of ICARDA tested under severe heat for grain yield and biomass. Correlation was measured between the BLUE for grain yield recorded along the Senegal River and the haplotype score. Accuracy, sensitivity, and specificity where determined using only the top 20 and worst 20 lines. AX-95260810 and AX-94432276 tag QTL.ICD.Heat.08, AX-95182463 tags QTL.ICD.Heat.09, AX-94408589 tags QTL.ICD.Heat.10.

4. Discussion

4.1. Evaluation of the Phenotypic Performance of Yield and Yield Components under Normal and Heat Stress Conditions

Several studies reported that wheat plants are very sensitive to elevated temperatures during flowering and grain filling phases [9,55,56], due to a reduction in seed development and fertility [56–58]. This study evaluated a set of durum wheat genotypes derived from a global collection for GY and yield components under heat and normal conditions. The genetic and phenotypic diversity shown by this set together with its relatively similar flowering time, promote it as an ideal panel to test heat tolerance. Further, the plastic tunnel method deployed here allowed to increase the temperatures well above 21 °C, the value that defines the absence of the stress [9]. A similar methodology was also successfully deployed by Corbellini et al. [54] to study the effect of heat shock proteins on technological quality characteristics. Compared to timely vs. delayed sowing experiments to simulate heat stress, the use of the plastic tunnel method avoids incurring false discovery due to changes in the phenological behavior of plants.

In the present study, a short and severe episode of heat stress was applied from the beginning of heading to the early dough stage, and resulted in 54% reduction in grain yield. This was in agreement with the study conducted by Ugarte et al. [59] that found a reduction of up to 52% when thermal treatment was applied via transparent chambers. Interestingly, our stress treatment caused an average temperature increase of 10 °C, which caused an average GY reduction of 5.4% for each 1 °C raise. This value is well within the 4.1% to 6.4% interval suggested by Liu et al. [10] for 1 °C raise in temperatures. GNSpk was the most affected trait (−61%) with the highest positive correlation to GY. This is in good agreement with previous studies that have shown that seed setting is the most sensitive parameter to heat stress, with a noticeable influence on yield [28,60–62]. Still, its non-linear relationship to yield confirms the complexity of the trait. Biom and HI were also found to have an influence on yield [63,64] with different relationships based on the occurrence of the stress. The presence of dissimilarities of the associations between the two treatments indicates clearly that there is a trade-off among the yield components as previously reported by Sukumaran et al. [65] for grain weight and grain number. Variation of one of the yield components affect the others positively or negatively. Compared to the simple regression, the additive model allowed to reveal the complexity of the relationship between GY and yield related traits.

The stress index SSI was developed by Fisher and Maurer [43] and modified by Nachit and Ouassou [44] as a useful indicator and a good parameter for selection. It measures the severity of the heat stress [66,67] and was also used in earlier studies in wheat to seek heat tolerant genotypes [23,68,69]. The TOL index is instead useful for selecting against yield depression, and it was used in several studies for heat or drought tolerance in wheat [27,44,67,70]. Improving heat tolerance should not be based on the use of these criterions alone as was suggested by Clarke et al. [71]. It is important to select simultaneously for good yield performance coupled with good adaptability (SSI < 1) and stability (low TOL) [44]. In that sense, the accessions Kunmiki, Berghouata1, Margherita2, and IDON37-141 originated from ICARDA durum wheat program, and Ourgh, a Moroccan cultivar, have been identified as high yielding genotypes that also show good heat stress tolerance based on the two indices.

4.2. *Dissection of Heat-Specific QTLs Associated with Yield-Related Traits and Stress Tolerance Indices*

The significant correlation identified between yield and its components were not linear in nature, and tend to change their mode of action based on the occurrence of the stress. Therefore, several physiological processes are simultaneously involved in protecting the wheat plant from the heat stress [72], and there is value in dissecting it into its genetic components. In this study GWAS was used to identify the genetic regions controlling the response of the various traits. To prevent the confounding effect that phenology-related loci might have [73], MTAs were identified for DTH and removed from downstream analysis. Additionally, flowering time was used as covariate in all analyses for the other traits. Very few MTAs for DTH were observed either in normal or stressed conditions due to the synchronized flowering of the entries used in this study. This indicated the absence of confounding effects between the two trials. i.e., almost all the accessions were exposed to the same conditions in each developmental phase [74] before imposing the stress.

Out of 12 QTLs identified, three occurred only when the heat stress was imposed, including indices. These three main genomic regions occurred on chromosomes 1AL, 5BL, and 6BS, and were considered as QTLs controlling heat tolerance. These three loci were confirmed by mean of haplotype analysis on a larger panel of modern lines (208 entries) field tested under severe heat along the Senegal River valley [35], to confirm that the presence of the positive alleles at all three loci provided a significant GY advantage of +182 kg ha^{-1} (+8%). The QTL on the long arm of chromosome 1A controlled GNSpk, HI, and TOL-GY, and it explained up to 27% of the phenotypic variation. In a study with double haploid population of bread wheat, Heidari et al. [75] identified a major QTL on the same chromosome (1A), influencing grain number per spike, grain weight per spike, and spikes/m^2. However, their phenotypic assessment was not performed under heat stress, the marker systems used was different compared to our study and the locus was identified in the short arm of chromosome 1A. Therefore, it is quite difficult to align the results from that study to the current one. Another study had previously reported many MTAs on chromosome 1A detected for yield components under heat stress, but all were found to have a pleiotropic relationship with days to heading and were also located on the short arm of 1A [26], instead of 1AL found here. A heat-specific QTL was also detected on the same chromosome in the short arm for spikelet compactness and leaf rolling in bread wheat [76]. An earlier study identified a QTL on 1AS for yield but associated with different stress conditions [77]. To the best of our knowledge, this is the first time that this region on 1AL is presented as associated to GNSpk, HI, and TOL-GY in durum wheat under heat stress conditions. The second major QTL region was detected on the long arm of chromosome 5B and found to be associated with GNSpk and the two indices SSI-GY and TOL-GY, contributing to 37% of the phenotypic variation. A region in the short arm of the same chromosome has been previously reported to be associated with grain number per square meter in bread wheat [76], and controlling thousand grain weight in durum wheat [27] under combined drought and heat stress. Shirdelmoghanloo et al. [25] and Acuna-Galindo et al. [78] reported loci for grain weight and other important traits on chromosome 5B under heat and non-heat conditions in hexaploid wheat. On the other hand, the same chromosome has been previously suggested to carry heat-specific QTLs for yield per se in bread wheat [26]. Sukumaran et al. [27] identified markers for heat susceptibility (HSI or SSI) and tolerance (TOL) indices for yield and grain number per square meter on the short arm of the chromosome 5B. Mason et al. [64] also detected QTL for HSI for kernel number on 5BL in bread wheat. The genomic region identified in this study on 5BL is likely to be a new QTL since no information has been reported earlier for this locus associated to GNSpk, SSI-GY, and TOL-GY in durum wheat and specific to heat stress, but we cannot exclude that it overlaps with previously reported QTLs. A third heat-responsive locus was identified on the short arm of chromosome 6B related to GY, SSI-GY, GNspk, HI, and Biom accounting for 36% of the phenotypic variance. An earlier study on bread wheat identified a locus on chromosome 6BS underpinning chlorophyll loss rates and heat susceptibility index for grain weight and chlorophyll loss rates under heat-stress conditions [25]. Under post-anthesis high temperatures stress, Vijayalakshmi et al. [20] reported a QTL on the short arm of chromosome 6B for senescence related traits in hexaploid wheat. McIntyre et al. [79] and Pinto et al. [21] reported QTLs on chromosome 6BL that were associated with many important traits (grain number per square meter and grain yield and water-soluble carbohydrate content) related to drought and heat tolerance. Ogbonnaya et al. [26] found a locus on the short arm of chromosome 6B for grain yield under heat stress in bread wheat. These previously reported QTLs in 6B could overlap with the one identified in this study, but they were either identified not in association with heat tolerance or detected in hexaploidy wheat. Therefore, this region is also assumed to have been reported for the first time here in relationship to heat tolerance for durum wheat. This locus affects multiple traits (GY, GNspk, HI, Biom, and two heat susceptibility indexes) and hence it is of good importance for deployment in breeding. The principal breeding objective is to develop varieties with high grain yield and stability when exposed to different stresses. However, grain yield is a complex trait controlled by many genes and strongly influenced by the environment [80–86]. Therefore, a good understanding of traits and underlying loci associated with tolerance to elevated temperatures is of a great importance for breeding new heat tolerant cultivars [87].

4.3. *Pyramiding Heat-Tolerant QTLS via MAS*

Three loci on chromosomes 1AL, 5BS, and 6BS showed an additive nature by means of haplotype analysis (Figure 5), revealing that only the combination of all three positive alleles generated a true yield advantage. Among the most heat tolerant elite lines identified here 'Kunmiki', 'Berghouata1', and 'Ourgh' confirmed to harbor the positive alleles for all three loci. This prompts their use in crossing schemes to pyramid the positive alleles, as well as the deployment of simple marker system to conduct MAS.

Axiom to KASP marker conversion and validation was attempted for 20 MTAs. Eleven KASP markers generated polymorphic haplotypes in an independent set of ICARDA elite lines. Four revealed a significant ($p < 0.05$) correlation to GY and biomass assessed under severe heat along the Senegal River Valley (Figure 6). In particular, AX-95182463 tags QTL.ICD.Heat.09 located on chromosome 5B and it revealed good correlation and sensitivity, but lacks in accuracy and specificity, and it is hence protected from Type II errors, but prone to Type I, with several elite lines wrongly identified as carrying the positive alleles. AX-95260810 tags QTL.ICD.Heat.08, linked to the two stress tolerance indices for GY (SSI-GY and TOL-GY) located on chromosome 5A. AX-94408589 tags QTL.ICD.Heat.10 located on chromosome 6B, and associated to several traits GNspk, HI, Biom, SSI-GY. In these two cases, the KASP markers explained 15% and 33% of the phenotypic variation of an independent validation set, with 100% and 90% specificity, and 74% and 65% accuracy, but medium sensitivity (43% and 40%). As such, these markers are protected against Type I errors (no false positive), but prone to Type II errors, with several elite lines identified as not carrying the positive allele while instead being tolerant to heat. Hence, while all converted KASP markers are prone to different types of errors, these three markers can be considered as validated and ready to be deployed in breeding. The combination of the three might represent a more stringent approach to protect against both types of errors. An additional nine QTLs were identified in this study, and their KASP conversion and validation are still ongoing and will require better targeted efforts to be achieved.

5. Conclusions

Heat stress causes a complex cascade of negative effects on the wheat plant, resulting in drastic reductions in grain yield. The deployment of heat tolerant varieties that will benefit greatly farmers requires first to enhance our understanding of this mechanism and loci governing it. Our study combined a discovery phase with a core set tested over two field seasons in Morocco under artificial heat-treatment with plastic tunnels, followed by a different confirmation set of germplasm grown for two seasons in Kaedi, Mauritania under severe natural heat, and completed with one final validation set tested one season in Kaedi. Our results confirmed that spike fertility (GNSpk) and maintenance of green leaves (Biom) are the most critical traits to drive tolerance to this stress, and hence should be the primary targets of durum wheat breeders. Further, the deployment of plastic tunnels proved to be a strategic methodology to study this stress and reveal its mechanisms without affecting the phenology of the plant. In addition, 12 loci were identified as responsible for controlling the main heat tolerance traits. Among these, three were activated only when the stress occurred and hence represent ideal targets for breeding. Two of these were validated into a KASP marker and are now ready for deployment via MAS, especially if associated with a third, also validated, KASP. Finally, three ICARDA elite lines and one Moroccan cultivar were confirmed as tolerant to heat, with high grain yield, and carrying positive alleles for three main QTLs. These are freely available and should be incorporated as crossing parents by other breeding programs. Altogether, this study has confirmed the key traits for heat tolerance as well as a new methodology to study it in durum wheat, it has revealed the main loci controlling these traits and proceeded to validate three of them for MAS, and it has also provided freely available elite lines to breed new cultivars better adapted to the stress.

Supplementary Materials: Table S1: List of durum wheat genotypes evaluated under plastic tunnel-mediated heat stress in the present study, Table S2: Markers associated with days to heading (DTH) under heat stress and normal conditions, Table S3: Sequence information of the KASP markers, Table S4: Pearson correlation matrix between all the measured traits under heat conditions (upper part) and normal (lower part) conditions. GY—Grain yield; Biom—Biomass; HI—Harvest index; Spkm^2—Spikes per square meter: GNspk—Grain number per spike; TKW—Thousand kernel weight; DTH—Days to heading. *, ** Significant at the 0.05 and 0.01 probability levels, respectively, Table S5: Correlation (r), linear regression estimated via ordinary least squares (OLS) and flexible regression estimated via regression additive model. (**a**) Under heat stress. (**b**) Under normal conditions, Table S6: Regression matrix between the haplotype of the peak markers for the 13 identified QTLs. *, significant loci similarity at $p < 0.05$ consistent with homeologous relationship; **, significant loci identity ($p < 0.01$) consistent with wrongly assigned genomic position, Figure S1: Plots of the additive regression model showing GNspk, biom, TKW, spkm^2 and HI as the spline function of the target trait grain yield (GY). (**a**) Under heat stress. (**b**) Under normal conditions

Author Contributions: F.M.B., M.N. and K.E.H. conceived and designed the study. K.E.H. and F.M.B. performed the field experiment. A.T.S. performed the field experiment in the Senegal river. A.A. contributed in the genotyping. K.E.H. and F.M.B. analyzed the data. K.E.H. Wrote the original draft. K.E.H., B.B., A.F.M., A.A., M.N., and F.M.B. wrote or reviewed the manuscript. All authors read and approved the final manuscript.

Acknowledgments: The authors wish to acknowledge the technical assistance provided by A. Rached and all ICARDA durum wheat program staff in handling field activities.

References

1. Hays, D.; Mason, E.; Hwa, D.J.; Menz, M.; Reynolds, M. Expression quantitative trait loci mapping heat tolerance during reproductive development in wheat (*T. aestivum* L.). In *Wheat Production in Stressed Environment*; Buck, H.T., Nisi, J.E., Salomon, N., Eds.; Springer: Amsterdam, The Netherlands, 2007; pp. 373–382.
2. Saini, H.; Aspinall, D. Abnormal sporogenesis in wheat (*Triticum aestivum* L.) induced by short periods of high temperature. *Ann. Bot.* **1982**, *49*, 835–846. [CrossRef]
3. Plaut, Z.; Butow, B.J.; Blumenthal, C.S.; Wrigley, C.W. Transport of dry matter into developing wheat kernels and its contribution to grain yield under post-anthesis water deficit and elevated temperature. *Field Crops Res.* **2004**, *86*, 185–198. [CrossRef]
4. Hunt, L.A.; van der Poorten, G.; Pararajasingham, S. Post-anthesis temperature effects on duration and rate of grain filling in some winter and spring wheats. *Can. J. Plant Sci.* **1991**, *71*, 609–617. [CrossRef]
5. Wollenweber, B.; Porter, J.R.; Schellberg, J. Lack of interaction between extreme high-temperature events at vegetative and reproductive growth stages in wheat. *J. Agron. Crop Sci.* **2003**, *189*, 142–150. [CrossRef]
6. Dias, A.S.; Lidon, F.C. Bread and durum wheat tolerance under heat stress: A synoptical overview. *J. Food Agric.* **2010**, *22*, 412–436. [CrossRef]
7. Dolferus, R.; Ji, X.; Richards, R.A. Abiotic stress and control of grain number in cereals. *Plant Sci.* **2011**, *181*, 331–341. [CrossRef] [PubMed]
8. Porter, J.R.; Gawith, M. Temperatures and the growth and development of wheat: A review. *Eur. J. Agron.* **1999**, *10*, 23–36. [CrossRef]
9. Farooq, M.; Bramley, H.; Palta, J.A.; Siddique, K.H.M. Heat stress in wheat during reproductive and grain-filling phases. *Crit. Rev. Plant Sci.* **2011**, *30*, 1–17. [CrossRef]
10. Liu, B.; Asseng, S.; Muller, C.; Ewert, F.; Elliott, J.; Lobell, D.B.; Rosenzweig, C. Similar estimates of temperature impacts on global wheat yield by three independent methods. *Nat. Clim. Chang.* **2016**, *6*, 1130–1136. [CrossRef]
11. Lillemo, M.; van Ginkel, M.; Trethowan, R.M.; Hernandez, E.; Crossa, J. Differential adaptation of CIMMYT bread wheat to global high temperature environments. *Crop Sci.* **2005**, *45*, 2443–2453. [CrossRef]
12. Hansen, J.; Sato, M.; Ruedy, R. Perception of climate change. *Proc. Natl. Acad. Sci. USA* **2012**, *109*, 2415–2423. [CrossRef] [PubMed]
13. United Nations Economic and Social Commission for Western Asia (ESCWA). *Arab Climate Change Assessment Report—Main Report*; E/ESCWA/SDPD/2017/RICCAR/Report; United Nations Publication: Beirut, Lebanon, 2017.
14. Ayeneh, A.; van Ginkel, M.; Reynolds, M.P.; Ammar, K. Comparison of leaf, spike, peduncle, and canopy temperature depression in wheat under heat stress. *Field Crops Res.* **2002**, *79*, 173–184. [CrossRef]
15. Yang, J.; Sears, R.G.; Gill, B.S.; Paulsen, G.M. Quantitative and molecular characterization of heat tolerance in hexaploid wheat. *Euphytica* **2002**, *126*, 275–282. [CrossRef]
16. Nachit, M.M.; Elouafi, I. Durum adaptation in the Mediterranean dryland breeding: Breeding, stress physiology, and molecular markers. *Crop Sci. Soc. Am. Am. Soc. Agron. CSSA* **2004**, *32*, 203–218.
17. Reynolds, M.P.; Hobbs, P.R.; Braun, H.J. Challenges to international wheat improvement. *J. Agric. Sci.* **2007**, *145*, 223–227. [CrossRef]
18. Tuberosa, R. Phenotyping for drought tolerance of crops in the genomics era. *Front. Physiol.* **2012**, *3*, 347. [CrossRef] [PubMed]
19. Reynolds, M.; Langridge, P. Physiological breeding. *Curr. Opin. Plant Biol.* **2016**, *31*, 162–171. [CrossRef] [PubMed]
20. Vijayalakshmi, K.; Fritz, A.K.; Paulsen, G.M.; Bai, G.; Pandravada, S.; Gill, B.S. Modeling and mapping QTL for senescence-related traits in winter wheat under high temperature. *Mol. Breed.* **2010**, *26*, 163–175. [CrossRef]

21. Pinto, R.S.; Reynolds, M.P.; Mathews, K.L.; McIntyre, C.L.; Olivares-Villegas, J.J.; Chapman, S.C. Heat and drought adaptive QTL in a wheat population designed to minimize confounding agronomic effects. *Theor. Appl. Genet.* **2010**, *121*, 1001–1021. [CrossRef]
22. Bennett, D.; Reynolds, M.; Mullan, D.; Izanloo, A.; Kuchel, H.; Langridge, P.; Schnurbusch, T. Detection of two major grain yield QTL in bread wheat (*Triticum aestivum* L.) under heat, drought and high yield potential environments. *Theor. Appl. Genet.* **2012**, *125*, 1473–1485. [CrossRef]
23. Paliwal, R.; Röder, M.S.; Kumar, U.; Srivastava, J.P.; Joshi, A.K. QTL mapping of terminal heat tolerance in hexaploid wheat (*T. aestivum* L.). *Theor. Appl. Genet.* **2012**, *125*, 561–575. [CrossRef] [PubMed]
24. Talukder, S.K.; Babar, M.A.; Vijayalakshmi, K.; Poland, J.; Prasad, P.V.V.; Bowden, R.; Fritz, A. Mapping QTL for the traits associated with heat tolerance in wheat (*Triticum aestivum* L.). *BMC Genet.* **2014**, *15*, 97. [CrossRef] [PubMed]
25. Shirdelmoghanloo, H.; Taylor, J.D.; Lohraseb, I.; Rabie, H.; Brien, C.; Timmins, A.; Martin, P.; Mather, D.E.; Emebiri, L.; Collins, N.C. A QTL on the short arm of wheat (*Triticum aestivum* L.) chromosome 3B affects the stability of grain weight in plants exposed to a brief heat shock early in grain filling. *BMC Plant Biol.* **2016**, *16*, 100. [CrossRef] [PubMed]
26. Ogbonnaya, F.C.; Rasheed, A.; Okechukwu, E.C.; Jighly, A.; Makdis, F.; Wuletaw, T.; Hagras, A.; Uguru, M.I.; Agbo, C.U. Genome-wide association study for agronomic and physiological traits in spring wheat evaluated in a range of heat prone environments. *Theor. Appl. Genet.* **2017**, *130*, 1819–1835. [CrossRef] [PubMed]
27. Sukumaran, S.; Reynolds, M.P.; Sansaloni, C. Genome-Wide Association Analyses Identify QTL Hotspots for Yield and Component Traits in Durum Wheat Grown under Yield Potential, Drought, and Heat Stress Environments. *Front. Plant Sci.* **2018**, *9*, 81. [CrossRef] [PubMed]
28. Sall, A.T.; Cisse, M.; Gueye, H.; Kabbaj, H.; Ndoye, I.; Maltouf, A.F.; El-Mourid, M.; Ortiz, R.; Bassi, F.M. Heat tolerance of durum wheat (*triticum durum* desf.) elite germplasm tested along the Senegal river. *J. Agric. Sci.* **2018**, *10*, 217–233. [CrossRef]
29. Pradhan, G.P.; Prasad, P.V.V.; Fritz, A.; Kirkham, M.B.; Gill, B.S. Effects of drought and high temperature stress on synthetic hexaploid wheat. *Funct. Plant Biol.* **2012**, *39*, 190–198. [CrossRef]
30. Weldearegay, D.F.; Yan, F.; Jiang, D.; Liu, F. Independent and combined effects of soil warming and drought stress during anthesis on seed set and grain yield in two spring wheat varieties. *J. Agron. Crop Sci.* **2012**, *198*, 245–253. [CrossRef]
31. Jha, U.C.; Bohra, A.; Singh, N.P. Heat stress in crop plants: Its nature, impacts and integrated breeding strategies to improve heat tolerance. *Plant Breed.* **2014**, *133*, 679–701. [CrossRef]
32. Mishra, S.C.; Singh, S.K.; Patil, R.; Bhusal, N.; Malik, A.; Sareen, S. Breeding for heat tolerance in wheat. In *Recent Trends on Production Strategies of Wheat in India*; Shukla, R.S., Mishra, P.C., Chatrath, R., Gupta, R.K., Tomar, S.S., Sharma, I., Eds.; JNKVV, Jabalpur & ICAR-IIWBR: Karnal, Indi, 2014; pp. 15–29.
33. Ni, A.; Li, H.; Zhao, Y.; Peng, H.; Hu, Z.; Xin, M.; Sun, Q. Genetic improvement of heat tolerance in wheat: Recent progress in understanding the underlying molecular mechanisms. *Crop J.* **2017**, *6*, 32–41. [CrossRef]
34. Kabbaj, H.; Sall, A.T.; Al-Abdallat, A.; Geleta, M.; Amri, A.; Filali-Maltouf, A.; Bassi, F.M. Genetic diversity within a global panel of durum wheat (*Triticum durum* L.) landraces and modern germplasm reveals the history of alleles exchange. *Front. Plant Sci.* **2017**, *8*, 1277. [CrossRef] [PubMed]
35. Sall, A.T.; Bassi, F.M.; Cisse, M.; Gueye, H.; Ndoye, I.; Filali-Maltouf, A.; Ortiz, R. Durum wheat breeding: In the heat of the Senegal river. *Agriculture* **2018**, *8*, 99. [CrossRef]
36. Bates, D.; Maechler, M.; Bolker, B.; Walker, S. Fitting linear mixed-effects models using lme4. *J. Stat. Softw.* **2015**, *67*, 1–48. [CrossRef]
37. R Development Core Team. *R: A Language and Environment for Statistical Computing*; R Foundation For Statistical Computing: Vienna, Austria, 2016.
38. DeLacy, I.H.; Basford, K.E.; Cooper, M.; Bull, J.K.; McLaren, C.G. Analysis of multi-environment trials: An historical perspective. In *Plant Adaptation and Crop Improvement*; Cooper, M., Hammer, G.L., Eds.; CAB International: Oxfordshire, UK, 1996.
39. Wickham, H. *ggplot2: Elegant Graphics for Data Analysis*; Springer: New York, NY, USA, 2009.
40. Wei, T.; Simko, V. *R package "corrplot": Visualization of a Correlation Matrix (Version 0.84)*; R Foundation for Statistical Computing: Vienna, Austria, 2017.

41. Wood, S.N. *Generalized Additive Models: An Introduction with R*, 2nd ed.; Chapman & Hall, CRC: Boca Raton, Florida, 2017; ISBN 9781498728331.
42. Wood, S.N.; Pya, N.; Saefken, B. Smoothing parameter and model selection for general smooth models. *J. Am. Stat. Assoc.* **2016**, *111*, 1548–1575. [CrossRef]
43. Fisher, R.A.; Maurer, R. Drought resistance in spring wheat cultivars I. grain yield responses. *Aust. J. Agric. Res.* **1978**, *29*, 897–917. [CrossRef]
44. Nachit, M.M.; Ouassou, A. Association of yield potential, drought tolerance and stability of yield in *Triticum turgidum* var *durum*. In Proceedings of the 7th International Wheat Symposium, Cambridge, UK, 13–19 July 1988; pp. 867–870.
45. Rosielle, A.A.; Hamblin, J. Theoretical aspects of selection for yield in stress and non-stress environments. *Crop Sci.* **1981**, *21*, 943–946. [CrossRef]
46. Bivand, R.; Ono, H.; Dunlap, R.; Stigler, M. R Package Classint: Choose Univariate Class Intervals. 2018. Available online: http://cran.r-project.org/package=classInt (accessed on 10 September 2019).
47. Lenth, R.; Love, J.; Maxime, H. Estimated Marginal Means, aka Least-Squares Means. 2017. Available online: http://cran.r-project.org/package=emmeans (accessed on 14 June 2019).
48. Maccaferri, M.; Harris, N.S.; Twardziok, S.O.; Pasam, R.K.; Gundlach, H.; Spannagl, M.; Himmelbach, A. Durum wheat genome highlights past domestication signatures and future improvement targets. *Nat. Genet.* **2019**, *51*, 885–895. [CrossRef] [PubMed]
49. Bradbury, P.J.; Zhang, Z.; Kroon, D.E.; Casstevens, T.M.; Ramdoss, Y.; Buckler, E.S. TASSEL: Software for association mapping of complex traits in diverse samples. *Bioinformatics* **2007**, *23*, 2633–2635. [CrossRef]
50. Jujumaan. Estimating and Plotting the Decay of Linkage Disequilibrium. Available online: https://jujumaan.com/2017/07/15/linkage-disequilibrium-decay-plot/ (accessed on 15 July 2017).
51. Bassi, F.; Brahmi, H.; Sabraoui, A.; Amri, A.; Nsarellah, N.; Nachit, M.M.; Al-Abdallat, A.; Chen, M.S.; Lazraq, A.; El Bouhssini, M. Genetic identification of loci for Hessian fly resistance in durum wheat. *Mol. Breed.* **2019**, *39*, 24. [CrossRef]
52. Sukumaran, S.; Xiang, W.; Bean, S.R.; Pedersen, J.F.; Kresovich, S.; Tuinstra, M.R.; Yu, J. Association mapping for grain quality in a diverse sorghum collection. *Plant Genom. J.* **2012**, *5*, 126–135. [CrossRef]
53. Duggal, P.; Gillanders, E.M.; Holmes, T.N.; Bailey-Wilson, J.E. Establishing an adjusted p-value threshold to control the family-wide type 1 error in genome wide association studies. *BMC Genom.* **2008**, *9*, 516. [CrossRef] [PubMed]
54. Pearson, K. Notes on regression and inheritance in the case of two parents. *Proc. R. Soc. Lond.* **1985**, *58*, 240–242.
55. Sofield, I.; Evans, L.T.; Cook, M.G.; Wardlaw, I.F. Factors influencing the rate and duration of grain filling in wheat. *Aust. J. Plant Physiol.* **1977**, *4*, 785–797. [CrossRef]
56. Gooding, M.J.; Ellis, R.H.; Shewry, P.R.; Schofield, J.D. Effects of restricted water availability and increased temperature on the grain filling, drying and quality of winter wheat. *J. Cereal Sci.* **2003**, *37*, 295–309. [CrossRef]
57. Corbellini, M.; Canevara, M.G.; Mazza, L.; Ciaffi, M.; Lafiandra, D.; Tozzi, L.; Borghi, B. Effect of the duration and intensity of heat shock during grain filling on dry matter and protein accumulation, technological quality and protein composition in bread and durum wheat. *Aust. J. Plant Physiol.* **1997**, *24*, 245–250. [CrossRef]
58. Corbellini, M.; Mazza, L.; Ciaffi, M.; Lafiandra, D.; Borghi, B. Effect of heat shock during grain filling on protein composition and technological quality of wheats. *Euphytica* **1998**, *100*, 147–154. [CrossRef]
59. Ugarte, C.; Calderini, D.F.; Slafer, G.A. Grain weight and grain number responsiveness to pre-anthesis temperature in wheat, barley and triticale. *Field Crops Res.* **2006**, *100*, 240–248. [CrossRef]
60. Nachit, M.M. Durum breeding for Mediterranean drylands of North Africa and West Asia. In *Durum Wheat: Challenges and Opportunities*; Rajaram, S., Saari, E., Hettel, G.P., Eds.; Wheat Special Report; CIMMYT: Ciudad Obregon, Mexico, 1992; Volume 9.
61. Barnabas, B.; Jager, K.; Feher, A. The effect of drought and heat stress on reproductive processes in cereals. *Plant Cell Environ.* **2008**, *31*, 11–38. [CrossRef]
62. Reynolds, M.; Foulkes, J.; Furbank, R.; Griffiths, S.; King, J.; Murchie, E.; Parry, M.; Slafer, G. Achieving yield gains in wheat. *Plant Cell Environ.* **2012**, *35*, 1799–1823. [CrossRef]
63. Masoni, A.; Ercoli, L.; Mariotti, M.; Arduini, I. Post-anthesis accumulation and remobilization of dry matter, nitrogen and phosphorus in durum wheat as affected by soil type. *Eur. J. Agron.* **2007**, *26*, 179–186. [CrossRef]

64. Prasad, P.V.V.; Pisipati, S.R.; Mutava, R.N.; Tuinstra, M.R. Sensitivity of grain sorghum to high temperature stress during reproductive development. *Crop Sci.* **2008**, *48*, 1911–1917. [CrossRef]
65. Sukumaran, S.; Lopes, M.; Dreisigacker, S.; Reynolds, M. Genetic analysis of multi-environmental spring wheat trials identifies genomic regions for locus-specific trade-offs for grain weight and grain number. *Theor. Appl. Genet* **2017**, *131*, 1–14. [CrossRef]
66. Sio-Se Mardeh, A.; Ahmadi, A.; Poustini, K.; Mohammadi, V. Evaluation of drought resistance indices under various environmental conditions. *Field Crops Res.* **2006**, *98*, 222–229. [CrossRef]
67. Dodig, D.; Zoric, M.; Kandic, V.; Perovic, D.; Surlan-Momirovic, G. Comparison of responses to drought stress of 100 wheat accessions and landraces to identify opportunities for improving wheat drought resistance. *Plant Breed.* **2012**, *131*, 369–379. [CrossRef]
68. Mason, R.E.; Mondal, S.; Beecher, F.W.; Pacheco, A.; Jampala, B.; Ibrahim, A.M.H.; Hays, D.B. QTL associated with heat susceptibility index in wheat (*Triticum aestivum* L.) under short-term reproductive stage heat stress. *Euphytica* **2010**, *174*, 423–436. [CrossRef]
69. Mason, R.E.; Mondal, S.; Beecher, F.W.; Hays, D.B. Genetic loci linking improved heat tolerance in wheat (*Triticum aestivum* L.) to lower leaf and spike temperature under controlled conditions. *Euphytica* **2011**, *180*, 181–194. [CrossRef]
70. Orcen, N.; Altinbas, M. Use of some stress tolerance indices for late drought in spring wheat. *Fresenius Environ. Bull.* **2014**, *24*.
71. Clarke, J.M.; Depauw, R.M.; Townley-Smith, T.F. Evaluation of methods for quantification of drought tolerance in wheat. *Crop Sci.* **1992**, *32*, 723–728. [CrossRef]
72. Rekika, D.; Kara, Y.; Souyris, I.; Nachit, M.M.; Asbati, A.; Monneveux, P. The tolerance of PSII to high temperatures in durum wheat (*T. Turgidum conv. Durum*): Genetic variation and relationship with yield under heat stress. *Cereal Res. Commun.* **2000**, *28*, 395–402.
73. Lopes, M.S.; Reynolds, M.P.; Jalal-Kamali, M.R.; Moussa, M.; Feltaous, Y.; Tahir, I.S.A.; Barma, N.; Vargas, M.; Mannes, Y.; Baum, M. The yield correlations of selectable physiological traits in a population of advanced spring wheat lines grown in warm and drought environments. *Field Crops. Res.* **2012**, *128*, 129–136. [CrossRef]
74. Reynolds, M.; Tuberosa, R. Translational research impacting on crop productivity in drought-prone environments. *Curr. Opin. Plant Biol.* **2008**, *11*, 171–179. [CrossRef] [PubMed]
75. Heidari, B.; Sayed-Tabatabaei, B.E.; Saeidi, G.; Kearsey, M.; Suenaga, K. Mapping QTL for grain yield, yield components, and spike features in a doubled haploid population of bread wheat. *Genome* **2011**, *54*, 517–527. [CrossRef] [PubMed]
76. Tahmasebi, S.; Heidari, B.; Pakniyat, H.; McIntyre, C.L. Mapping QTLs associated with agronomic and physiological traits under terminal drought and heat stress conditions in wheat (*Triticum aestivum* L.). *Genome* **2017**, *60*, 26–45. [CrossRef] [PubMed]
77. Quarrie, S.A.; Steed, A.; Calestani, C.; Semikhodskii, A.; Lebreton, C.; Chinoy, C.; Schondelmaier, J. A high-density genetic map of hexaploid wheat (*Triticum aestivum* L.) from the cross Chinese Spring x SQ1 and its use to compare QTL for grain yield across a range of environments. *Theor. Appl. Genet.* **2005**, *110*, 865–880. [CrossRef] [PubMed]
78. Acuna-Galindo, M.A.; Mason, R.E.; Subrahmanyam, N.K.; Hays, D. Meta-analysis of wheat QTL regions associated with adaptation to drought and heat stress. *Crop Sci.* **2014**, *55*, 477–492. [CrossRef]
79. McIntyre, C.L.; Mathews, K.L.; Rattey, A.; Drenth, J.; Ghaderi, M.; Reynolds, M.; Chapman, C.S.; Shorter, R. Molecular detection of genomic regions associated with grain yield and yield components in an elite bread wheat cross evaluated under irrigated and rainfed conditions. *Theor. Appl. Genet.* **2010**, *120*, 527–541. [CrossRef] [PubMed]
80. Araus, J.L.; Bort, J.; Steduto, P.; Villegas, D.; Royo, C. Breeding cereals for Mediterranean conditions: Ecophysiological clues for biotechnology application. *Ann. Appl. Biol.* **2003**, *142*, 129–141. [CrossRef]
81. Slafer, G.A.; Araus, J.L.; Royo, C.; García Del Moral, L.F. Promising eco-physiological traits for genetic improvement of cereal yields in Mediterranean environments. *Ann. Appl. Biol.* **2005**, *146*, 61–70. [CrossRef]
82. McCartney, C.A.; Somers, D.J.; Humphreys, D.G.; Lukow, O.; Ames, N.; Noll, J.; Cloutier, S.; McCallum, B.D. Mapping quantitative trait loci controlling agronomic traits in the spring wheat cross RL4452 9 'AC Domain'. *Genome* **2005**, *48*, 870–883. [CrossRef]
83. Collins, N.C.; Tardieu, F.; Tuberosa, R. Quantitative trait loci and crop performance under abiotic stress: Where do we stand? *Plant Physiol.* **2008**, *147*, 469–486. [CrossRef]

84. Rebetzke, G.J.; van Herwaarden, A.F.; Jenkins, C.; Weiss, M.; Lewis, D.; Ruuska, S.; Tabe, L.; Fettell, N.A.; Richards, R.A. Quantitative trait loci for water soluble carbohydrates and associations with agronomic traits in wheat. *Aust. J. Agric. Res.* **2008**, *59*, 891–905. [CrossRef]
85. Zaim, M.; El Hassouni, K.; Gamba, F.; Filali-Maltouf, A.; Belkadi, B.; Ayed, S.; Amri, A.; Nachit, M.; Taghouti, M.; Bassi, F. Wide crosses of durum wheat (*Triticum durum* Desf.) reveal good disease resistance, yield stability, and industrial quality across Mediterranean sites. *Field Crops Res.* **2017**, *214*, 219–227. [CrossRef]
86. Bassi, F.; Sanchez-Garcia, M. Adaptation and stability analysis of ICARDA durum wheat elites across 18 countries. *Crop Sci.* **2017**, *57*, 2419–2430. [CrossRef]
87. Graziani, M.; Maccaferri, M.; Royo, C.; Salvatorelli, F.; Tuberosa, R. QTL dissection of yield components and morpho-physiological traits in a durum wheat elite population tested in contrasting thermo-pluviometric conditions. *Crop Pasture Sci.* **2014**, *65*, 80–95. [CrossRef]

7

Relationships between Iraqi Rice Varieties at the Nuclear and Plastid Genome Levels

Hayba Badro, Agnelo Furtado and Robert Henry *

Queensland Alliance for Agriculture and Food Innovation, University of Queensland, Brisbane, QLD 4072, Australia; haybaq@yahoo.com (H.B.); a.furtado@uq.edu.au (A.F.)

* Correspondence: robert.henry@uq.edu.au

Abstract: Due to the importance of the rice crop in Iraq, this study was conducted to determine the origin of the major varieties and understand the evolutionary relationships between Iraqi rice varieties and other Asian rice accessions that could be significant in the improvement of this crop. Five varieties of *Oryza sativa* were obtained from Baghdad/Iraq, and the whole genomic DNA was sequenced, among these varieties, Amber33, Furat, Yasmin, Buhooth1 and Amber al-Baraka. Raw sequence reads of 33 domesticated Asian rice accessions were obtained from the Sequence Read Archive (SRA-NCBI). The sequence of the whole chloroplast-genome was assembled while only the sequence of 916 concatenated nuclear-genes was assembled. The phylogenetic analysis of both chloroplast and nuclear genomes showed that two main clusters, Indica and Japonica, and further five sub-clusters based upon their ecotype, *indica, aus, tropical-japonica, temperate-japonica* and *basmati* were created; moreover, Amber33, Furat, Yasmin and Buhooth1 belonged to the *basmati, indica* and *japonica* ecotypes, respectively, where Amber33 was placed in the *basmati* group as a sister of cultivars from Pakistan and India. This confirms the traditional story that Amber was transferred by a group of people who had migrated from India and settled in southern Iraq a long time ago.

Keywords: rice (*Oryza sativa*); evolutionary relationships; chloroplast genome; nuclear genome; phylogeny

---

1. Introduction

Rice is grown in a wide range of environments worldwide, however, most of the world's rice is cultivated and consumed in Asia [1–3]. Iraq has favorable agricultural conditions for rice cultivation, where rice is a staple food for the majority of the Iraqi people [4]. In Iraq, rice grows as a summer crop, and there are a number of traditional, introduced and improved rice varieties that are cultivated in the central and southern region, as well as in the valleys of northern Iraq [1].

The variety Amber is the most important local Iraqi rice variety and is characterised by high quality in terms of taste (aromatic character) [1]. It has been cultivated in central and southern Iraq, especially in the marshes, for a long time. Anecdotal evidence suggests that Amber was introduced to the marshlands of southern Iraq when water buffalo breeding was introduced to the region by a foreign group from the south Asia, probably from the Indian subcontinent. This popular view was reported in a study by Al-Zahery et al. [5], that highlighted the paternal and maternal origin of the human population in the marsh areas, and observed marginal influences of Indian origin on the gene pool of an autochthonous population of the region. A number of rice varieties have also been introduced to Iraq since the middle of the last century to improve rice productivity [6]. IR8 was the first variety introduced in 1968 by the International Rice Research Institute (IRRI) (Philippines), it has high yield potential but the grain quality has not been high compared to Amber. Since aroma is one of the key traits in determining grain quality in rice [7], Amber became a control variety in the central and south regions of Iraq to assess the grain quality of introduced varieties [1]. Accordingly, Furat and Yasmin were also introduced from Vietnam to Iraq in the late 20th century because they are aromatic, tolerant to limited water, highly productive, and have high grain quality [4]. An understanding of the origin of local Iraqi rice and the genetic relationships between Iraqi rice and Asian domesticated rice will effectively guide Iraqi rice breeding (the aim of the current study). However, few studies have investigated Iraqi rice in general, and the origin and the evolution of Iraqi varieties, especially Amber, in particular [4,8,9].

Each living organism is the consequence of an evolutionary process [10]; therefore, it is imperative to enrich our perception of the evolutionary history of organisms and the relationships among them to guide their genetic improvement. Methods of determining evolutionary history (Phylogeny) have undergone many stages of development. Morphological markers maybe influenced by environmental factors and growth practices. More recent methods have used molecular markers which are independent of environmental factors [11], including techniques such as RFLP, AFLP, RAPD, SSR and ISSR along with morphological markers, to study phylogenetic relationships [12]. The development of high-throughput sequencing technology has revolutionised the study of genetics and evolutionary relationships. Most recently, through next-generation sequencing (NGS), whole-genome sequencing and re-sequencing have become available, so the investigation of the entire genome, rather than targeting precise regions, is now a real opportunity [13–15].

Every plant cell has three genomes—nuclear, chloroplast, and mitochondrial—that may differ in evolutionary history. The chloroplast genome is a maternal genome which is highly-conserved and not involved in recombination, therefore, it is the most commonly used tool to determine the origin and the evolutionary relationships among plant species [16–19]. However, sometimes, evolutionary analysis based on the chloroplast genome must be supported by nuclear genome-based analysis to achieve the most reliable results because the chloroplast genome can only represent the maternal evolutionary history with a slow evolutionary rate [20,21]. Phylogenetic analysis using the nuclear genome can deliver inconsistent trees due to recombination that may confuse phylogenetic resolution. However, this analysis provides greater insights into evolutionary relationships. Several studies have strongly suggested applying this analysis along with chloroplast phylogenetic analysis [18,19,22]. Many studies have applied phylogenetic analysis at both genome levels [23–25], and the results of most of these studies showed that the nuclear genome followed a different evolutionary history pattern to that of the chloroplast genome.

We reported the whole chloroplast genome sequences for Iraqi rice and compared them with the whole chloroplast sequences of other domesticated Asian rice varieties. This provided an important tool for estimating genetic distance and determining evolutionary relationships between rice accessions; the nuclear genomes also provided further information on the relationships between the varieties studied. The study aimed to determine the origin and evolution of Iraqi rice, especially Amber33.

2. Results

2.1. DNA Sequencing and Data Processing

The sequencing process of the five Iraqi varieties (Table 1) generated about 51 Gb of data containing 337 million of 151-bp paired-end reads. The minimum and the maximum number of reads were about 58 and 93 million reads with sequence depth ranging between 23× and 38× for Buhooth1 and Furat, in turn. When raw data was trimmed at the quality limit of 0.01, an average of 15% of the reads' length and 9% of the number of reads were removed, thus the number of reads and data coverage reduced to the range between 53 and 86 million, and 18× and 30×, respectively (Table S1). In terms of downloaded data (Table 2), the average length of raw reads was 83-bp, and the minimum and the maximum number of reads ranged between 43 and 117 million reads while the sequence coverage fluctuated between 10× and 26×. Finally, the number of reads and the data coverage of each of the data sets were assessed after trimming the raw reads at the quality limit of 0.01 (Table S1).

Table 1. The Iraqi plant materials used in this study.

Varieties	History	Varietal Group	BioProject ID	BioSample Accessions
Amber33	Local (Iraq)	Aromatic, medium grain type	PRJNA576935	SAMN13014963
Furat	Introduced from (Vietnam) in 1996	Aromatic, medium grain type	PRJNA576935	SAMN13014964
Yasmin	Introduced from (Vietnam) in 1998	Aromatic, medium grain type	PRJNA576935	SAMN13014965
Buhooth1	Improved	Non-Aromatic, long grain type	PRJNA576935	SAMN13014966
Amber al-Baraka	Improved	Aromatic, long grain type	PRJNA576935	SAMN13014967

Table 2. Summary of data downloaded for sequence comparisons.

No	Sample Unique ID	Project Accession	Species	Country of Origin	Ecotype *	Alignment Name (in Figures 1 and 2) *
1	B243	ERP005654	*O. sativa*	China	Aus	Ch(Aus)B243
2	CX165	ERP005654	*O. sativa*	China	TmpJ	Ch(TmpJ)CX165
3	CX352	ERP005654	*O. sativa*	China	TrpJ	Ch(TrpJ)CX352
4	CX10	ERP005654	*O. sativa*	China	In	Ch(In)CX10
5	CX368	ERP005654	*O. sativa*	India	Aus	India(Aus)CX368
6	IRIS_313–10670	ERP005654	*O. sativa*	India	Bas	India(Bas)IRIS_313-10670
7	IRIS_313–11153	ERP005654	*O. sativa*	India	TmpJ	India(TmpJ)IRIS_313-11153
8	IRIS_313–11479	ERP005654	*O. sativa*	India	TrpJ	India(TrpJ)IRIS_313-11479
9	IRIS_313–11152	ERP005654	*O. sativa*	India	In	India(In)IRIS_313-11152
10	CX129	ERP005654	*O. sativa*	Indonesia	TrpJ	Indo(TrpJ)CX129
11	CX25	ERP005654	*O. sativa*	Indonesia	In	Indo(In)CX25
12	CX104	ERP005654	*O. sativa*	Iran	Bas	Iran(Bas)CX104
13	CX227	ERP005654	*O. sativa*	Japan	Aus	Jap(Aus)CX227
14	CX140	ERP005654	*O. sativa*	Japan	TmpJ	Jap(TmpJ)CX140
15	IRIS_313–10073	ERP005654	*O. sativa*	Japan	TrpJ	Jap(TrpJ)IRIS_313-10073
16	IRIS_313–10549	ERP005654	*O. sativa*	Pakistan	Aus	Pak(Aus)IRIS_313-10549
17	IRIS_313–11021	ERP005654	*O. sativa*	Pakistan	Bas	Pak(Bas)IRIS_313-11021
18	IRIS_313–11026	ERP005654	*O. sativa*	Pakistan	Bas	Pak(Bas)IRIS_313-11026
19	IRIS_313–8656	ERP005654	*O. sativa*	Pakistan	Bas	Pak(Bas)IRIS_313-8656
20	IRIS_313–11829	ERP005654	*O. sativa*	Pakistan	TmpJ	Pak(TmpJ)IRIS_313-11829
21	IRIS_313–10380	ERP005654	*O. sativa*	Philippines	Aus	Phil(Aus)IRIS_313-10380
22	CX59	ERP005654	*O. sativa*	Philippines	Bas	Phil(Bas)CX59
23	IRIS_313–10373	ERP005654	*O. sativa*	Philippines	TmpJ	Phil(TmpJ)IRIS_313-10373
24	CX243	ERP005654	*O. sativa*	Philippines	TrpJ	Phil(TrpJ)CX243
25	IRIS_313–9505	ERP005654	*O. sativa*	Philippines	In	Phil(In)IRIS_313-9505
26	CX126	ERP005654	*O. sativa*	Philippines	In	Phil(In)CX126
27	IRIS_313–10718	ERP005654	*O. sativa*	Sri Lanka	Aus	SriL(Aus)IRIS_313-10718
28	IRIS_313–9949	ERP005654	*O. sativa*	Sri Lanka	TrpJ	Sril(TrpJ)IRIS_313-9949
29	IRIS_313–11248	ERP005654	*O. sativa*	Thailand	TrpJ	Thai(TrpJ)IRIS_313-11248
30	CX106	ERP005654	*O. sativa*	Vietnam	TrpJ	Viet(TrpJ)CX106
31	B009	ERP005654	*O. sativa*	Vietnam	In	Viet(In)B009
32	CX37	ERP005654	*O. sativa*	Vietnam	In	Viet(In)CX37
33	O.glaberrima-PRJNA13765	SRP038750	*O. glaberrima*	-	-	*O. glaberrima*

32 domesticated Asian rice accessions and one domesticated African rice as an out-group downloaded from SAR-NCBI: their unique ID, species, country of origin, and ecotype was from the study of [26]; the alignment names were generated in this study. * In: *indica* subpopulation, TrpJ: *tropical japonica* subpopulation, TmpJ: *temperate japonica* subpopulation, Aus: *aus* population, Bas: *basmati* population, Ch: China, Indo: Indonesia, Jap: Japan, Pak: Pakistan, Phil: Philippines, Sril: SriLanka, Viet: Vietnam.

2.2. *Chloroplast Genome Assembly*

Mapping all varieties against the reference, *O. sativa* sub sp. japonica *Nipponbare* "GenBank: GU592207.1", under three various fraction settings clarified the most accurate and reliable mapping setting. The number of mismatches and gaps of each variety was virtually stable in all different settings (Table S2). Indeed, this stability confirms that most of these variations were produced from actual differences between the sequences of samples and reference, not due to using different settings; based on that, setting number two (length fraction (LF) of 1 and similarity fraction (SF) of 0.8) was applied to the other steps of assembly, Improvement process (Imp). Moreover, three different settings of Word "W" and Bubble "B" size in *de novo* assembly generated a satisfactory number of contigs that cover the whole chloroplast genome area, around five large chloroplast-contigs with a length of more than 12 kb produced from each setting. Subsequently, four main regions of the chloroplasts, large single copy

(LSC), inverted repeat A (IR A), small single copy (SSC) and inverted repeat B (IR B), were assembled successfully for all 38 varieties through a *de novo* assembly pipeline. The lengths of these regions were about 80 kb for LSC, 12 kb for SSC and 20 kb for IR A and IR B (Figure S1 shows only Iraqi rice varieties). In manual-curation, the comparison between both sub-approaches of the chloroplast genome assembly pipeline showed no significant differences in terms of the number of variations; however, any minor conflicts were resolved by reference to the reads (Table S3 shows only Iraqi rice varieties). The minimum and maximum lengths of the whole chloroplast for all Iraqi varieties and downloaded accessions were 134,259 and 134,556 bp, respectively, while the coverages ranged from 839× to up to 11,466×, and the average coverage was 3818× (Table 3).

Table 3. The results of the chloroplast and nuclear genome assembly.

Varieties	Chloroplast Genome		Length of Nuclear Genome (bp)
	Length of Genome (bp)	Coverage (×)	
Amber33	134,536	2909	616,371
Furat	134,500	7819	616,190
Yasmin	134,502	5495	616,301
Buhooth1	134,550	4759	616,393
Amber al-Baraka	134,493	3651	616,310
B243	134,497	2203	616,278
CX165	134,542	8818	616,377
CX352	134,553	5305	616,324
CX10	134,503	11,466	616,274
CX368	134,504	3674	616,236
IRIS_313–10670	134,535	1669	616,369
IRIS_313–11153	134,551	1639	616,360
IRIS_313–11479	134,259	2726	616,367
IRIS_313–11152	134,503	1413	616,271
CX129	134,535	6076	616 337
CX25	134,503	5830	616,210
CX104	134,532	6784	616,348
CX227	134,504	4267	616,314
CX140	134,547	5185	616,393
IRIS_313–10073	134,556	2036	616,355
IRIS_313–10549	134,495	1636	616 324
IRIS_313–11021	134,531	1978	616,383
IRIS_313–11026	134,532	1723	616,358
IRIS_313–8656	134,532	2334	616 380
IRIS_313–11829	134,539	4164	616,389
IRIS_313–10380	134,496	1857	616,331
CX59	134,536	5913	616,370
IRIS_313–10373	134,551	1464	616,363
CX243	134,556	4375	616,362
IRIS_313–9505	134,503	968	616,283
CX126	134,503	3510	616,220
IRIS_313–10718	134,531	2332	616,324
IRIS_313–9949	134,532	3041	616 385
IRIS_313–11248	134,413	974	616,336
CX106	134,529	6145	616,339
B009	134,528	839	616,231
CX37	134,503	4836	616,258
O.glaberrima-PRJNA13765	134,542	2567	616,099

The table includes the length of the chloroplast genome, the number of bases of mapped reads, and the coverage of assembled chloroplast genome for five Iraqi varieties and 32 domesticated Asian accessions and one domesticated African rice as an out-group downloaded from SAR-NCBI. This table also shows the length of the nuclear genome.

2.3. Phylogenetic Analysis of the Chloroplast Genome

Two phylogenetic approaches were used to analyse the multiple alignments of thirty-nine chloroplast genomes which had a total length of 134,535 bp. Although the result of both phylogenetic methods showed some minor alterations at the end of some subclades, the content of the main clades and subclades, which followed their ecotype classifications, were identical (Figure 1). Phylogenetic analysis of the chloroplast genome divided the thirty-nine rice accessions into two main clades, an Indica clade and a Japonica clade. The Indica clade (In) included most individuals under *indica* (6 accessions) and *aus* (5 accessions) ecotypes except two individuals, B009 and IRIS_313-10718. The Japonica clade contained two subclades, a main Japonica clade and a Basmati clade; the first subclade which was the main Japonica clade (Jap) included all *japonica* individuals (13 accessions) from the two subpopulations of *japonica* ecotype, *tropical* and *temperate*, while the second subclade, the Basmati clade (Bas), involved all individuals of *basmati* ecotype (6 accessions) and the excluded individuals from the first clade (Indica). Additionally, the Iraqi varieties were distributed as following: Furat, Yasmin and Amber al-Baraka into the Indica clade whereas Amber33 and Buhooth1 into the Japonica clade. Buhooth1 was close to accessions from *tropical japonica* ecotype more than accessions under *temperate japonica* ecotype, and interestingly, Amber33 was located within the Basmati subclade.

The multiple alignments of chloroplast genomes comprised 134,535 bp, the number of identical sites was 134,270 characters (99.8%) while the number of variable bases among all the accessions totaled 265 (0.2%). These 265 variable bases were sorted into 85 variation positions which were in turn grouped into four types of polymorphisms including single nucleotide polymorphism (SNP), multi nucleotide polymorphism (MNP), insertions (Ins) and deletions (Del) (Table 4). The most abundant polymorphism types among all accessions were SNPs. Out of 85 polymorphisms, 83%, 12% and 5% were located in the four main regions of the chloroplast genome, LSC, SSC and IR A and B, respectively (Table S4).

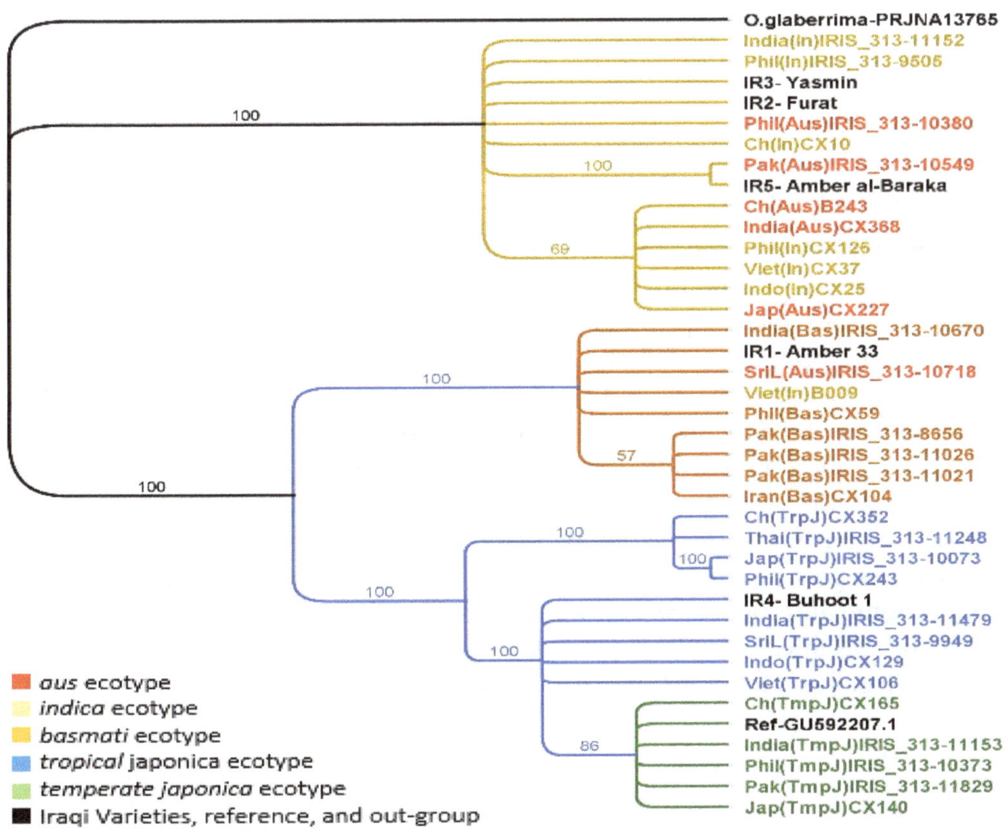

Figure 1. Phylogenetic relationships among chloroplast genomes of thirty-nine rice accessions. Tree topology based on MrBayes software (branch labels represent probability percentage).

Considering the variations identified, all thirty-nine rice accessions were sorted into three main groups: (1) Indica, (2) Japonica and (3) Basmati. As expected, the highest number of polymorphisms among the species studied (255 bases in 76 variant positions) was found in the Indica group, 11 accessions and 3 Iraqi varieties; within 76 variants, there was only one variation (1-bp deletion at position of 75990 bp) between *indica* and *aus* accessions. While the second largest number of variations (55 bases within 21 variant positions) was within the Basmati group, 8 accessions and one Iraqi variety. Part of the Basmati group, 4 accessions, showed unique polymorphisms (2 variable bases (SNPs) within 2 variant positions), three accessions were from Pakistan IRIS_313–8656, IRIS_313–11026, and IRIS_313–11021) and one from Iran (CX104). As expected, the Japonica group, 13 accessions along with the reference (*O. sativa* sub sp. japonica Nipponbare "GenBank: GU592207.1") and one Iraqi variety, possessed the lowest number of polymorphisms (13 bases within 10 variant positions) (Table S4). Most of the polymorphisms in the Japonica group belonged to only four accessions from *tropical japonica* (TrpJ) subpopulation, including CX352, IRIS_313–10073, CX243, and IRIS_313–11248.

Furthermore, a heat-map was drawn according to the number of variable bases (Table S5); in this map, the two main clusters, Indica and Japonica, were clearly distinguished, whereas the Basmati group was comprised within the Japonica group. Within the Japonica group two individuals, 24:IRIS_313–11479 and 27:IRIS_313–11248, clearly showed the greatest distances among the rice accessions. This cluster surprisingly also included two individuals, 33:IRIS_313–10718 and 34:B009, from the *aus* and *indica* ecotype, respectively. There were no variable bases between a number of pairs (dark red in Table S5) such as (3:IRIS_313–11152 and 4:IRIS_313–9505), (9:CX126 and 11:CX37), (9:CX126 and 13:CX25), (17:IRIS_313–10073, and 18:CX243), (19:Ref-GU592207.1 and 20:IRIS_313–11153), (20:IRIS_313–11153 and 21:IRIS_313–10373); and (30:IRIS_313–8656, 31:IRIS_313–11026 and 37:CX104); whereas the highest number of variable bases, 260 bases, was found between (14:CX227 and 24:IRIS_313–11479) (dark green in Table S5). The smallest number of variable bases between Iraqi varieties and other domesticated rice accessions were 1, 3, 1, 6 and 4 bases, those bases were between Iraqi varieties: Amber33, Furat, Yasmin, Buhooth1, and Amber al-Baraka, and the following accessions: 28:IRIS_313-10670, 12:CX10, 3:IRIS_313–11152, 20:IRIS_313–11153 and 5:IRIS_313–10549, respectively (Table S5).

2.4. *Phylogenetic Analysis of the Nuclear Genome*

Within a group of thirty-nine rice accessions, the multiple alignment of 916 concatenated nuclear genes was 621,012 bp in length; the minimum and maximum lengths were 616,099 and 616,393 bp, respectively (Table 3). The nuclear phylogenies using two different methods showed that the two main clusters, Indica and Japonica, and further five sub-clusters were based upon their ecotype, *indica*, *aus*, *tropical japonica*, *temperate japonica* and *basmati* (Figure 2). Unlike the results of the chloroplast phylogeny, the accessions of *indica*, and *aus* ecotypes were represented by two well-resolved subclades within the Indica clade. The Iraqi varieties, Furat and Yasmin, were found in the *indica* subclade while the rest of the Iraqi collection was grouped in the Japonica clade, where Amber33 acted as a sister to all the *basmati* varieties within the *basmati* subcluster, which included all accessions with the *basmati* ecotype. Buhooth1 was part of the *temperate japonica* subcluster that comprised accessions from the *temperate japonica* ecotype and the reference, *O.s japonica* cv. Nipponbare. Amber al-Baraka was a sister to both the Indica and Japonica clades; however, Geneious Tree Builder showed that it was close to the Indica clade, while MrBayes suggested that Amber al-Baraka was closer to the Japonica clade.

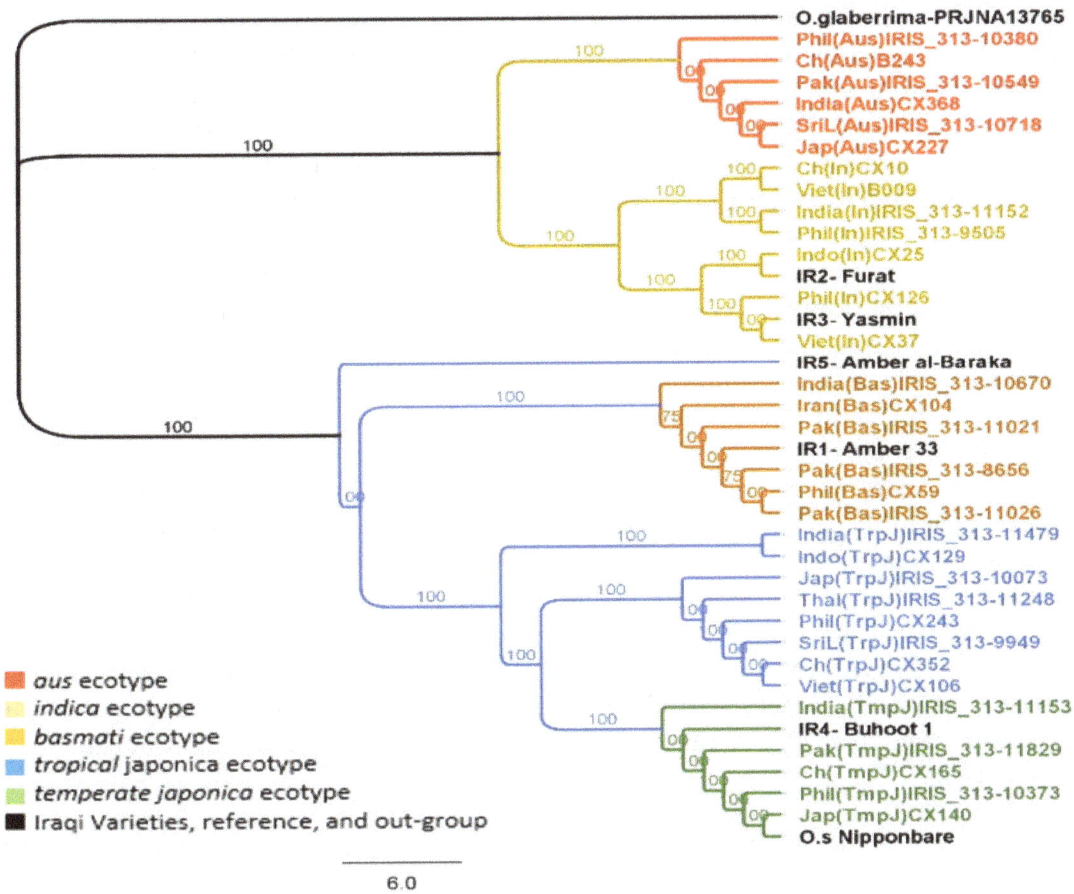

Figure 2. Evolutionary relationships among the multiple alignment of 916 concatenated nuclear genes of domesticated rice. Tree topology based on MrBayes software (branch labels represent probability percentage).

3. Discussion

Rice phylogeny has been extensively studied as a better understanding of the evolutionary relationships among rice species is critical for rice breeding programmes as well as comparative genomics studies. Recent advances in next-generation DNA sequencing (NGS) have improved the phylogenetic reconstruction of any plant species including *Oryza*. In this study, both plastid and nuclear genomes were assembled using NGS reads (whole genome DNA sequencing) to identify the phylogenetic relationships among Iraqi rice varieties and other accessions. According to Sims et al. [27], the accuracy of a genome assembly using NGS reads depends on many factors including sequencing depth (coverage) and the accuracy of the assembly pipeline. Therefore, even after trimming, the sequence coverage of the sequenced and downloaded accessions (Table S1) was enough to ensure coverage of all the chloroplast and most of the nuclear genome, thereby guaranteeing a high-quality assembly.

A dual pipeline was applied to the assembly of the chloroplast genome in this study; this pipeline consisted of two procedures, mapping assembly (MA), and *de novo* assembly (*d*A). The comparison between the sequence of *de novo* and mapping consensus showed no significant differences in terms of the number of variations. Interestingly, the variety Yasmin showed no difference in both approaches with regard to a number of variations (Table S3), but the length of the consensuses was different; this observation indicates that even when the number of copies of an insertion or deletion was similar, the number of bases that were inserted or deleted was diverse. Therefore, in agreement with an earlier study [28], a manual-curation step was critical in resolving any conflicts by reference to the reads. A pipeline of nuclear genes assembly was also developed in this study. This pipeline involved multiple tools on the CLC Genomics Workbench, unlike a previous study [25] that used different software packages to assemble the nuclear genes for phylogenetic analysis at the nuclear genome level. The number of genes selected to represent the nuclear genome in the phylogenetic analysis was only 916 genes with a length of 621,012 bp, considerably lower than that reported previously [25].

Phylogenetic analysis of the chloroplast genome sorted the thirty-nine rice accessions into two main clades, an Indica clade and a Japonica clade (Figure 1). The Indica clade (In) included most individuals under the *indica* and the *aus* ecotypes except for two accessions. Accessions from *indica* and *aus* ecotypes were not clearly distinct but were placed together in one clade; this was confirmed by the results of genetic polymorphism analysis that showed only one variation (1-bp deletion at the position of 75,990 bp; Table S4) between the *indica* and *aus* accessions. The Japonica clade (Jap) contained two subclades, the main Japonica clade (Jap) which included all individuals from the two *japonica* subpopulations, and the Basmati clade (Bas) that included all basmati accessions as well as the individuals excluded from the first clade (Indica). Moreover, the presence of accessions from *aus* ecotype in the Indica clade as well as in the Basmati subclade within Japonica clade agrees with earlier outcomes [18] which indicated that the two different ecotypes, *indica* and *japonica*, might be involved in the origins of the maternal genome in two Korean *aus* landrace rices. This also agrees with the conclusion made by Civáň et al. [29]. which suggested that *aromatic* rice resulted from a hybridization between *japonica* and *aus*. Analysis of genetic polymorphisms at the chloroplast genome level revealed that the most abundant variation types were SNPs, 57% of 85 variants (Table 4). This analysis also showed 255 nucleotide differences within 76 variant positions between the *O.sativa* spp. *indica* and the *japonica* reference (GU592207) in agreement with the previous studies of Brozynska et al. [22] and Wambugu et al. [28].

Table 4. Summary of the number and types of variants in the chloroplast-genomes of thirty-nine domesticated rice-accessions.

Group \ Variation Type	SNPs	MNPs	Del	Ins	Total
Indica	37	5	10	9	61
Japonica	1	0	1	1	3
Basmati	4	1	0	1	6
Indica & Basmati	3	2	3	0	8
Indica and Basmati and Japonica	3	0	1	3	7
Total	48	8	15	14	85

At the nuclear genome level, the phylogenetic analysis using two different approaches sorted accessions from *indica* and *aus* ecotypes into two completely independent subclades within the Indica clade, unlike the result of the chloroplast phylogeny, whilst the second clade was a Japonica clade which included three sub-clusters *tropical japonica*, *temperate japonica*, and *basmati* (Figure 2). Accordingly, the findings of the evolutionary relationship based on nuclear and chloroplast data in the current study aligned with an earlier study by Garris et al. [30] which reported that the closest evolutionary relationships were between *indica* and *aus* groups, and among the *tropical japonica*, *temperate japonica*, and *aromatic* groups. In general, in the present study, the phylogenetic analysis at both genome levels, chloroplast and nuclear, showed relatively comparable evolutionary history patterns with insignificant differences at the end of clades, unlike other studies that recorded significant differences in evolutionary history pattern using both chloroplast and nuclear genomes (regardless of plant materials) [23–25,29]. Furthermore, the phylogenetic trees of both genomes, chloroplast and nuclear, constructed using different methodologies, were highly compatible. However, Amber al-Baraka showed slightly different relationships at the level of the nuclear genome according to the method used; where Geneious Tree Builder software placed Amber al-Baraka close to the Indica clade whereas Amber al-Baraka was closer to the Japonica clade and distant from the Indica clade by MrBayes software. This was unexpected and requires further investigation.

The phylogenetic analysis of both the chloroplast and nuclear genomes indicated that Amber33, Furat and Yasmin, and Buhooth1 belonged to *basmati*, *indica* and *japonica* ecotypes, respectively. Our results supported that Buhooth1 is an improved cultivar, where the nuclear phylogenies showed a divergent relationship to those deduced from the chloroplast genomes, analogous to *temperate japonica* subpopulation and *tropical japonica* subpopulation, respectively. Furat and Yasmin were introduced to Iraq from Vietnam [4], this was obvious by the results of phylogenetic analysis of the nuclear genome, but their chloroplast genome was closely related to accessions from China, India and Philippines. This may be explained by the breeding history of the genotype.

In this study, Amber33, which is local Iraqi variety, was placed in the *basmati* ecotype group as a sister of cultivars from Pakistan and India by analysing the evolutionary relationship at both levels of the genome. Based on distance analysis, the number of differences in the chloroplast genome between Amber33 and all accessions within the Basmati subclade was in the following order: 28:IRIS_313–10670 O (1 bp), 35:CX59 (1 bp), 30:IRIS_313–8656 (3 bp), 31:IRIS_313–11026 (3 bp), 37:CX104 (3 bp), 33:IRIS_313–10718 (3 bp), 32:IRIS_313–11021 (5 bp), 34:B009 (6 bp) (Table S5); it can be accordingly concluded that Amber33 is closely related to accession from India which is visibly reflected in the observed phylogenetic tree (Figure 1). This confirms the popular tradition that says that the Amber variety was transferred by a group of people who had migrated from India (the Southeast) and settled in southern Iraq a long time ago.

Recently, the term Basmati has been used to indicate a long-grain and high-quality rice, but this name originally refers to *aromatic* rice because it was derived from the Sanskrit words "Vas" and "Matup" which stand for "aroma" and "ingrained from the beginning", respectively, and then both words were combined making 'Vasmati' which changed to become 'Basmati' later on [31,32]. Therefore, the presence of Amber33 within the Basmati subcluster does not necessarily mean that it is a long grain cultivar; indeed, it is an aromatic medium-grain cultivar. Furthermore, Basmati is a group that can be described basically as the fifth isozyme group identified by Glaszmann [33], and it is closer to the *japonica* group than the *indica* [7,30,34]; this group is also phenotypically diverse as it includes both long or medium grain, and aromatic or nonaromatic varieties [7]. In many studies, this group is also known as the "*aromatic*" subpopulation [7,33], but most of the time it is known as "Group V" to avoid confusion. In this study, we refer to this group as "Basmati" according to Wang et al.'s study [26] which is the information resource of the downloaded accessions.

4. Conclusions

In the present study, we have assembled the whole chloroplast genome and the nuclear genome of the five Iraqi rice varieties, together with thirty-three domesticated Asian rice, to find the origin of Iraqi varieties, especially Amber33, and to gain insight into the evolutionary relations between Iraqi and domesticated Asian rice varieties. Our results suggest that the possibility of an Indian and/or Pakistani origin for Amber33; to evaluate this hypothesis, further historical biogeographical analyses are required. Moreover, further study on the chloroplast and nuclear genome in Iraqi rice varieties are required to determine the functional genome annotations that might be useful for future rice breeding programmes in Iraq.

5. Materials and Methods

5.1. Plant Materials

A total of five varieties of *Oryza sativa* were provided and tested by the Office of Agricultural Research, and Directorate of Seed Testing and Certification, Ministry of Agriculture, Baghdad, IRAQ, respectively. Among these varieties, one variety, Amber33, is local and one of the most highly valued varieties in Iraq because of its fragrance, and two varieties, Furat and Yasmin, were introduced from Vietnam; however, they are successfully cultivated in the central and southern regions of Iraq; while the other two, Buhooth1 and Amber al-Baraka, are improved varieties [4]. The plant materials used in this chapter are described in detail in Table 1.

5.2. *Seed Germination and Growth*

About 15 seeds of each individual, a total of 75 seeds, were first dehusked, and then placed in a container with plenty of liquid fertilizer, Flowfeed EX7, that was diluted to half concentration (full concentration is 0.5 g/1L) to break the dormancy phase; this method was the non-heat treatment method. Once the radicle emerged, the germinated seeds were transferred to a petri dish covered with a layer of tissue that was saturated with liquid fertilizer, and planted within three days. All the germination and planting processes were carried out under extremely restricted quarantine conditions in quarantine facilities.

5.3. *DNA Extraction and Sequencing*

After harvesting leaves tissues, total genomic DNA was extracted individually using the modified CTAB protocol described by Furtado [35] with slight modifications. The modifications that were made can be summarised as the following: the mixture of ground plant tissue and nuclear extraction buffer was incubated at 65 °C for 60 min with periodic mixing by inverting the tubes every 5 min; as well as the speed and time of centrifuge were increased to 4000× *g* and 7 min, respectively, after the steps of protein denaturation and DNA precipitation. However, the most vital modification in the DNA extraction procedure was the exclusion of the mixture of phenol:chloroform:isoamyl alcohol (25:24:1). The quality of DNA was assessed by NanoDrop™ 8000 Spectrophotometers (Thermo Scientific, http://www.nanodrop.com) while the DNA quantity was estimated by agarose gel electrophoresis (1%, 120 V for 1 h) based on Furtado's study [35].

The whole genomic DNA of Iraqi rice varieties was sequenced by preparing and indexing five PCR-free libraries separately (one library for each variety), then pooling them together and sequencing over a half lane of an Illumina HiSeq 4000 flow-cell at MACROGEN (Seoul, Korea; http://dna.macrogen.com).

5.4. *Data Downloaded for Sequence Comparisons*

Raw sequence reads of 33 domesticated rice accessions were sourced from the Sequence Read Archive (SRA)-NCBI website (https://www.ncbi.nlm.nih.gov/sra) using "Download/Search for Reads in SRA" tool on CLC Genomics Workbench version 11.0.1 (CLC Bio, a QIAGEN Company, Aarhus, Denmark; www.clcbio.com). All of the species, except one, were Asian rice (*O. sativa*) relatives. *O. glaberrima*, an African rice, was included as an out-group. All related information such as the sample unique ID, project accession, species, country of origin, and ecotype was obtained from an earlier study [26], as shown in Table 2.

5.5. *Data Processing*

The raw reads of both sequenced and downloaded data were subjected to quality control (QC) analysis using the "Create Sequencing QC Report" tool in the CLC Genomics Workbench, which was used to verify the integrity of the data and determine the appropriate trimming score. The low-quality reads were trimmed at a quality limit of 0.01 and a minimum PHRED score of 25 "Trim Sequences" tool on the CLC.

5.6. *Chloroplast Genome Assembly*

A chloroplast genome of the domesticated rice was assembled and validated using a dual pipeline approach: (1) mapping assembly (MA), and (2) *de novo* assembly (*d*A) [36]. In the mapping assembly (MA) pipeline, the trimmed reads were mapped against the reference, which is *O. sativa* sub sp. japonica Nipponbare "GenBank: GU592207.1", using "Map reads to reference" tool at three various fraction settings of length-fraction and similarity-fraction (1) 0.8 and 0.8, (2) 1 and 0.8, and (3) 1 and 0.9, this step was known as "R". Additionally, in an attempt to mend the Cp map, two tools, "InDels and Structural Variants" and "Local Realignment", were applied. This step was named "S". All the analyses of mapping assembly were performed on the CLC Genomics Workbench 11.0.1.

In the *de novo* assembly pipeline, the Fast "F" model was used with combinations of Word "W" and Bubble "B" settings. Contigs generated by *de novo* were blasted against the Cp reference *O. sativa* sub sp. japonica Nipponbare "GenBank: GU592207.1" to select the Cp-exclusive contigs, and they were then updated using the "Map Reads to Contigs" tool on the CLC Genomics Workbench 11.0.1. Lastly, the updated contigs were aligned to a reference sequence to recognise overlaps and gaps using Clone Manager Professional 9.0 (www.scied.com). When non-overlapping contigs were produced, supplemental *de novo* assembly was conducted at various W-and B-settings to plug all gaps by creating additional contigs, and then all the overlapping contigs were subjected to the further analysis.

An additional improvement process was performed on both the mapping and *de novo* assembly pipelines. The improvement (Imp) process was similar to the mapping assembly (MA) pipeline, repeated twice, Imp-1 and Imp-2, with one difference, the consensus generated from each process would be a reference for the following process. The sequences of both improved Cp consensus generated by the mapping and *de novo* improvement processes were compared to identify all mismatches and then were manually corrected by reference to the reads; this step was named "manual-curation" (Figure S2). Eventually, the Cp sequence of each variety was ready for the phylogenetic analysis.

5.7. Phylogenetic Analysis

The consensus chloroplast sequences of the Iraqi rice and the other domesticated rice accessions were used to perform a phylogenetic analysis using Geneious software version 9.1.8 (https://www.geneious.com). The multiple alignment was conducted using the plugin MAFFT Alignment [37] with default parameters; subsequently, to analyse evolutionary relationships; the phylogenetic tree was constructed through software that roots the constructed tree based on the outgroup method: MrBayes [38], and Geneious Tree Builder. The distance between the chloroplast genomes of Iraqi and comparative rice was determined by detecting all the variants using the "variant/SNP detection" tool on Geneious software and then counting the differences (number of bases which are not identical), one of the outputs of the phylogenetic tree construction process.

5.8. Phylogenetic Analysis of the Nuclear Genome

An evolutionary relationship analysis at the level of the nuclear genome was undertaken using the CLC Genomics Workbench 11.0.1 and Geneious software version 9.1.8; this analysis started with the nuclear genome assembly (NGA) pipeline (Figure S3). In NGA pipeline, the "Map Reads to Reference" tool was used to map the trimmed reads of the Iraqi rice (Table 1), and the domesticated rice accessions from Asia and Africa (Table 2) against the reference, which is *O. sativa* sub-spp. Japonica cv Nipponbare "GenBank: IRGSP1.0", applying the following setting: length-fraction of 1 and similarity-fraction of 0.8. After mapping, the consensus sequence of a whole genome for each variety was extracted using the "Extract Consensus Sequence" tool, and from that, the genome and coding sequence (CDS) tracks were generated by the "Convert to Tracks" tool. By investigating the CDS tracks for all varieties, a subset of 916 genes was identified in all varieties, and then the nucleotide sequences of 916 CDS were separately extracted from the genomes using the "Extract Annotations" tool. At the final stage of the nuclear genome assembly (NGA), all the nucleotide sequences of the 916 CDS selected from each genome were concatenated into a super-matrix of 621,012 bp by the "Join Sequences" tool. The super-matrices of all varieties were then aligned using multiple alignments MAFFT [37] on Geneious at default parameters; the alignment output was used in the following phylogenetic inference. A phylogenetic tree was constructed and rooted using the outgroup methods which are MrBayes [38], and Geneious Tree Builder (https://www.geneious.com); the default tree search settings were applied for both methods.

Supplementary Materials:
Figure S1: The results of *de novo* Assembly on Clone Manager Professional 9.0 software; Figure S2: Illustration of the Chloroplast Genome Assembly Pipeline; Figure S3: Illustration of the Nuclear Genome Assembly Pipeline. Table S1: Summary of the output of sequencing and downloading (Raw Data) and trimming processes; Table S2: Summary of Mapping Assembly process using three different setting of Length fraction and Similarity Fraction; Table S3: Comparison between Mapping and *de novo* assembly in the number of variations in the chloroplast-genome; Table S4: Details of the polymorphisms identified in aligned chloroplast-genomes using the "variant/SNP detection" tool; Table S5: Distance matrix corresponding to the number of non-identical bases in the sequences of domesticated-rice chloroplast-genomes.

Author Contributions: Conceptualization, R.H. and H.B.; Methodology, H.B and A.F.; Software, A.F. and H.B.; Validation, H.B.; Formal Analysis, H.B.; Investigation, H.B.; Resources, H.B.; Data Curation, H.B.; Writing-Original Draft Preparation, H.B.; Writing-Review & Editing, R.H., A.F. and H.B.; Visualization, H.B.; Supervision, R.H. and A.F.; Project Administration, A.F.; Funding Acquisition, H.B.

Acknowledgments: We would like to acknowledge the HCED Iraq program for providing PhD scholarship and sincerely thank the Office of Agricultural Research and Directorate of Seed Testing and Certification (Ministry of Agriculture, Baghdad, IRAQ) for providing us with the seeds of five Iraqi varieties for phylogenetic analysis. We acknowledge the University of Queensland Research Computing Centre (UQ-RCC) for providing all the computing resources.

References

1. Chakravarty, H. *Plant Wealth of Iraq: A Dictionary of Economic Plants*; Botany Directorate, Ministry of Agriculture & Agrarian Reform: Baghdad, Iraq, 1976.
2. Chakravarthi, B.K.; Naravaneni, R. SSR marker based DNA fingerprinting and diversity study in rice (*Oryza sativa* L.). *Afr. J. Biotechnol.* **2006**, *5*, 684–688.
3. Rabbani, M.A.; Pervaiz, Z.H.; Masood, M.S. Genetic diversity analysis of traditional and improved cultivars of Pakistani rice (*Oryza sativa* L.) using RAPD markers. *Electron. J. Biotechnol.* **2008**, *11*, 52–61. [CrossRef]
4. Younan, H.Q.; Al-Kazaz, A.A.; Sulaiman, B.K. Investigation of Genetic Diversity and Relationships among a Set of Rice Varieties in Iraq Using Random Amplified Polymorphic DNA (RAPD) Analysis. *Jordan J. Biol. Sci.* **2011**, *4*, 249–256.
5. Al-Zahery, N.; Pala, M.; Battaglia, V.; Grugni, V.; Hamod, M.A.; Kashani, B.H.; Olivieri, A.; Torroni, A.; Santachiara-Benerecetti, A.S.; Semino, O. In search of the genetic footprints of Sumerians: A survey of Y-chromosome and mtDNA variation in the Marsh Arabs of Iraq. *BMC Evol. Biol.* **2011**, *11*, 288. [CrossRef] [PubMed]
6. Chao, L. *Rice Production*; Report No. 1081 to the Government of Iraq; FAO: Rome, Italy, 1959.
7. Kovach, M.J.; Calingacion, M.N.; Fitzgerald, M.A.; Mccouch, S.R. The origin and evolution of fragrance in rice (*Oryza sativa* L.). *Proc. Natl. Acad. Sci. USA* **2009**, *106*, 14444–14449. [CrossRef] [PubMed]
8. Al-Judy, N. Detecting of DNA Fingerprints and Genetic Relationship Analysis in Local and Improved Rice (*Oryza sativa* L.) Varieties in Iraq. Ph.D. Thesis, College of Science-Baghdad University, Baghdad, Iraq, 2004.
9. Al-Kazaz, A.K.A. Sequence Variation and Phylogenetic Relationships Among Ten Iraqi Rice Varieties Using RM171 Marker. *Iraqi J. Sci.* **2014**, *55*, 145–150.
10. Patwardhan, A.; Ray, S.; Roy, A. Molecular Markers in Phylogenetic Studies—A Review. *J. Phylogenet. Evol. Biol.* **2014**, *2*, 131.
11. Liu, X.; Zhang, H. Advance of molecular marker application in the tobacco research. *Afr. J. Biotechnol.* **2008**, *7*, 4827–4831.
12. Sotowa, M.; Ootsuka, K.; Kobayashi, Y.; Hao, Y.; Tanaka, K.; Ichitani, K.; Flowers, J.; Purugganan, M.; Nakamura, I.; Sato, Y.-I.; et al. Molecular relationships between Australian annual wild rice, *Oryza meridionalis* and two related perennial forms. *Rice* **2013**, *6*, 26. [CrossRef]
13. Nock, C.J.; Waters, D.L.E.; Edwards, M.A.; Bowen, S.G.; Rice, N.; Cordeiro, G.M.; Henry, R.J. Chloroplast genome sequences from total DNA for plant identification. *Plant Biotechnol. J.* **2011**, *9*, 328–333. [CrossRef]

14. Straub, S.C.K.; Parks, M.; Weitemier, K.; Fishbein, M.; Cronn, R.C.; Liston, A. Navigating the tip of the genomic iceberg: Next-generation sequencing for plant systematics. *Am. J. Bot.* **2012**, *99*, 349–364. [CrossRef] [PubMed]
15. Mcpherson, H.; Van Der Merwe, M.; Delaney, S.K.; Edwards, M.A.; Henry, R.J.; Mcintosh, E.; Rymer, P.D.; Milner, M.L.; Siow, J.; Rossetto, M. Capturing chloroplast variation for molecular ecology studies: A simple next generation sequencing approach applied to a rainforest tree. *BMC Ecol.* **2013**, *13*, 8. [CrossRef] [PubMed]
16. Waters, D.L.E.; Nock, C.J.; Ishikawa, R.; Rice, N.; Henry, R.J. Chloroplast genome sequence confirms distinctness of Australian and Asian wild rice. *Ecol. Evol.* **2012**, *2*, 211–217. [CrossRef] [PubMed]
17. Yang, J.-B.; Tang, M.; Li, H.-T.; Zhang, Z.-R.; Li, D.-Z. Complete chloroplast genome of the genus Cymbidium: Lights into the species identification, phylogenetic implications and population genetic analyses. *BMC Evol. Biol.* **2013**, *13*, 84. [CrossRef] [PubMed]
18. Tong, W.; He, Q.; Wang, X.Q.; Yoon, M.Y.; Ra, W.H.; Li, F.; Yu, J.; Oo, W.H.; Min, S.K.; Choi, B.W.; et al. A chloroplast variation map generated using whole genome re-sequencing of Korean landrace rice reveals phylogenetic relationships among *Oryza* sativa subspecies. *Biol. J. Linn. Soc.* **2015**, *115*, 940–952. [CrossRef]
19. Tong, W.; Kim, T.-S.; Park, Y.-J. Rice Chloroplast Genome Variation Architecture and Phylogenetic Dissection in Diverse *Oryza* Species Assessed by Whole-Genome Resequencing. *Rice* **2016**, *9*, 57. [CrossRef]
20. Parks, M.; Cronn, R.; Liston, A. Increasing phylogenetic resolution at low taxonomic levels using massively parallel sequencing of chloroplast genomes. *BMC Biol.* **2009**, *7*, 84. [CrossRef]
21. Moore, M.J.; Pamela, S.S.; Charles, D.B.; Burleigh, J.G.; Douglas, E.S. Phylogenetic analysis of 83 plastid genes further resolves the early diversification of eudicots. *Proc. Natl. Acad. Sci. USA* **2010**, *107*, 4623–4628. [CrossRef]
22. Brozynska, M.; Omar, E.S.; Furtado, A.; Crayn, D.; Simon, B.; Ishikawa, R.; Henry, R.J. Chloroplast Genome of Novel Rice Germplasm Identified in Northern Australia. *Trop. Plant Biol.* **2014**, *7*, 111–120. [CrossRef]
23. Huang, X.; Kurata, N.; Wang, Z.X.; Wang, A.; Zhao, Q.; Zhao, Y.; Liu, K.; Lu, H.; Li, W.; Guo, Y.; et al. A map of rice genome variation reveals the origin of cultivated rice. *Nature* **2012**, *490*, 497–501. [CrossRef]
24. Kim, H.; Jeong, E.; Ahn, S.-N.; Doyle, J.; Singh, N.; Greenberg, A.; Won, Y.; Mccouch, S. Nuclear and chloroplast diversity and phenotypic distribution of rice (*Oryza sativa* L.) germplasm from the democratic people's republic of Korea (DPRK North Korea). *Rice* **2014**, *7*, 7. [CrossRef] [PubMed]
25. Brozynska, M.; Copetti, D.; Furtado, A.; Wing, R.A.; Crayn, D.; Fox, G.; Ishikawa, R.; Henry, R.J. Sequencing of Australian wild rice genomes reveals ancestral relationships with domesticated rice. *Plant Biotechnol. J.* **2017**, *15*, 765–774. [CrossRef] [PubMed]
26. Wang, W.; Mauleon, R.; Hu, Z.; Chebotarov, D.; Tai, S.; Wu, Z.; Li, M.; Zheng, T.; Fuentes, R.R.; Zhang, F.; et al. Genomic variation in 3,010 diverse accessions of Asian cultivated rice. *Nature* **2018**, *557*, 43–49. [CrossRef] [PubMed]
27. Sims, D.; Ian, S.; Nicholas, E.I.; Andreas, H.; Chris, P.P. Sequencing depth and coverage: Key considerations in genomic analyses. *Nat. Rev. Genet.* **2014**, *15*, 121. [CrossRef]
28. Wambugu, P.W.; Brozynska, M.; Furtado, A.; Waters, D.L.; Henry, R.J. Relationships of wild and domesticated rices (*Oryza* AA genome species) based upon whole chloroplast genome sequences. *Sci. Rep.* **2015**, *5*, 13957. [CrossRef]
29. Civáň, P.; Craig, H.; Cox, C.J.; Brown, T.A. Three geographically separate domestications of Asian rice. *Nat. Plants* **2015**, *1*, 15164. [CrossRef]
30. Garris, A.J.; Tai, T.H.; Coburn, J.; Kresovich, S.; Mccouch, S. Genetic Structure and Diversity in *Oryza sativa* L. *Genetics* **2005**, *169*, 1631–1638. [CrossRef]
31. Ahuja, S.C.; Panwar, D.V.S.; Uma, A.; Gupta, K.R. *Basmati Rice: The Scented Pearl, Hisar, Directorate of Publications*; CCS Haryana Agricultural University: Hisar, India, 1995.
32. Akram, M. Aromatic rices of Pakistan: A review. *Pak. J. Agric. Res.* **2009**, *22*, 154–160.
33. Glaszmann, J.C. Isozymes and classification of Asian rice varieties. *Tag. Theor. Appl. Genet. Theor. Angew. Genet.* **1987**, *74*, 21–30. [CrossRef]
34. Caicedo, A.L.; Williamson, S.H.; Hernandez, R.D.; Boyko, A.; Fledel-Alon, A.; York, T.L.; Polato, N.R.; Olsen, K.M.; Nielsen, R.; McCouch, S.R.; et al. Genome-wide patterns of nucleotide polymorphism in domesticated rice. *PLoS Genet.* **2007**, *3*, e163. [CrossRef]
35. Furtado, A. DNA extraction from vegetative tissue for next-generation sequencing. *Methods Mol. Biol.* **2014**, *1099*, 1–5. [PubMed]

36. Moner, A.M.; Furtado, A.; Henry, R.J. Chloroplast phylogeography of AA genome rice species. *Mol. Phylogenet. Evol.* **2018**, *127*, 475–487. [CrossRef] [PubMed]
37. Katoh, K.; Misawa, K.; Kuma, K.; Miyata, T. MAFFT: A novel method for rapid multiple sequence alignment based on fast Fourier transform. *Nucleic Acids Res.* **2002**, *30*, 3059–3066. [CrossRef] [PubMed]
38. Huelsenbeck, J.P.; Ronquist, F. MRBAYES: Bayesian inference of phylogenetic trees. *Bioinformatics* **2001**, *17*, 754–755. [CrossRef]

8

Durum Wheat Seminal Root Traits within Modern and Landrace Germplasm in Algeria

Ridha Boudiar [1,2,*], Juan M. González [3], Abdelhamid Mekhlouf [1], Ana M. Casas [4] and Ernesto Igartua [4,*]

1 Laboratoire d'Amélioration et de Développement de la Production Végétale et Animale (LADPVA), University of Ferhat ABBAS (UFAS-Sétif1), Setif 19000, Algeria; mekhloufa@yahoo.fr
2 Biotechnology Research Center (CRBt), UV 03 BP E73, Nouvelle Ville Ali Mendjli, Constantine 25016, Algeria
3 Department of Biomedicine and Biotechnology, University of Alcalá, 28805 Alcalá de Henares, Spain; juanm.gonzalez@uah.es
4 Estación Experimental de Aula Dei, EEAD-CSIC, Avenida Montañana 1005, 50059 Zaragoza, Spain; acasas@eead.csic.es
* Correspondence: boudiarreda@yahoo.fr (R.B.); igartua@eead.csic.es (E.I.);

Abstract: Seminal roots are known to play an important role in crop performance, particularly under drought conditions. A set of 37 durum wheat cultivars and local landraces was screened for variation in architecture and size of seminal roots using a laboratory setting, with a filter paper method combined with image processing by SmartRoot software. Significant genetic variability was detected for all root and shoot traits assessed. Four rooting patterns were identified, with landraces showing overall steeper angle and higher root length, in comparison with cultivars, which presented a wider root angle and shorter root length. Some traits revealed trends dependent on the genotypes' year of release, like increased seminal root angle and reduced root size (length, surface, and volume) over time. We confirm the presence of a remarkable diversity of root traits in durum wheat whose relationship with adult root features and agronomic performance should be explored.

Keywords: proxy traits; genetic resources; root screening; root architecture

---

1. Introduction

The root system of wheat includes two main types, seminal (embryonic) and nodal roots, also known as the crown or adventitious roots [1,2]. Both types of roots play a crucial role in plant growth and are active throughout the whole plant life. Seminal roots, however, could be more important under specific circumstances, like drought conditions, as they penetrate deeper into the soil layers than nodal roots, making water in deep layers accessible to the plant [3–5]. Seminal roots also play a capital role during crop establishment, as they are the only roots existing before the emergence of the fourth leaf. Seminal roots include one primary root, two pairs of symmetric roots at each side, and, at times, a sixth central root [6].

The main features of root systems are encompassed under two categories, root system architecture (RSA) and morphology. RSA is related to the whole, or a large subset, of the root system, and may be described as topological or geometric measures of the root shape. Root morphology, as defined by J. Lynch, refers to "the surface features of a single root axis as an organ, including characteristics of the epidermis such as root hairs, root diameter, the root cap, the pattern of appearance of daughter roots, undulations of the root axis, and cortical senescence" [7]. The traits often used to describe wheat roots are total root length, root surface area, root volume, root angle, number of roots, and root diameter [8–10].

Roots are difficult to measure readily in natural conditions. Root trait determination has become accessible through the development of phenotyping methods in artificial systems, for instance, gel chambers [11], rolled germination paper [12], clear pots and growth pouches [13], "Termita" chamber and Whatman paper system [14], or growth pouches system [15]. Seminal roots can be phenotyped early and easily compared to the root system of mature plants [16,17], and for this reason, they have been proposed as good candidates to act as proxy traits in wheat [18] and maize [19,20]. Nevertheless, phenotyping these traits could be of interest only if they are useful to predict root growth and functioning in adult plants [21,22]. Indeed, several studies have found useful associations with traits in adult plants of wheat species [23–26]. For instance, the seminal root angle was correlated with nodal root angle [5,27], and with grain yield under drought conditions [28]. The seminal root number was correlated with thousand kernel weight (TKW) under stress, while the primary root length at the seedling stage was correlated with TKW under wetter conditions [25]. A steeper angle between the outermost roots and a higher root number in wheat seedlings have been linked to a more compact root system with more roots at depth in wheat [11,24,29].

Genotypic variation in root architecture has been reported within genotypes of different crop species [30–32], including wheat [13,21,25,33]. The presence of variation for the trait of interest is an essential requirement to improve the adaptability of crops under changing environmental conditions [34].

Local landraces are considered well adapted to the region where they were grown and contain large genetic diversity useful to improve crops like durum wheat [35]. These landraces were replaced by high yielding but more uniform semi-dwarf cultivars, better adapted to modern agriculture. However, scientists are convinced that local landraces still constitute a genetic resource useful to improve commercially valuable traits [36]. It is assumed that root traits enhance response to drought stress [37], but the realization of their contribution to superior grain yield depends on the type of drought and the agro-ecological conditions [38]. A deep rooting ideotype ("steep, cheap, and deep") was proposed by [39] to optimize water and N acquisition, building on the assumption that deeper rooting genotypes will use water that is beyond reach for shallower rooting genotypes. Modern breeding has caused some shifts in the root system architecture of durum wheat, from shallower and densely rooted systems in landraces of Mediterranean origin to deeper and more evenly distributed systems throughout the soil depth in cultivars worldwide [5].

The current study aims at evaluating the diversity of seminal root traits, including root angle and depth, during early growth of a set of durum wheat genotypes, consisting of modern cultivars and local landraces which are representative of the germplasm adapted to the mostly semi-arid conditions of Algerian cereal-growing regions before and after the advent of modern breeding. The study aims to reveal morphological diversity that could have agronomic relevance and, therefore, interest breeders.

2. Materials and Methods

2.1. Plant Material

We studied thirty-seven genotypes (landraces and modern cultivars), representative of durum wheat (*Triticum turgidum* ssp. *durum* Desf.) grown in Algeria. Geographical origins were varied (Algeria, France, Italy, Spain, Tunisia), and included genotypes produced at international breeding programs addressing semi-arid areas, namely the International Maize and Wheat Improvement Center (CYMMIT), the International Center for Agricultural Research in the Dry Area (ICARDA) and the Arab Center for the Studies of Arid zones and Drylands (ACSAD). These genotypes are representative of different periods of agriculture in Algeria, before and after the Green Revolution (Table 1).

Table 1. Name, type of cultivar, origin, and year of release of 37 genotypes of durum wheat used in the experiment.

N°	Genotype	Type/Pedigree	Origin	Year of Release
1	Beliouni	Landrace	Algeria	1958
2	Bidi 17	Landrace	Algeria	1930
3	Djenah Khotifa	Landrace	North Africa	1955
4	Gloire de Montgolfier	Landrace	Algeria	1960
5	Guemgoum R' khem	Landrace	Algeria	1960
6	Hedba 3	Landrace	Algeria	1921
7	Langlois	Landrace	Algeria	1930
8	Mohammed Ben Bachir (MBB)	Landrace	Algeria	1930
9	Montpellier	Landrace	Algeria	1965
10	Oued Zenati 368	Landrace	Algeria	1936
11	Acsad 65	Gerardo-vz-469/3/Jori-1// Nd-61-130/Leeds	ACSAD	1984
12	Altar 84	Ruff/Flamingo,mex// Mexicali-75/3/Shearwater	CYMMIT	1984
13	Ammar 6	Lgt3/4/Bicre/3/Ch1// Gaviota/Starke	ICARDA	2010
14	Bousselem	Heider//Martes/ Huevos de oro	ICARDA	2007
15	Boutaleb	Hedba 3/Ofanto	Algeria	2013
16	Capeiti	Eiti*6/Senatore-Cappelli	Italy	1940
17	Chen's	Shearwater(sib)/(sib)Yavaros-79	CYMMIT	1983
18	Ciccio	Appulo/Valnova(f6)// (f5)Valforte/Patrizio	Italy	1996
19	Cirta	Hedba-3/Gerardo-vz-619	Algeria	2000
20	Core	Platani/Gianni	Italy	2008
21	GTA Dur	Crane/4/Polonicum PI185309//T.glutin enano/2* Tc60/3/Gll	CIMMYT	1972
22	INRAT 69	Mahmoudi/(bd-2777)Kyperounda	Tunisia	1969
23	Korifla	Durum-dwarf-s-15/Crane//Geier	ICARDA	1987
24	Mansourah	Bread wheat/MBB	Algeria	2012
25	Massinissa	Ofanto/Bousselem	Algeria	2012
26	Megress	Ofanto/Waha//MBB	Algeria	2007
27	Mexicali 75	Gerardo-vz-469/3/ Jori(sib)//Nd-61-130/Leeds	CIMMYT	1975
28	Ofanto	Ademelio/Appulo	Italy	1990
29	Oued El Berd	Gta dur/Ofanto	Algeria	2013
30	Polonicum	Triticum polinicum/Zenati boulette 1953-58	France	1973
31	Sahell	Cit"s"/4/Tace/4*tc//2*zb/ wls/3/aa"s"/5/Ruff"s"/Albe"s"	CYMMIT	1977
32	Simeto	Capeiti-8/Valnova	Italy	1988
33	Sitifis	Bousselam/Ofanto	Algeria	2011
34	Vitron	Turkey77/3/Jori/Anhinga//Flamingo	Spain	1987
35	Waha	Plc/Ruff//Gta's/3/ Rolette	ICARDA	1986
36	Wahbi	Bidi 17/Waha//Bidi 17	Algeria	2002
37	ZB × Fg	Zb/fg"s" lk/3/ko 120/4/Ward cs 10604	Algeria	1983

*: Backcross.

2.2. Root Phenotyping

2.2.1. Preparation of Seeds

Twelve seeds of uniform size and healthy aspects were visually selected from each genotype and surface sterilized in a sodium hypochlorite solution (1.25% + one detergent drop, Mistol Henkel Iberica®). Seeds of each genotype were soaked and shaken in the solution for 15–20 min. Then, they were rinsed four times with sterile deionized water, in sterile conditions. Twelve seeds of each genotype were placed in Petri dishes, each with two filter papers soaked with 4 mL of sterile water. Then the Petri dishes were placed in a dark room at 4 °C for four days, and then at 22 °C/18 °C in a growth chamber with a 12 h light/darkness photoperiod for about 16 h.

Finally, the pre-germinated grains were transferred to the rhizo-slide system, described in detail in the next section and Figure S1. The experiment was carried out at the Laboratory of Cellular Biology and Genetics, Department of Biomedicine and Biotechnology of the University of Alcalá, Spain.

2.2.2. The Rhizo-Slide System

The rhizo-slide system was constructed as a sandwich made with glass plate, black cardboard, filter paper, and a black plastic sheet. Sheets of A4-size black cardboard (180 g/m^2, www.liderpapel.com) and filter papers were previously sterilized in an autoclave and then soaked in the nutritive solution Aniol [40]. The nutritive solution was prepared by dissolving 0.5550 g of Ca Cl_2, 0.8215 g of KNO_3, 0.6352 g of $MgCl_2 \cdot 6H_2O$, 0.0165 g of $(NH_4)_2SO$, 0.0400 g of NH_4NO_3 in 100 mL of distilled water, to which 500 µL/L of Plant Preservative Mixture (PPM^{TM}, Plant Cell Technology) at pH 5.8 was added. Each 8 mL was used to prepare 1 L of nutritive solution. Black cardboard with a nick made at the top center was placed on a glass plate with the same dimensions; then the pre-germinated grain (with embryonic part downward) was positioned just below the nick and covered by a filter paper. A black plastic sheet was used to cover the filter paper to ensure obscurity for roots, shifted ~2 cm upwards to allow better contact of the cardboard, and filter paper sheets with the nutritive solution. Two rhizo-slides were confronted to each other by the glass plate side, and the set was placed vertically in a glass box (internal dimensions of 32.2, 22, and 16 cm, length, width, and height) with two liters of the nutritive solution at the bottom, and then secured with two paper clips. Each glass box held 6 glass plates with two rhizoslides each, for a total of 12 seedlings, consisting of two genotypes, 6 seedlings for each (Figure S1). In total, each genotype was replicated 12 times. More details on the system are found in Ruiz et al. (2018) [25].

Once placed in the rhizoslides and the glass boxes, the seedlings were grown in a growth chamber for 7 days at 22/18 °C and 12/12 h photoperiod, day/night. The 37 genotypes were processed in batches of 6. Pre-germinated seeds of each 6 genotypes were placed into six glass boxes, each holding 6 seeds of two different genotypes. A complete batch comprised six boxes, three glass boxes prepared each Monday, and three each Thursday, every week. In total, 7 batches (14 runs) were performed until the experiment was completed (accounting for some seedlings that had to be replicated for various reasons). The set of genotypes for each run was selected randomly.

During the experiment, the boxes were replenished with distilled water every two days, to refill to the initial solution level. At the same time, to minimize seedling failure, each single seedling received 10 mL of the nutritive solution, applied with a pipette, near each seed. On the eighth day, the rhizo-slides were opened and shoots were immediately collected. The fresh roots were scanned using a Canon "LiDE210" scanner at 300 ppi to capture the first image then overlapped roots were manually separated and a second scan was done. The individual plant shoot dry weight (SDW) was obtained after oven-drying at 80 °C for six hours.

2.2.3. Image Analysis

The two images of a rhizo-slide were analyzed using SmartRoot software v.3.32 [41] plugin for ImageJ1.46R (http://imagej.nih.gov/ij/download.html). The first image was used to measure only root angles and the second one to assess the other root traits using manual and semi-automatic SmartRoot procedures. Each root of the seedling was traced, semi-automatically, and then SmartRoot automatically generated the corresponding traits. In total, ten variables from the Smartroot output were recorded for each seedling: total root length (TRL), primary root length (PRL), mean length of the other seminal roots (MRL), total root surface area (Surface), mean root diameter (Diameter), total root volume (Volume), root number (RN), and shoot dry weight (SDW). The root angle was determined for each root with respect to the vertical (90°). From this determination, we extracted the maximum vertical angle (MVA) represented by the root growing with the steepest angle, the least vertical angle (LVA) represented by the root growing with the widest angle, and mean vertical angle (MRA) of all the roots, for each seedling.

2.3. Statistical Analysis

The experiment was considered a completely randomized design, with 12 replicates per genotype. Statistical analyses were performed using the REML (Restricted maximum likelihood) procedure with Genstat 18 [42]. Genotypes were considered as fixed factors and replications were considered as a random factor. The "Genotype" factor (n-1 degrees of freedom) was broken down into a single degree of freedom comparison of landraces vs. cultivars (named "Type" effect), and a "within type" factor (n-2) which corresponds to the variation of genotypes within each type. Multiple means separation was carried out using LSD at 0.05 level, for variables in which the F-value for "Genotypes" was significant. A principal component analysis (PCA) and a hierarchical cluster analysis (HC) were performed using the R package FactoMineR [43]. The hcut function was used for tree cutting levels truncation. The R package Factoextra [44] was employed for extracting and visualizing the results. Broad-sense heritability (h^2) was calculated on an entry mean basis using the REML procedure, as follows:

$$h^2 = \sigma^2_g/(\sigma^2_g + (\sigma^2_e/r)) \quad (1)$$

where σ^2_g is the genotypic variance, σ^2_e is the error variance and r is the number of replications.

3. Results

3.1. Genotypic Variability

We found remarkable genetic variability for all measured traits, as revealed by the highly significant differences among genotypes in the analyses of variance (Table 2). Significant differences were also found in the "type" comparison for most traits, except for Diameter and RN (Table 2). For the other traits, the mean squares for type were 4 to 12 times larger than those for genotypes.

The means of landraces showed higher or equal mean values compared to cultivars for all traits, except root angle (MRA, LVA, and MVA), which was higher in cultivars (Figure 1). It is worth mentioning that the landrace group presented higher root depth (PRL) than the cultivars.

All traits but SDW were root-related traits so, henceforward all the traits will be referred to generally as root traits unless stated otherwise. All traits (except RN) showed a near-normal distribution (Figure 1) which denotes their polygenic control. A wide range of phenotypic values was observed for most traits (Table 2). The landrace group showed a larger range of variation for TRL, Surface, Volume, and SDW than the cultivars. For the other traits, the cultivars had higher ranges of variation (Tables S1 and S2).

The coefficients of variation (CV) ranged from small values like 5.19 (Diameter) to 24.60 (MRA, Table 2). The exception was the large CV found for MVA, 59.82. When calculated separately for landraces and cultivars, slightly higher CV for most traits were found in landraces compared to cultivars (Table S2). All the traits exhibited high broad sense heritability (h^2), ranging from 0.80 for MVA to 0.98 for MRA (Table 2).

Table 2. Descriptive statistics, broad sense heritability (h^2), ANOVA summary, and correlation coefficients for the root traits assessed in 37 durum wheat cultivars.

	Descriptive Statistics					ANOVA		Correlation									
Traits	Min	Mean	Max	CV	h^2	Genotype	Type	TRL	Surface	Volume	Diameter	PRL	SDW	MRA	LVA	MVA	RN
TRL (cm)	54.28	98.49	137.22	13.51	0.90	***	***	1	***	***	ns	***	***	***	**	***	***
Surface (cm^2)	8.25	16.77	24.82	14.74	0.90	***	***	0.95	1	***	***	***	***	***	ns	***	***
Volume (cm^3)	0.0962	0.2326	0.3721	17.03	0.90	***	***	0.83	0.97	1	***	***	***	***	ns	**	***
Diameter (cm)	0.0447	0.0538	0.0620	5.19	0.87	***	ns	0.08	0.38	0.59	1	***	***	***	*	**	***
PRL (cm)	14.32	26.51	32.06	10.76	0.90	***	***	0.55	0.55	0.52	0.17	1	***	ns	**	***	ns
SDW (g)	5.70	14.72	23.50	18.66	0.94	***	***	0.55	0.63	0.64	0.40	0.58	1	ns	**	*	ns
MRA (°)	1.20	30.36	45.69	24.60	0.98	***	***	−0.39	−0.32	−0.23	0.20	−0.01	0.00	1	***	***	ns
LVA (°)	20.40	42.71	61.47	17.81	0.92	***	***	−0.28	−0.23	−0.18	0.12	0.08	0.01	0.76	1	***	ns
MVA (°)	0.00	14.10	43.49	59.82	0.80	***	***	−0.25	−0.20	−0.15	0.12	−0.27	−0.13	0.62	0.25	1	**
RN (no.)	4.00	5.32	6.00	9.13	0.87	***	ns	0.37	0.32	0.26	−0.22	−0.01	0.06	−0.09	0.02	−0.14	1

*, **, ***: sources of variation in the analyses of variance or correlation coefficients significant at $p < 0.05$, 0.01, and 0.001, respectively. TRL: Total root length, Surface: total root surface area, Volume: Total root volume, Diameter: mean root diameter, PRL: Primary root length, SDW: Shoot dry weight, MRA: Mean root angle, LVA: Least vertical angle, MVA: Maximum vertical angle, RN: Root number.

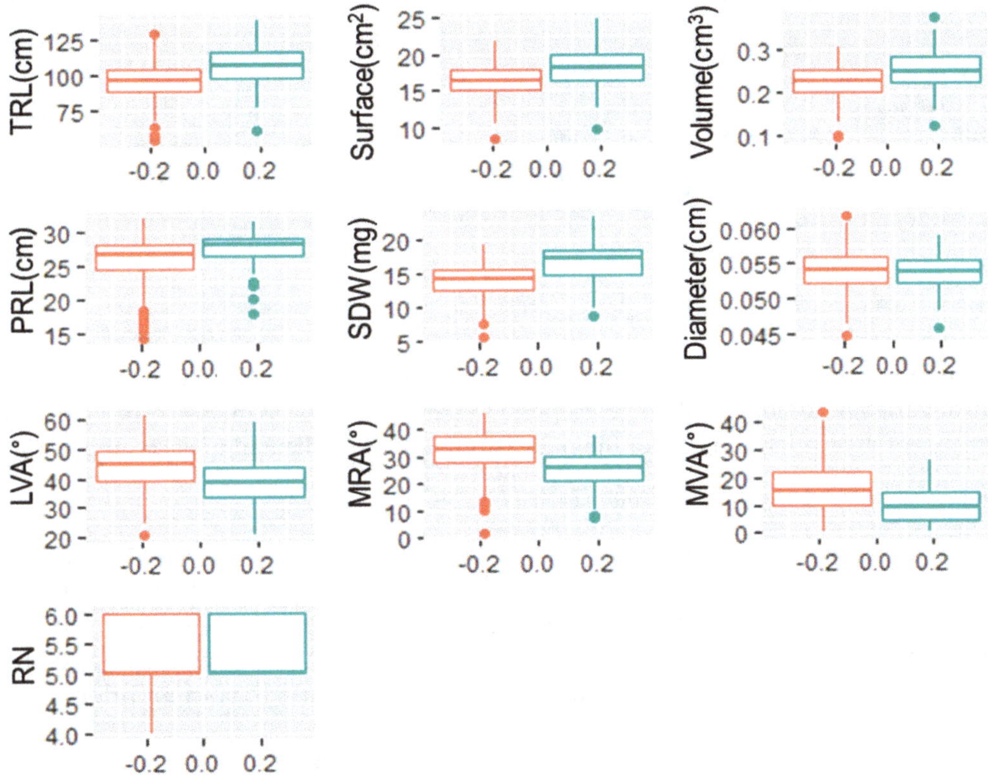

Figure 1. Boxplots for root traits for the cultivar (red) and landrace (blue) groups. Horizontal lines splitting the boxes indicate the median values; box limits indicate the 25th and 75th percentiles; whiskers extend 1.5 times the interquartile range from the 25th and 75th percentiles; outliers are represented by dots. Variable names coded as in Table 2. Genotype mean values are in Table S1.

3.2. Relationships between Traits

Highly significant correlations were found between most traits (Table 2). TRL, Surface and Volume were highly and positively correlated among them. There were moderate positive correlations between TRL, surface, and Volume, with PRL, SDW, RN, and negative ones with root angle variables (seedlings with higher TRL, Surface, and Volume tended to have steeper root angles). Seedlings with higher RN tended to have roots with thinner root diameter, indicating that there could be some kind of compensation between these traits (more roots with a finer diameter and vice versa). Interestingly, seedlings with higher primary root length produced more shoot biomass. Performing correlations between traits within each group (cultivars and landraces) showed, in general, similar patterns to the correlations performed for the entire dataset (Table S2). The moderate relationship of PRL with MVA and RN disappeared in the landrace group, compared to the cultivars and the whole dataset (Table S2).

3.3. Time Trends of Root Traits

When the genotypic means were plotted against year of release of the genotypes, different trends were observed (Figure 2 and Figure S2), in which, all the traits presented significant regression coefficients except Diameter, RN and PRL (Table S3). This trend was largely influenced by the comparison of landraces vs. cultivars because landraces are older. The trend was positive or negative depending on the trait. Overall, cultivars reduced their seminal root length and developed a shallower root angle compared to landraces (Figure 2). The root surface and volume of root presented the same trend as root length, as they were highly correlated, as mentioned above. MVA and LVA showed the same trend as MRA. No substantial variation was observed for RN, Diameter, and PRL. Regarding the shoot, a remarkable and steady reduction in SDW over the years was detected (Figure 1 and Figure S2).

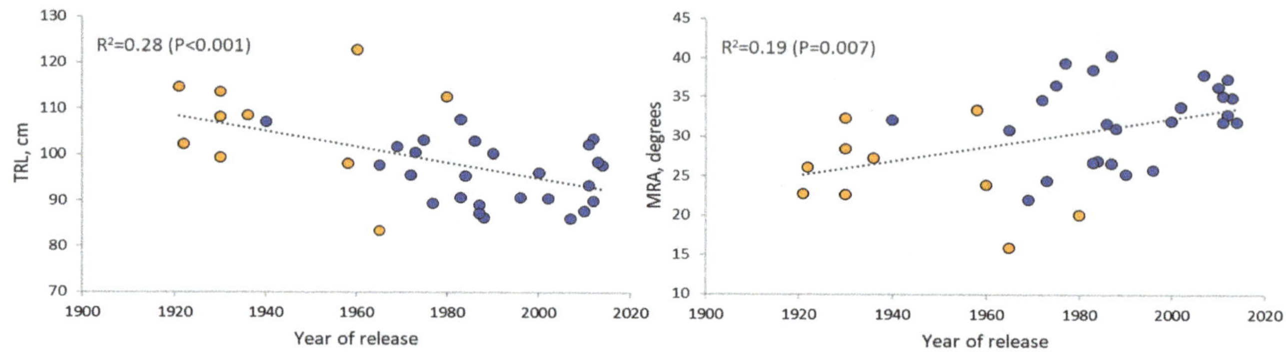

Figure 2. Time trends for total root length (TRL) and mean root angle (MRA) in seedlings of 37 durum wheat varieties. Yellow symbols correspond to landraces; blue symbols correspond to cultivars. The coefficients of determination (R^2) of the regression lines are indicated in each graph.

3.4. Grouping of Genotypes According to Root Traits

The first two principal components explained 69.63% of the total variation (Figure 3A). The first component (46%) was most related to Surface, TRL, Volume, and SDW, with the respective contributions of 20.19, 19.04, 18.16, and 11.51 (Table S4). MRA, LVA, and Diameter had the highest loadings for the second component (PC2). Correlations between these traits are discussed above (Table 2). Thus, the first axis (PC1) was related to root size traits and the second one to root architecture traits.

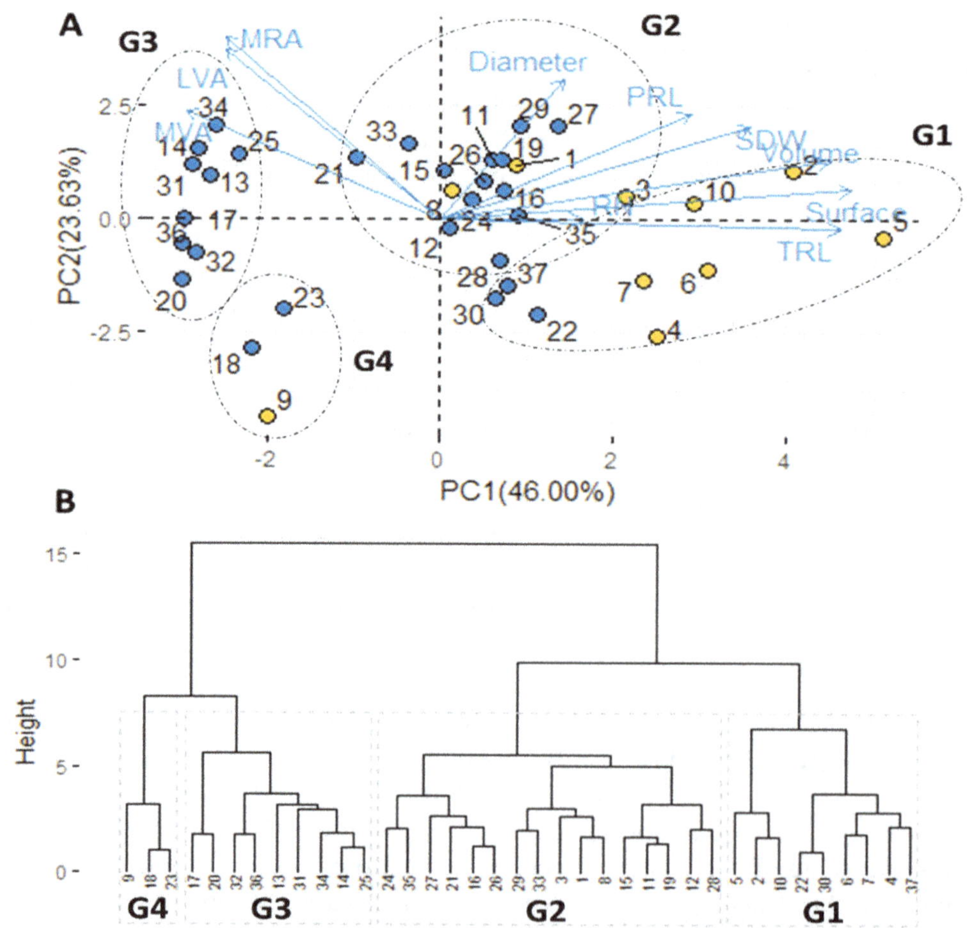

Figure 3. Biplot of the first two principal components (**A**) and dendrogram resulting from hierarchical clustering (**B**) based on seedling traits for 37 durum wheat genotypes. Ellipses in (**A**) encompass the individuals according to the clustering presented in (**B**). Yellow symbols correspond to landraces; blue symbols correspond to cultivars. Genotypes coded with numbers as in Table 1.

Genotypes were better distributed along with the first component, as a result of the contrasting position between landraces, many with large positive scores on PC1 (due to their higher root size and shoot weight) and the cultivars, with lower positive or negative scores in PC1, so the discrimination between these two groups was clear (Figure 3B). From the hierarchical classification, which was carried out based on the original data, four groups were created (G1 to G4) (Figure 3B). G1 was mostly formed by landraces. G2 was the largest one and was constituted by cultivars, and two landraces. This group was at a central position in the biplot graph (Figure 3B), presenting close to average values for most traits. G3 was located on the negative side of PC1, contrasting with G1 by having a relative smaller root size. Finally, the last group (G4) was formed by only three genotypes depicted on the negative quadrant, for both PC1 and PC2, having smaller values for both classes of root traits; fine, steeper root angle and reduced root traits related to biomass. This group included landrace Montpellier (genotype 9), which showed a special root system architecture compared to other landraces, with steeper root angle, and lower SDW, closer to two cultivars from Italy and ICARDA.

4. Discussion

The durum wheat collection used in this study was assembled to explore the seminal root variability present in a set of genotypes cultivated in Algeria, with a historical perspective on the possible changes caused by modern breeding. The method chosen enabled data acquisition and processing of 444 single plants, by one person, in two months. Its performance could be easily expanded by increasing the number of boxes and operators. Therefore, it is amenable to the scale needed for the type of studies carried out in plant genetics and breeding. Root number together with root length, the main results of this type of experiment, describe how extensively the seminal axes can potentially explore the rooting volume. These easily measurable traits at an early stage can have agronomic implications. For example, root spread angle is an additional feature whose variation can influence how crops cope with water-limited conditions and/or other environmental constraints, such as high pH, toxic ions, or low nutrient availability [45,46]. The root angular spread at an early growth stage can be used to predict the partitioning of root biomass in the soil profile at the adult plant stage [5,27,28], a feature relevant for water use efficiency in wheat [21,47]. Therefore, artificial systems are efficient at revealing phenotypic (and presumably genetic) variability, but its implications on agronomic performance must be validated later under field conditions.

4.1. Large Genotypic Variation for Seminal Root Traits

An overview of the results found in different studies sheds more light on the actual genetic variation available for seminal root traits, better than any single study. Differences among studies may be partly due to slight differences in the experimental methods, but also to the size and scope of the genetic material used. Nevertheless, some meaningful conclusions can be derived.

We found significant genetic variation for all traits. We found a range of values for the least vertical root angle (LVA) from 20.40° to 61.47°. Multiplying these values by two (range from 40.80° to 122.94°) allows the comparison of our study with others, in which the values of the total opening of the angle of the root system was reported. Our range was superior to those found by others in durum [47] and bread wheat [24,29]. Our wheat genotypes displayed similar low ranges of variation in mean root number as in similar studies in durum [47,48], with a slightly higher mean. In our genotypes, the sixth root was present in about a third of all genotypes, with no significant differences between landraces and cultivars. This is a similar proportion than found in a study of Mediterranean and North-American elite material [47], with the striking difference that in the former study they reported almost absence of the sixth seminal root in native Mediterranean materials [47]. Neither sample of landrace materials was large enough to derive definitive conclusions from these studies, but at least we can say that Algerian landraces are not more likely to lack the sixth seminal root than modern cultivars.

Based on the coefficients of variation, overall, landraces showed higher slightly variability for most traits, especially for root angle, even though the sample size was lower than for cultivars. Previous reports indicate that native Mediterranean landraces are likely to provide additional genetic variability for root architecture [46], particularly in wheat accessions that experienced long-term natural selection in drought-prone environments [49], and in barley [50]. Overall, the Algerian landraces showed sizeable genetic variation for most traits, indicating that they harbor relevant root morphology variation that should be further investigated by geneticists and breeders.

4.2. Classification of Durum Wheat Genotypes According to Root Morphology

Overall, genotypes with higher root length tended to have larger root number, as found in a previous study [47], and a narrower root angle. Other authors [24,51] found no correlation between root angle and root number. Sanguinetti et al. [47] also found no correlation of root angle with other traits and suggested that the root angle was controlled by an independent set of genes. In our study, however, given the negative correlation between MRA and root size traits, we cannot rule out that these two traits are controlled by the same set of genes.

We found that higher root length and Diameter were associated with higher SDW (r of 0.55 and 0.40, respectively, Table 2), suggesting a size effect that affected the whole plant. Rather similar observations were done in the Spanish core collection of tetraploid wheat, but the plant size effect was visible for subsp. *dicoccon* and *turgidum*, but not for *durum* [25]. Correlation between root length and volume and SDW was also found in hexaploid wheat [52]. We found no correlation between RN and MRA, in agreement with previous studies [24,51]. It seems that an overall plant size effect that affects harmonically roots and shoots is common in wheat species.

Our genotypes displayed different seminal root system patterns, from vigorous and steep to a small and shallow root system. These root patterns may be related to phylogenetic relationships, regional origin, and functional plant adaptation to different environments, as indicated in previous studies [53]. There were differences in the length of the seminal roots of single plants. This was made evident by calculating the difference between the length of the primary root (PRL), and the average of the rest (MRL). G2 and G3 had a higher difference between PRL and the mean length of other roots (MRL), compared to groups G1 and G4, which had roots with more similar lengths (Table S5). G2 genotypes combined a significantly longer primary root (Table S5) with the largest difference between it and the other seminal roots (together with G3). This rooting pattern, based on dissimilar growth of the roots, could have an impact on overall soil exploring capacity that should be explored further, particularly its usefulness in semi-arid environments, to access to stored water at deep layers at critical periods (flowering and grain filling), while keeping enough shallow roots to take advantage of in-season precipitations.

Two groups (G1 and G3) showed the highest contrast in the multivariate analysis (Figure 3). G1, with a majority of landraces, displayed a vigorous seminal root system, in contrast with G3, formed entirely by cultivars with small root systems. Our finding was in agreement with the study of a collection of 160-durum wheat landraces [33] in terms of larger seminal root size. This study found that landraces coming from the eastern Mediterranean region (Turkey), the driest and warmest areas considered in the study, showed the largest seminal root size and widest root angle compared to landraces from eastern Balkan countries. The authors claimed that these differences were due to the adaptations of landraces to the contrasting environmental conditions of these two regions. The larger root size and wider root angle from Turkish landraces would allow better exploration of the full soil profile and better water capture. Among the four groups found in this study, no one combined the highest MRA and TRL, comparable to Turkish landraces. Therefore, there could be room for improvement for the root systems of durum wheat for Algeria. Crosses to combine these traits in a single genotype should be devised, and Turkish landraces could be tested in Algerian conditions, to assess their potential.

In our germplasm, the landraces showed on an average narrower angle and higher root size. Previous studies on Mediterranean durum wheat [33] found that the genotypes with the narrowest angle came from the western Mediterranean region and that they also had heavier grains [54,55].

Additionally, it was reported that *Triticum turgidum* subsp. *dicoccon* landraces coming from cooler and wetter zones had shallower seminal root systems than those from warmer and drier areas [25]. The subsp. *durum* landraces, developed in warmer and drier areas, tended to have larger and steeper root patterns than landraces coming from cooler and wetter zones. Accordingly, the root system architecture of the Algerian landraces would indicate adaptation to a warm and dry environment. Other studies have found different root morphologies in apparent adaptation to stressful conditions. For instance, the drought-tolerant bread wheat cv. SeriM82 has a compact root system [21], associated with a limited water use early in the season, facilitating access to stored water later in the reproductive phase. Contrary to our landraces, SeriM82 exhibited less vigorous shoot growth. In contrast with our findings, a study of bread wheat germplasm grown historically in the semi-arid northwestern of China [56] found that breeding caused a narrowing of the seminal root angle, reduced root number, and increase of primary seminal root length. In that study "newer cultivars produced higher yields than older ones only at the higher sowing density, showing that increased yield results from changes in competitive behavior." This view was confirmed and expanded later [57], confirming that the advantage of new Chinese wheat cultivars came from the attenuation of inter-plant competition and increased plasticity in root morphology. A seminal root architecture with fewer, longer seminal roots with narrower root angle, would overlap less with neighbors, leading to less competition between individuals [58], and these trends agree with the hypothesis of weakening of "selfish" traits [59].

The shift in root morphology observed in Algeria in the step from landraces to modern cultivars does not conform to the scenario described in those works. There was a reduction of overall root length and volume after the advent of modern breeding, which could be consistent with the reduction of inter-plant competition but combined with the widening of the root angle, which does not bode well with that hypothesis. It seems that wheat breeding may have resulted in different trends for root morphology in different parts of the world. This could be the result of the adaptation of Algerian landraces to agronomic conditions different from current agriculture. The difference in rooting patterns between landraces from different geographical areas and cultivars may lie in the agronomic environments in which they were developed. In general, modern durum wheat cultivars were bred under high plant densities [57], whereas landraces were grown in stands with density adapted to the environment. The morphology of Algerian landraces (long seminal roots growing in steep angles) conforms to the "steep, cheap, and deep root ideotype" [60], and could be the result of adaptation to accessing water in deep soil layers. Further studies with adult plants are needed to evaluate if root features of seedlings are maintained when the competition between individuals for root growth is increased (as the seminal and nodal roots require more space and resources than just the seminal roots of the seedlings). A shovelomics experiment is being carried out with the same genotypes, which could elucidate this issue at least for some measurable traits like root angle.

The high SDW of our durum landraces compared to cultivars could be related to the lack of dwarfing genes in the landraces. This hypothesis was already confirmed previously for bread and durum wheat for some height reducing genes [61], which reduced the first seedling leaf growth in *Rht* genotypes compared with the corresponding tall wheat lines.

4.3. Conclusions and Perspectives

We have found wide genetic variability in a collection of durum wheat genotypes cultivated in Algeria and unraveled a possible historic trend that sheds light on the outcomes of modern breeding. An important issue is to what extent this variability found at the seedling stage can reflect the variability in the field with the same genetic material, more precisely, which traits can be consistent across plant phases (seedling and adult plant), enabling the selection at the early seedling stage. If this relationship is not found, then the room for the testing of seminal root traits is very limited. Experiments to evaluate this relationship are ongoing.

Overall, landraces showed a larger root size and steeper root angle. These two traits could be involved in the adaptation of landraces to water-stressed environments. The dwarfing genes seem to influence biomass partitioning; screening the current germplasm for these genes would elucidate this issue. The root size and shape in our data indicated some independence that would open opportunities to design cultivars with the desired combinations of traits.

Overall, the current genotypes present a diverse root system architecture, from compact deep-rooting to wide shallow one. This opens the opportunity to test the four different root ideotypes found (G1–G4) for functional implications under water and nutrient-limited environments. Based on the above results, we hypothesize that root architecture difference between cultivars, landraces (or steep deep vs. shallow root systems) may result in different strategies of adaptation to the availability of water and nutrients over the soil profile.

Supplementary Materials:
Table S1. Mean values and standard errors (12 replications) of the seminal root traits for the 37 durum wheat genotypes. Table S2: Descriptive statistics and correlation coefficients for seedling traits, calculated separately for the cultivar and landrace groups. Table S3: Results of linear regressions of traits over years of release. Table S4: Contribution (%) of the traits to the first two principal components (PC1, PC2), as represented in Figure 3A. Table S5: Means comparison between groups formed by hierarchical clustering for the traits assessed. Figure S1. (A) One-week-old durum wheat seedlings in the rhizoslide system; (B) pictures of two genotypes contrasting for root angle, landrace Gloire de Montgolfier (left) and cultivar Oued El Berd (right); (C) schematic representation of a glass box, holding 6 glass plates, each holding two rhizoslides. Figure S2: Time trend of seminal root traits over the year of release.

Author Contributions: Conceptualization, E.I.; methodology, J.M.G.; formal analysis, R.B. and E.I.; investigation, R.B. and E.I.; resources, A.M.; software, J.M.G.; supervision, J.M.G., A.M., and E.I.; data curation, R.B.; visualization, J.M.G.; writing—original draft preparation, R.B. and J.M.G.; writing—review and editing, R.B., J.M.G., A.M., A.M.C., and E.I.; project administration, A.M.C.; funding acquisition, R.B.; A.M.; A.M.C., and E.I. All authors have read and agreed to the published version of the manuscript.

Acknowledgments: R.B. is a recipient a PhD grant in the framework of Programme National Exceptionel (PNE 2018/2019) funded by Algerian Ministry of Higher Education and Scientific Research.

References

1. Chochois, V.; Vogel, J.P.; Rebetzke, G.J.; Watt, M. Variation in adult plant phenotypes and partitioning among seed and stem-borne roots across *Brachypodium distachyon* accessions to exploit in breeding cereals for well-watered and drought environments. *Plant Physiol.* **2015**, *168*, 953–967. [CrossRef]
2. Sinha, S.K.; Rani, M.; Kumar, A.; Kumar, S.; Venkatesh, K.; Mandal, P.K. Natural variation in root system architecture in diverse wheat genotypes grown under different nitrate conditions and root growth media. *Theor. Exp. Plant Physiol.* **2018**, *30*, 223–234. [CrossRef]
3. Araki, H.; Iijima, M. Deep rooting in winter wheat: Rooting nodes of deep roots in two cultivars with deep and shallow root systems. *Plant Prod. Sci.* **2001**, *4*, 215–219. [CrossRef] [PubMed]
4. Manske, G.G.B.; Vlek, P.L.G. Root Architecture-Wheat as a Model Plant. In *Plant Roots: The Hidden Half*, 3rd ed.; Waisel, Y., Eshel, A., Beeckman, T., Kafkafi, U., Eds.; Marcel Dekker: New York, NY, USA, 2002; pp. 249–259.
5. Maccaferri, M.; El-Feki, W.; Nazemi, G.; Salvi, S.; Canè, M.A.; Colalongo, M.C.; Stefanelli, S.; Tuberosa, R. Prioritizing quantitative trait loci for root system architecture in tetraploid wheat. *J. Exp. Bot.* **2016**, *67*, 1161–1178. [CrossRef] [PubMed]
6. Esau, K. *Plant Anatomy*, 2nd ed.; John Wiley: New York, NY, USA, 1965; 767p.
7. Lynch, J. Root architecture and plant productivity. *Plant Physiol.* **1995**, *109*, 7–13. [CrossRef] [PubMed]
8. Ahmadi, J.; Pour-Aboughadareh, A.; Fabriki-Ourang, S.; Mehrabi, A.A.; Siddique, K.H.M. Screening wheat germplasm for seedling root architectural traits under contrasting water regimes: Potential sources of variability for drought adaptation. *Arch. Agron. Soil Sci.* **2018**, *64*, 1351–1365. [CrossRef]
9. York, L.M.; Slack, S.; Bennett, M.J.; Foulkes, M.J. Wheat shovelomics I: A field phenotyping approach for characterizing the structure and function of root systems in tillering species. *BioRxiv* **2018**, 280875. [CrossRef]

10. Nguyen, V.L.; Stangoulis, J. Variation in root system architecture and morphology of two wheat genotypes is a predictor of their tolerance to phosphorus deficiency. *Acta Physiol. Plant.* **2019**, *41*, 109. [CrossRef]
11. Bengough, A.G.; Gordon, D.C.; Al-Menaie, H.; Ellis, R.P.; Allan, D.; Keith, R.; Thomas, W.T.B.; Forster, B.P. Gel observation chamber for rapid screening of root traits in cereal seedlings. *Plant Soil* **2004**, *262*, 63–70. [CrossRef]
12. Watt, M.; Moosavi, S.; Cunningham, S.C.; Kirkegaard, J.A.; Rebetzke, G.J.; Richards, R.A. A rapid, controlled-environment seedling root screen for wheat correlates well with rooting depths at vegetative, but not reproductive, stages at two field sites. *Ann. Bot.* **2013**, *112*, 447–455. [CrossRef]
13. Richard, C.A.; Hickey, L.T.; Fletcher, S.; Jennings, R.; Chenu, K.; Christopher, J.T. High-throughput phenotyping of seminal root traits in wheat. *Plant Methods* **2015**, *11*, 13. [CrossRef] [PubMed]
14. González, J.M.; Friero, E.; Selfa, L.; Froilán, S.; Jouve, N.A. Comparative study of root system architecture in seedlings of *Brachypodium* spp. using three plant growth supports. *Cereal Res. Commun.* **2016**, *44*, 69–78. [CrossRef]
15. Adeleke, E.; Millas, R.; McNeal, W.; Faris, J.; Taheri, A. Variation analysis of root system development in wheat seedlings using root phenotyping system. *Agronomy* **2020**, *10*, 206. [CrossRef]
16. Richard, C.; Christopher, J.; Chenu, K.; Borrell, A.; Christopher, M.; Hickey, L. Selection in early generations to shift allele frequency for seminal root angle in wheat. *Plant Genome* **2018**, *11*, 170071. [CrossRef] [PubMed]
17. El Hassouni, K.; Alahmad, S.; Belkadi, B.; Filali-Maltouf, A.; Hickey, L.T.; Bassi, F.M. Root system architecture and its association with yield under different water regimes in durum wheat. *Crop Sci.* **2018**, *58*, 2331–2346. [CrossRef]
18. Bai, C.; Liang, Y.; Hawkesford, M.J. Identification of QTLs associated with seedling root traits and their correlation with plant height in wheat. *J. Exp. Bot.* **2013**, *64*, 1745–1753. [CrossRef]
19. Tuberosa, R.; Salvi, S.; Sanguineti, M.C.; Landi, P.; Maccaferri, M.; Conti, S. Mapping QTLs regulating morpho-physiological traits and yield: Case studies, shortcomings and perspectives in drought-stressed maize. *Ann. Bot.* **2002**, *89*, 941–963. [CrossRef]
20. Tuberosa, R.; Sanguineti, M.C.; Landi, P.; Giuliani, M.M.; Salvi, S.; Conti, S. Identification of QTLs for root characteristics in maize grown in hydroponics and analysis of their overlap with QTLs for grain yield in the field at two water regimes. *Plant Mol. Biol.* **2002**, *48*, 697–712. [CrossRef]
21. Manschadi, A.M.; Christopher, J.; De Voil, P.; Hammer, G.L. The role of root architectural traits in adaptation of wheat to water-limited environments. *Funct. Plant Biol.* **2006**, *33*, 823–837. [CrossRef]
22. Paez-Garcia, A.; Motes, C.M.; Scheible, W.R.; Chen, R.; Blancaflor, E.B.; Monteros, M.J. Root traits and phenotyping strategies for plant improvement. *Plants* **2015**, *4*, 334–355. [CrossRef]
23. Løes, A.K.; Gahoonia, T.S. Genetic variation in specific root length in Scandinavian wheat and barley accessions. *Euphytica* **2004**, *137*, 243–249. [CrossRef]
24. Manschadi, A.M.; Hammer, G.L.; Christopher, J.T.; Devoil, P. Genotypic variation in seedling root architectural traits and implications for drought adaptation in wheat (*Triticum aestivum* L.). *Plant Soil* **2008**, *303*, 115–129. [CrossRef]
25. Ruiz, M.; Giraldo, P.; González, J.M. Phenotypic variation in root architecture traits and their relationship with eco-geographical and agronomic features in a core collection of tetraploid wheat landraces (*Triticum turgidum* L.). *Euphytica* **2018**, *214*, 54. [CrossRef]
26. Li, T.; Ma, J.; Zou, Y.; Chen, G.; Ding, P.; Zhang, H.; Yang, C.; Mu, Y.; Tang, H.; Liu, Y.; et al. Quantitative trait loci for seeding root traits and the relationships between root and agronomic traits in common wheat. *Genome* **2020**, *63*, 27–36. [CrossRef] [PubMed]
27. Alahmad, S.; El Hassouni, K.; Bassi, F.M.; Dinglasan, E.; Youssef, C.; Quarry, G.; Able, J.A.; Christopher, J.A. Major root architecture QTL responding to water limitation in durum wheat. *Front. Plant Sci.* **2019**, *10*, 436. [CrossRef]
28. Ali, M.L.; Luetchens, J.; Nascimento, J.; Shaver, T.M.; Kruger, G.R.; Lorenz, A.J. Genetic variation in seminal and nodal root angle and their association with grain yield of maize under water-stressed field conditions. *Plant Soil* **2015**, *397*, 213–225. [CrossRef]
29. Nakamoto, T.; Oyanagi, A. The direction of growth of seminal roots of *Triticum aestivum* L. and experimental modification thereof. *Ann. Bot.* **1994**, *73*, 363–367. [CrossRef]
30. Masi, C.E.A.; Maranville, J.W. Evaluation of sorghum root branching using fractals. *J. Agric. Sci.* **1998**, *131*, 259–265. [CrossRef]

31. Liao, H.; Rubio, G.; Yan, X.; Cao, A.; Brown, K.M.; Lynch, J.P. Effect of phosphorus availability on basal root shallowness in common bean. *Plant Soil* **2001**, *232*, 69–79. [CrossRef]
32. Lynch, J.P.; Brown, K.M. Topsoil foraging–an architectural adaptation of plants to low phosphorus availability. *Plant Soil* **2001**, *237*, 225–237. [CrossRef]
33. Roselló, M.; Royo, C.; Sanchez-Garcia, M.; Soriano, J.M. Genetic dissection of the seminal root system architecture in Mediterranean durum wheat landraces by genome-wide association study. *Agronomy* **2019**, *9*, 364. [CrossRef]
34. El-Beltagy, A.; Madkour, M. Impact of climate change on arid lands agriculture. *Agric. Food Secur.* **2012**, *1*, 3. [CrossRef]
35. Nazco, R.; Villegas, D.; Ammar, K.; Pena, R.J.; Moragues, M.; Royo, C. Can Mediterranean durum wheat landraces contribute to improved grain quality attributes in modern cultivars? *Euphytica* **2012**, *185*, 1–17. [CrossRef]
36. Lopes, M.S.; El-Basyoni, I.; Baenziger, P.S.; Singh, S.; Royo, C.; Ozbek, K.; Aktas, H.; Ozer, E.; Ozdemir, F.; Manickavelu, A.; et al. Exploiting genetic diversity from landraces in wheat breeding for adaptation to climate change. *J. Exp. Bot.* **2015**, *66*, 3477–3486. [CrossRef]
37. Araujo, S.S.; Beebe, S.; Crespi, M.; Delbreil, B.; González, E.M.; Gruber, V.; Lejeune-Henaut, I.; Link, W.; Monteros, M.J.; Prats, E.; et al. Abiotic stress responses in legumes: Strategies used to cope with environmental challenges. *Crit. Rev. Plant Sci.* **2015**, *34*, 237–280. [CrossRef]
38. Rao, I.M.; Beebe, S.E.; Polania, J.; Grajales, M.; Cajiao, C.; Ricaurte, J.; García, R.; Rivera, M. Evidence for genotypic differences among elite lines of common bean in the ability to remobilize photosynthate to increase yield under drought. *J. Agric. Sci.* **2017**, *155*, 857–875. [CrossRef]
39. Lynch, J.P. Steep, cheap and deep: An ideotype to optimize water and N acquisition by maize root systems. *Ann. Bot.* **2013**, *112*, 347–357. [CrossRef]
40. Aniol, A. Induction of aluminum tolerance in wheat seedlings by low doses of aluminum in the nutrient solution. *Plant Physiol.* **1984**, *75*, 551–555. [CrossRef]
41. Lobet, G.; Pagès, L.; Draye, X. A novel image-analysis toolbox enabling quantitative analysis of root system architecture. *Plant Physiol.* **2011**, *157*, 29–39. [CrossRef]
42. Payne, R.W.; Murray, D.A.; Harding, S.A.; Baird, D.B.; Soutar, D.M. *GenStat for Windows Introduction*, 12th ed.; VSN International: Hemel Hempstead, UK, 2009.
43. Le, S.; Josse, J.; Husson, F. FactoMineR: An R Package for Multivariate Analysis. *J. Stat.* **2008**, *25*, 18.
44. Kassambara, A.; Mundt, F. Package 'factoextra'. *Extr. Vis. Results Multivar. Data Anal.* **2017**, *76*. Available online: http://www.sthda.com/english/rpkgs/factoextra (accessed on 2 March 2020).
45. Devaiah, B.N.; Nagarajan, V.K.; Raghothama, K.G. Phosphate homeostasis and root development in Arabidopsis is synchronized by the zinc finger transcription factor ZAT6. *Plant Physiol.* **2007**, *145*, 147–159. [CrossRef] [PubMed]
46. Da Silva, A.; Bruno, I.P.; Franzini, V.I.; Marcante, N.C.; Benitiz, L.; Muraoka, T. Phosphorus uptake efficiency, root morphology and architecture in Brazilian wheat cultivars. *J. Radioanal. Nucl. Chem.* **2016**, *307*, 1055–1063. [CrossRef]
47. Sanguineti, M.C.; Li, S.; Maccaferri, M.; Corneti, S.; Rotondo, F.; Chiari, T.; Tuberosa, R. Genetic dissection of seminal root architecture in elite durum wheat germplasm. *Ann. Appl. Biol.* **2007**, *151*, 291–305. [CrossRef]
48. Cane, M.A.; Maccaferri, M.; Nazemi, G.; Salvi, S.; Francia, R.; Colalongo, C.; Tuberosa, R. Association mapping for root architectural traits in durum wheat seedlings as related to agronomic performance. *Mol. Breed.* **2014**, *34*, 1629–1645. [CrossRef]
49. Robertson, B.M.; Waines, J.G.; Gill, B.S. Genetic variability for seedling root number in wild and domesticated wheats. *Crop Sci.* **1979**, *19*, 843–847. [CrossRef]
50. Grando, S.; Ceccarelli, S. Seminal root morphology and coleoptile length in wild (*Hordeum vulgare* ssp. *spontaneum) and cultivated (Hordeum vulgare ssp. vulgare) barley. Euphytica* **1995**, *86*, 73–80.
51. O'Brien, L. Genetic variability of root growth in wheat (*Triticum aestivum* L.). *Aust. J. Agric. Res.* **1979**, *30*, 587–595. [CrossRef]
52. Narayanan, S.; Mohan, A.; Gill, K.S.; Prasad, P.V. Variability of root traits in spring wheat germplasm. *PLoS ONE* **2014**, *9*, e100317. [CrossRef]
53. Bodner, G.; Leitner, D.; Nakhforoosh, A.; Sobotik, M.; Moder, K.; Kaul, H.P. A statistical approach to root system classification. *Front. Plant Sci.* **2013**, *4*, 292. [CrossRef]

54. Royo, C.; Nazco, R.; Villegas, D. The climate of the zone of origin of Mediterranean durum wheat (*Triticum durum* Desf.) landraces affects their agronomic performance. *Genet. Resour. Crop Evol.* **2014**, *61*, 1345–1358. [CrossRef]
55. Soriano, J.M.; Villegas, D.; Aranzana, M.J.; Del Moral, L.F.G.; Royo, C. Genetic structure of modern durum wheat cultivars and Mediterranean landraces matches with their agronomic performance. *PLoS ONE* **2016**, *11*, e0160983. [CrossRef] [PubMed]
56. Zhu, Y.H.; Weiner, J.; Yu, M.X.; Li, F.M. Evolutionary agroecology: Trends in root architecture during wheat breeding. *Evol. Appl.* **2019**, *12*, 733–743. [CrossRef] [PubMed]
57. Song, L.; Zhang, D.W.; Li, F.M.; Fan, X.W.; Ma, Q.; Turner, N.C. Drought stress: Soil water availability alters the inter- and intra-cultivar competition of three spring wheat cultivars bred in different eras. *J. Agron. Crop Sci.* **2010**, *196*, 323–335. [CrossRef]
58. De Parseval, H.; Barot, S.; Gignoux, J.; Lata, J.; Raynaud, X. Modelling facilitation or competition within a root system: Importance of the overlap of root depletion and accumulation zones. *Plant Soil* **2017**, *419*, 97–111. [CrossRef]
59. Weiner, J.; Du, Y.L.; Zhang, C.; Qin, X.L.; Li, F.M. Evolutionary agroecology: Individual fitness and population yield in wheat (*Triticum aestivum*). *Ecology* **2017**, *98*, 2261–2266. [CrossRef]
60. Kembel, S.W.; Cahill, J.F. Independent evolution of leaf and root traits within and among temperate grassland plant communities. *PLoS ONE* **2011**, *6*, e19992. [CrossRef]
61. Ellis, M.H.; Rebetzke, G.J.; Chandler, P.; Bonnett, D.; Spielmeyer, W.; Richards, R.A. The effect of different height reducing genes on the early growth of wheat. *Funct. Plant Biol.* **2004**, *31*, 583–589. [CrossRef]

9

The Development and Characterization of Near-Isogenic and Pyramided Lines Carrying Resistance Genes to Brown Planthopper with the Genetic Background of *japonica* Rice (*Oryza sativa* L.)

Cuong D. Nguyen [1,2], Holden Verdeprado [3], Demeter Zita [4], Sachiyo Sanada-Morimura [5], Masaya Matsumura [5], Parminder S. Virk [3,6], Darshan S. Brar [3,7], Finbarr G. Horgan [8], Hideshi Yasui [9] and Daisuke Fujita [1,3,4,9,*]

1 United Graduate School of Agricultural Sciences, Kagoshima University, Kagoshima 890-0065, Japan; cuongdhqt2000@gmail.com
2 College of Food Industry, 101B Le Huu Trac Street, Son Tra District, Da Nang City 550000, Vietnam
3 International Rice Research Institute, DAPO Box 7777, Metro Manila 1301, Philippines; h.verdeprado@irri.org (H.V.); p.virk@cgiar.org (P.S.V.); darshanbrar@pau.edu (D.S.B.)
4 Faculty of Agriculture, Saga University, Saga 840-8502, Japan; demeterzita2012@gmail.com
5 NARO Kyushu Okinawa Agricultural Research Center, 2421 Suya, Koshi, Kumamoto 861–1192, Japan; sanadas@affrc.go.jp (S.S.-M.); mmasa@affrc.go.jp (M.M.)
6 International Center for Tropical Agriculture, A.A, 6713 Cali, Colombia
7 School of Agricultural Biotechnology, Punjab Agricultural University, Ludhiana 141027, India
8 EcoLaVerna Integral Restoration Ecology, Bridestown, Kildinan, Co. Cork, T56 CD39, Ireland; f.g.horgan@gmail.com
9 Plant Breeding Laboratory, Graduate School, Kyushu University, Fukuoka 812-8581, Japan; hyasui@agr.kyushu-u.ac.jp
* Correspondence: dfujita@cc.saga-u.ac.jp.

Abstract: The brown planthopper (BPH: *Nilaparvata lugens* Stål.) is a major pest of rice, *Oryza sativa*, in Asia. Host plant resistance has tremendous potential to reduce the damage caused to rice by the planthopper. However, the effectiveness of resistance genes varies spatially and temporally according to BPH virulence. Understanding patterns in BPH virulence against resistance genes is necessary to efficiently and sustainably deploy resistant rice varieties. To survey BPH virulence patterns, seven near-isogenic lines (NILs), each with a single BPH resistance gene (*BPH2*-NIL, *BPH3*-NIL, *BPH17*-NIL, *BPH20*-NIL, *BPH21*-NIL, *BPH32*-NIL and *BPH17-ptb*-NIL) and fifteen pyramided lines (PYLs) carrying multiple resistance genes were developed with the genetic background of the *japonica* rice variety, Taichung 65 (T65), and assessed for resistance levels against two BPH populations (Hadano-66 and Koshi-2013 collected in Japan in 1966 and 2013, respectively). Many of the NILs and PYLs were resistant against the Hadano-66 population but were less effective against the Koshi-2013 population. Among PYLs, *BPH20*+*BPH32*-PYL and *BPH2*+*BPH3*+*BPH17*-PYL granted relatively high BPH resistance against Koshi-2013. The NILs and PYLs developed in this research will be useful to monitor BPH virulence prior to deploying resistant rice varieties and improve rice's resistance to BPH in the context of regionally increasing levels of virulence.

Keywords: rice (*Oryza sativa* L.); brown planthopper; near-isogenic lines; pyramided lines; resistance; virulence

1. Introduction

The brown planthopper (BPH: *Nilaparvata lugens* Stål.) is a major pest of rice (*Oryza sativa* L.) in tropical and subtropical Asia [1]. BPH damages rice by sucking phloem from the plants (mechanical damage) or by transmitting viruses such as rice grassy stunt virus (RGSV), rice ragged stunt phytoreovirus (RRSV) and rice wilted stunt virus (RWSV) [2,3]. At high planthopper densities, rice crops display patches of desiccated rice known as 'hopperburn.' Insecticides have been widely used to reduce BPH populations [4]. However, insecticides are damaging to human health and the environment, and are increasingly recognized as contributing to BPH outbreaks through physiological and ecological pest resurgence mechanisms [1]. Host plant resistance is considered a potentially effective alternative to harmful insecticides, that reduces BPH damage without detrimental effects on the natural enemies of BPH [5].

To date, more than 34 BPH resistance genes have been identified from rice cultivars and wild rice species. Seven genes: *BPH9*, *BPH14*, *BPH17*, *BPH18*, *BPH26*, *BPH29* and *BPH32* have been cloned and characterized for different BPH resistance levels [6–12]. Four gene clusters (chromosomal regions) strongly associated with BPH resistance have been identified. These occur on chromosomes 4S (short arm), 4L (long arm), 6S and 12 [3,13]. Four genes (*BPH12* from B14, *BPH15*, *BPH17* and *BPH20*) and six quantitative trait loci (QTLs) (*QBph4, QBPH4.1, QBPH4.2, QBph4.2 qBph4.3* and *qBph4.4*) for BPH resistance have been identified on chromosome 4S [13–20]. Among those QTLs, *QBph4* (6.7–6.9 Mbp) from IR02W101 (*Oryza officinalis*) and *QBph4.2* (6.6–6.9 Mbp) from IR65482-17 (*Oryza australiensis*) were identified at similar locations based on physical distance; *QBPH4.1* (5.8–7.8 Mbp) and *QBPH4.2* (15.2–17.2 Mbp) were identified from Rathu Heenati; *qBph4.3* (0.2–0.7 Mbp) and *qBph4.4* (0.7–13.1 Mbp) were detected in Salkathi. On chromosome 4L, five genes for BPH resistance—*BPH6*, *BPH12(t)* from GSK185-2, *BPH18(t)* from BPH2183, *BPH27* from GX2183 and *BPH27(t)* from Balamawee have been identified [21–25]. Six genes/QTLs have been detected on the short arm of chromosome 6: *BPH3, BPH4, BPH25, BPH32, Qbph6* and *qBPH6(t)* [12,26–30]. Eight genes have been identified on the long arm of chromosome 12—*BPH1, BPH2, BPH7, BPH9, BPH10* and *BPH18* from IR65482-7-216-1-2, and *BPH21* and *BPH26* [17,29,31–38]. The BPH resistance genes on chromosome 12 were classified into four allelic types based on their amino acid sequences and different resistance levels: type 1—*BPH1, BPH10, BPH18* and *BPH21*; type 2—*BPH2* and *BPH26*; type 3—*BPH7*; and type 4—*BPH9* [6]. A number of highly resistant rice breeding lines and varieties contain several genes with resistance to BPH and other phloem feeding Hemiptera. These include IR71033-121-15 introgressed from *Oryza minuta* carrying *BPH20, BPH21* and *qBPH6(t)* [17,39]; Rathu Heenati carrying *BPH3, BPH17, QBPH4.1* and *QBPH4.2* [8,13,27]; and PTB33 carrying *BPH2* and *BPH32* [12,40,41].

Monogenic resistance is vulnerable to rapid adaptation by BPH populations. Research indicates that BPH populations have sufficient genetic variability to enable them to overcome specific resistance genes when selected on a resistant host over multiple generations [4,42,43]. In the late 1970s, BPH populations adapted to varieties carrying the *BPH1* and/or *BPH2* genes after these were widely deployed in rice varieties across Asia [4,42,44]. A recent multi-national study has indicated that BPH populations across much of Asia have adapted to feed on rice carrying the *BPH1, BPH2, BPH5, BPH7, BPH8, BPH9, BPH10* and *BPH18* genes [45]. Under laboratory conditions, BPH populations continually reared for between seven to 15 generations on resistant rice varieties were capable of adapting to resistance from a range of genes, including *BPH1, BPH2, BPH3, BPH8, BPH9, BPH10* and *BPH32* [43,46–49]. Through adaptation to resistance genes, BPH acquires stronger virulence against resistance genes and BPH virulence remains stable for several decades [4,50]. Therefore, it is important to preserve the effects of resistance genes by preventing BPH adaptation.

To prevent further adaptation by BPH populations to available resistance genes, a strategy for deploying resistance based on insect virulence is necessary [4]. However, BPH virulence varies under different environments depending on the predominant rice cultivars, BPH migration routes, and the length of population exposure to different resistance genes [4]. Without exposing resistance genes to BPH populations under controlled conditions prior to deployment, the potential effectiveness of the resistance genes for target regions is difficult to predict. In previous studies, the virulence of BPH has been characterized using resistant varieties [45,48,51–53]. However, many BPH-resistant varieties have multiple resistance genes, such that the effects of any single resistance gene cannot be assessed using these varieties. In contrast, the effects of any single resistance gene may be revealed in detail by using near-isogenic lines (NILs) that carry the gene on the genetic background of a susceptible variety. Recently, more than 16 NILs with BPH resistance genes (*BPH3, BPH4, BPH6, BPH9, BPH10, BPH12, BPH14, BPH15, BPH17, BPH18, BPH20, BPH21, BPH25, BPH26, BPH30* and *BPH32*) have been developed on the genetic backgrounds of several *indica* and *japonica*-susceptible varieties. These NILs have been evaluated against different BPH populations from China, the Philippines and Japan [14,54–58].

Because the resistance of rice varieties carrying single genes is weaker and less durable (i.e., allowing rapid BPH adaptation) to BPH than varieties with multiple resistance genes, several researchers have proposed the pyramiding of two or more genes to enhance resistance levels and thereby avoid pest adaptation [59]. Combinations of multiple BPH resistance genes have been reported to increase levels of plant resistance to BPH. For example, a pyramided line (PYL) with *BPH14* and *BPH15* enhanced resistance against BPH from China compared to monogenic NILs with either *BPH14* or *BPH15* alone [60]. Similarly, the pyramided lines *BPH6* + *BPH12* PYL and *BPH3* + *BPH27* PYL exhibited greater resistance levels in bulk seedling tests than monogenic lines with each of the genes present alone [14,55], and *BPH17* + *BPH21* PYL had greater resistance against BPH in the Philippines than lines with either gene alone [57]. Pyramiding the *BPH25* and *BPH26* genes into a single rice line was reported to have positive epistatic effects against BPH populations collected in Vietnam, the Philippines and Japan [51,61]. Therefore, the development of rice varieties carrying multiple BPH resistance genes might be an effective way to enhance BPH resistance.

In this study, seven NILs with BPH resistance genes (*BPH2, BPH3, BPH17, BPH20, BPH21, BPH32* and *BPH17-ptb*) and a *japonica* rice genetic background were developed to evaluate the effects of different resistance genes on BPH populations. Based on the NILs developed, 15 pyramided lines (PYLs) carrying two or three resistance genes were developed to enhance levels of resistance against BPH. Additionally, using the NILs and PYLs we developed, the study compared resistance against two BPH populations collected in Japan: the first was collected in 1966 (before resistant varieties were widely released) and the second was collected in 2013 (recently migrated from China to Japan). Comparisons of the reactions by BPH from each population to the NILs and PYLs indicates the utility of resistance genes and their different combinations (some with epistatic effects) against modern BPH populations.

2. Results

2.1. Development of Seven NILs for BPH Resistance

Seven NILs with BPH resistance genes from three donor parents on the genetic background of Taichung 65 (T65) were developed through marker-assisted selection (MAS) and backcrossing (Tables 1 and 2). For three donor parents, IR71033-121-15 has *BPH20* and *BPH21*; Rathu Heenati contains *BPH3* and *BPH17*; and PTB33 carries *BPH2, BPH17-ptb* and *BPH32* based on previous studies. For PTB33, there has been no previous report of a BPH resistance gene on chromosome 4S. However, amino acid sequences for the *BPH17* locus in PTB33 were identical to those of Rathu Heenati [8]. Thus, we assume that PTB33 contains a gene for BPH resistance on chromosome 4S and tentatively named this as *BPH17-ptb*. The substituted chromosomal segments of the NILs were detected by polymorphic simple sequence repeat (SSR) markers that were equally distributed across the whole genome (Table 3; Figure 1). The genetic background of *BPH2*-NIL was analyzed using 203 polymorphic SSR markers. The ratio of substituted segments from PTB33 on *BPH2*-NIL was 9.1–14.8% (total 33.9–55.0 Mbp). One substituted segment with a size of 21.3–25.4 Mbp encompassing *BPH2* was detected between RM247 and RM5479 on chromosome 12. The other three segments were detected between RM5426 and RM248 on chromosome 7 with a size of 3.4–4.2 Mbp, between RM5688 and RM444 on chromosome 9 with a size of 4.2–9.2 Mbp and between RM7492 and RM216 on chromosome 10 with a size of 5.0–16.2 Mbp.

The genetic background of *BPH3*-NIL was confirmed using 195 polymorphic SSR markers, and the ratio of substituted segment from Rathu Heenati was 1.0–3.0% (total 3.8–11.3 Mbp). One segment with a size of 1.6–1.8 Mbp including *BPH3* was detected between MSSR1 and RM1369 on the short arm of chromosome 6. The other substituted segments were detected between RM1359 and RM1155 on chromosome 4 with a size of 0.5–4.2 Mbp and between RM1345 and RM3155 on chromosome 8 with a size of 1.8–4.9 Mbp.

The genetic background of *BPH17*-NIL was surveyed using 173 polymorphic SSR markers. The ratio of substituted segments was 1.0–4.8% (total 3.8–17.6 Mbp) containing one segment located between RM8213 and B40 on chromosome 4, including the *BPH17* region.

The genetic background of *BPH17-ptb*-NIL was analyzed using 229 polymorphic SSR markers, and the ratio of substituted segments from PTB33 on *BPH17-ptb*-NIL was 2.8–7.6% (total 10.5–28.1 Mbp). One substituted segment with 5.8–13.1 Mbp encompassing *BPH17-ptb* was detected between C61009 and B40 on chromosome 4. Two other substituted segments were detected at RM3126 on chromosome 3 and between RM7048 and RM6971 on chromosome 9 (4.7–11.5 Mbp).

The genetic background of *BPH20*-NIL was confirmed using 237 polymorphic SSR markers and the ratio of substituted segments of IR71033-121-15 was 5.6–9.6% (total 20.6–35.5 Mbp). One segment with a size of 13.8–19.9 Mb containing *BPH20* was detected between RM335 and RM5900 on chromosome 4. Two other substituted segments were detected between RM224 and RM5926 on chromosome 11 (1.5–7.7 Mbp) and between RM7315 and RM3103 on chromosome 12 (5.3–7.9 Mbp).

The genetic background of *BPH21*-NIL was surveyed using 229 polymorphic SSR markers, and the ratio of substituted segments from IR71033-121-15 was 7.1–11.6% (total 26.4–43.1 Mbp). One segment with a size of 22.6–23.7 Mbp, including *BPH21*, was detected between RM1880 and RM28493 on chromosome 12. Three other segments were detected between RM6841 and RM3348 on chromosome 5 (2.3–7.2 Mbp), around RM1328 on chromosome 9 and between RM224 and RM5926 on chromosome 11 (1.5–7.7 Mbp).

The genetic background of *BPH32*-NIL was confirmed using 233 polymorphic SSR markers. The ratio of substituted segments of PTB33 on *BPH32*-NIL was 1.9–4.1% (total 7.1–15.1 Mbp). One segment with a size of 1.6–3.2 Mbp containing *BPH32* was detected between RM6775 and RM190 on chromosome 6. Three other segments from the donor parent were detected between RM5755 and RM3280 on chromosome 3 with a size of 4.9–8.1 Mbp, between RM1306 and RM248 on chromosome 7 with a size of 0.4–3.2 Mbp and between RM5349 and RM5961 on chromosome 11 with a size of 0.2–0.6 Mbp.

Table 1. The SSR markers used for maker-assisted selection of nine genes for resistance to the brown planthopper.

Marker	Resistance Gene Tagged	Chromosome	Forward Primer Sequence (5′→3′)	Reverse Primer Sequence (5′→3′)	Physical Position (Mbp) *	Predicted Size (bp) **
RM1246 [a]	*BPH2, BPH21*	12	GGCTCACCTCGTTCTCGATCC	CATAAATAAATAGGGCGCCACACC	19.16	195
RM28493 [b]	*BPH2, BPH21*	12	ACCGTTAGATGACACAAGCAACG	GGTTAGCAAGACTGGAGGAGACG	23.28	259
RM508 [c]	*BPH3, BPH32*	6	AGAAGCCGGTTCATAGTTCATGC	ACCCGTGAACCACAAAGAACG	0.44	158
RM588 [c]	*BPH3, BPH32*	6	TCTTGCTGTGCTGTTAGTGTACG	GCAGGACATAAATACTAGGCATGG	1.61	97
RM8213 [a]	*BPH17, BPH17-ptb*	4	TGTTGGGTGGGTAAAGTAGATGC	CCCAGTGATACAAAGATGAGTTGG	4.42	178
MS10 [d]	*BPH17, BPH17-ptb, BPH20*	4	CAATACGAGAAGCCCCTCAC	CTGAAGGAACACGCGGTAGT	8.08	167
RM5900 [a]	*BPH20*	4	TTCTACGTTTGACCGTCA	TCTAGGAGCGTTTGTAGGAG	13.77	248
S00310 [e]	*BPH25*	6	CAACAAGATGGACGGCAAGG	TTGGAAGAAAAGGCAGGCAC	0.21	215
MSSR1 [e]	*BPH25*	6	CTAGCTGCTCTGCTCTGCTG	CGGCAATCTCTCCGAATC	0.22	114
RM309 [f]	*BPH26*	12	CACGCACCTTTCTGGCTTTCAGC	AGCAACCTCCGACGGGAGAAGG	21.52	177
RM28438 [b]	*BPH26*	12	GTTCGTGAGCCACAACAAATCC	GTTAAATGCTCCACCAAACACACC	22.59	216
InD14 [g]	*BPH26*	12	GGCCGAGTAGGATACTCTAGAAA	CTGCGAGAAAGGAGAGGTGG	22.87	387
RM28466 [b]	*BPH26*	12	CCGACGAAGAAGACGAGGAGTAGCC	AGGCCGGAGAGCAATCATGTCG	22.98	93
RM28481 [b]	*BPH26*	12	GTCAATTAACCATTGCCCATGC	TTCACGTGGGAACTACTCATGC	23.14	242
MSSR2 [e]	*BPH26*	12	CATGTCGAAGAGGTTGCAGA	GGTTTCATCCAAGTCCACGA	25.03	265

Primer sequence information was obtained from: [a]: McCouch et al. (2002) [62], [b]: International Rice Genome Sequencing Project (IRGSP) (2005) [63], [c]: Temnykh et al. (2001) [64], [d]: Rahman et al. (2009) [17], [e]: Myint et al. (2012) [29], [f]: Temnykh et al. (2000) [65] and [g]: Zhao et al. (2016) [6]. * The physical position of marker was the physical location of forward primer for each marker obtained from The Rice Annotation Project Database [66]. ** Physical distance and predicted size were estimated on the basis of the Nipponbare genome sequence.

Table 2. Details of the seven near-isogenic lines and 15 pyramided lines carrying brown planthopper resistance genes.

Entry	Gene (Donor Parent)						Generation
BPH2-NIL	***BPH2***	**(PTB33)**					BC_4F_3
BPH3-NIL	*BPH3*	(Rathu Heenati)					BC_4F_4
BPH17-NIL	*BPH17*	(Rathu Heenati)					BC_4F_4
BPH20-NIL	*BPH20*	(IR71033-121-15)					BC_4F_5
BPH21-NIL	*BPH21*	(IR71033-121-15)					BC_4F_5
BPH32-NIL	*BPH32*	(PTB33)					BC_4F_4
BPH17-ptb-NIL	*BPH17-ptb*	(PTB33)					BC_4F_3
BPH2+BPH17-PYL	*BPH2*	(PTB33)	*BPH17*	(Rathu Heenati)			BC_4F_3 equivalent
BPH2+BPH25-PYL	*BPH2*	(PTB33)	*BPH25*	(ADR52)			BC_4F_3 equivalent
BPH2+BPH32-PYL	*BPH2*	(PTB33)	*BPH32*	(PTB33)			BC_4F_3 equivalent
BPH2+BPH17-ptb-PYL	*BPH2*	(PTB33)	*BPH17-ptb*	(PTB33)			BC_4F_3 equivalent
BPH3+BPH17-PYL	*BPH3*	(Rathu Heenati)	*BPH17*	(Rathu Heenati)			BC_4F_4 equivalent
BPH17+BPH21-PYL	*BPH17*	(Rathu Heenati)	*BPH21*	(IR71033-121-15)			BC_4F_3 equivalent
BPH20+BPH21-PYL	*BPH20*	(IR71033-121-15)	*BPH21*	(IR71033-121-15)			BC_3F_8 equivalent
BPH20+BPH32-PYL	*BPH20*	(IR71033-121-15)	*BPH32*	(PTB33)			BC_4F_3 equivalent
BPH21+BPH25-PYL	*BPH21*	(IR71033-121-15)	*BPH25*	(ADR52)			BC_4F_3 equivalent
BPH21+BPH17-ptb-PYL	*BPH21*	(IR71033-121-15)	*BPH17-ptb*	(PTB33)			BC_4F_3 equivalent
BPH25+BPH17-ptb-PYL	*BPH25*	(ADR52)	*BPH17-ptb*	(PTB33)			BC_4F_3 equivalent
BPH32+BPH17-ptb-PYL	*BPH32*	(PTB33)	*BPH17-ptb*	(PTB33)			BC_3F_8 equivalent
BPH2+BPH3+BPH17-PYL	*BPH2*	(PTB33)	*BPH3*	(Rathu Heenati)	*BPH17*	(Rathu Heenati)	BC_4F_3 equivalent
BPH2+BPH32+BPH17-ptb-PYL	*BPH2*	(PTB33)	*BPH32*	(PTB33)	*BPH17-ptb*	(PTB33)	BC_4F_3 equivalent
BPH20+BPH21+BPH32-PYL	*BPH20*	(IR71033-121-15)	*BPH21*	(IR71033-121-15)	*BPH32*	(PTB33)	BC_4F_3 equivalent

Table 3. Background survey analysis of seven near-isogenic lines using SSR polymorphic markers.

NIL	Donor	No. of SSR Markers			Genome Ratio (%)		Total Physical Distance of Donor Segment (Mbp) *	
		T65	Donor	Total	T65	Donor		
bph2-NIL	PTB33	183	20	203	85.2–90.9	9.1–14.8	33.9	55.0
Bph3-NIL	Rathu Heenati	181	14	195	97.0–99.0	1.0–3.0	3.8	11.3
Bph17-NIL	Rathu Heenati	170	3	173	95.2–99.0	1.0–4.8	3.8	17.6
BPH17-ptb-NIL	PTB33	219	10	229	92.4–97.2	2.8–7.6	10.5	28.1
Bph20-NIL	IR71033-121-15	224	13	237	90.4–94.4	5.6–9.6	20.6	35.5
Bph21-NIL	IR71033-121-15	210	19	229	88.4–92.9	7.1–11.6	26.4	43.1
BPH32-NIL	PTB33	220	13	233	95.9–98.1	1.9–4.1	7.1	15.1

* The minimum physical distance of donor segment was calculated by the distance between two markers delimiting the substituted segment and the maximum amount was calculated by two flanking markers of substituted segments.

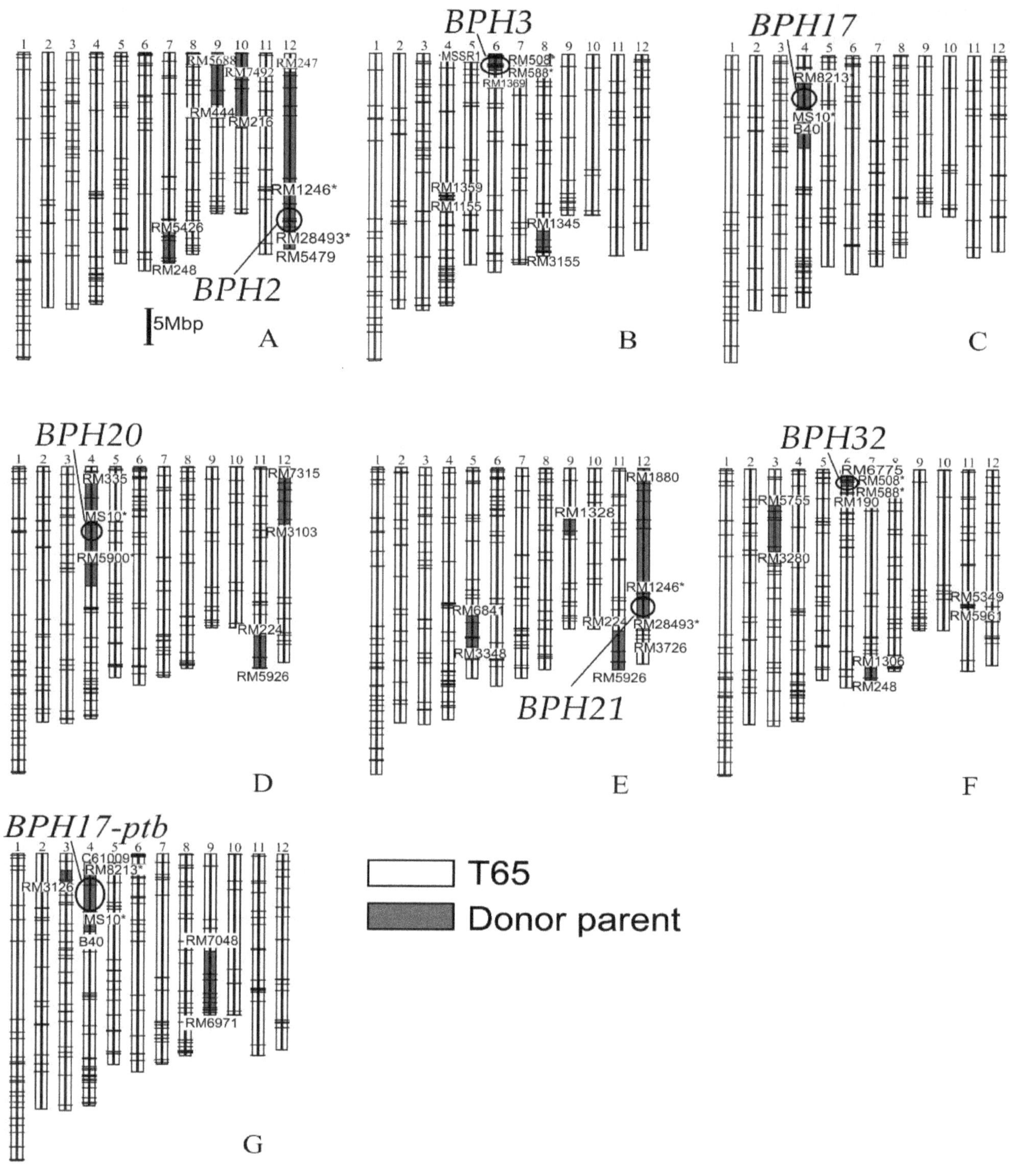

Figure 1. Graphical genotypes of *BPH2*-NIL (**A**), *BPH3*-NIL (**B**), *BPH17*-NIL (**C**), *BPH20*-NIL (**D**), *BPH21*-NIL (**E**), *BPH32*-NIL (**F**) and *BPH17-ptb*-NIL (**G**). The 12 bars indicate 12 chromosomes of rice. Horizontal lines across the chromosomes show the positions of polymorphic SSR markers. Circles indicate the approximate positions of brown planthopper resistant genes. The asterisks (*) indicate SSR markers that were used for marker-assisted selection.

2.2. Development of 15 PYLs Carrying Two or Three BPH Resistance Genes

Twelve PYLs carrying two BPH resistance genes (BPH2 + BPH17-PYL, BPH2 + BPH25-PYL, BPH2 + BPH32-PYL, BPH2 + BPH17-ptb-PYL, BPH3 + BPH17-PYL, BPH17 + BPH21-PYL, BPH20 + BPH21-PYL, BPH20 + BPH32-PYL, BPH21 + BPH25-PYL, BPH21 + BPH17-ptb-PYL, BPH25 + BPH17-ptb-PYL and BPH32 + BPH17-ptb-PYL) and three PYLs containing three BPH resistance genes (BPH2 + BPH3 + BPH17-PYL, BPH2 + BPH32 + BPH17-ptb-PYL and BPH20 + BPH21 + BPH32-PYL) were developed using NILs and PYLs with BPH resistance gene(s) (Table 2). The PYLs were confirmed for resistance genes through foreground selection using flanking SSR markers tightly linked to each resistance gene. Most of PYLs were selected from the BC_4F_3 equivalent generation, except BPH3 + BPH17-PYL from the BC_4F_4 equivalent generation, BPH20+BPH21-PYL from the BC_3F_8 generation and BPH32 + BPH17-ptb-PYL from the BC_3F_8 generation.

2.3. Comparison of Resistance Levels against Hadano-66 by Modified Seedbox Screening Test (MSST)

T65 was highly damaged (damage score (DS) = 8.2) by the Hadano-66 population (Figure 2A). The DSs of the donor parents were significantly lower (0.7 for IR71033-121-15, 0.7 for PTB33 and 0.2 for Rathu Heenati) than that of T65. The donor parents also had higher levels of resistance compared with their respective NILs and PYLs. Among the NILs, *BPH2*-NIL (DS: 3.0) and *BPH17*-NIL (3.2) showed the highest resistance levels. The other NILs *BPH3*-NIL (6.0), *BPH20*-NIL (6.0), *BPH21*-NIL (6.5), *BPH25*-NIL (6.7), *BPH26*-NIL (4.8), *BPH32*-NIL (6.7) and *BPH17-ptb*-NIL (5.7), had lower DSs than T65's but were not significantly different from the T65. Damage scores across the 15 PYLs ranged from 2.3 to 6.0. Among PYLs, the DSs of 10 PYLs—*BPH2* + *BPH17*-PYL (2.7), *BPH2* + *BPH25*-PYL (2.5), *BPH2* + *BPH32*-PYL (3.0), *BPH2* + *BPH17-ptb*-PYL (3.0), *BPH17* + *BPH21*-PYL (2.3), *BPH20* + *BPH21*-PYL (2.3), *BPH21* + *BPH25*-PYL (3.3), *BPH21* + *BPH17-ptb*-PYL (2.7), *BPH2* + *BPH3* + *BPH17*-PYL (3.0) and *BPH20* + *BPH21* + *BPH32*-PYL (2.3), were equal to or less than 3.3, while the DSs of five PYLs—*BPH3* + *BPH17*-PYL (5.0), *BPH20* + *BPH32*-PYL (5.3), *BPH25* + *BPH17-ptb*-PYL (6.0), *BPH32* + *BPH17-ptb* (5.0) and *BPH2* + *BPH32* + *BPH17-ptb*-PYL (4.3), were more than 4.3. Although the DSs between NILs and PYLs were not significantly different, the resistance levels of the PYLs tended to be higher than those of the NILs.

Additionally, fresh biomass reduction rates (FBRRs) of the NILs and PYLs were calculated as an indicator of resistance (Figure 2B). T65 had the highest FBRR (89.0%) and was significantly different from the donor parents: IR71033-121-15 (35.7%), PTB33 (39.2%) and Rathu Heenati (20.4%). Among the NILs, *BPH17*-NIL (58.7%) had the lowest FBRR and was significantly different from T65. The other NILs, *BPH2*-NIL (68.6%), *BPH3*-NIL (82.4%), *BPH20*-NIL (77.3%), *BPH21*-NIL (84.3%), *BPH25*-NIL (85.3%), *BPH26*-NIL (73.8%), *BPH32*-NIL (86.7%) and *BPH17-ptb*-NIL (77.6%), had lower FBRRs than T65; however, the differences were not significant. The FBRRs of four PYLs—*BPH2* + *BPH32*-PYL (59.1%), *BPH2* + *BPH17-ptb*-PYL (56.7%), *BPH21* + *BPH17-ptb*-PYL (50.1%) and *BPH2* + *BPH3* + *BPH17*-PYL (57.6%), were less than 60%. The FBRRs of five PYLs—*BPH2* + *BPH17*-PYL (64.5%), *BPH2* + *BPH25*-PYL (68.4%), *BPH20* + *BPH21*-PYL (62.3%), *BPH2* + *BPH32* + *BPH17-ptb*-PYL (64.0%) and *BPH20* + *BPH21* + *BPH32*-PYL (63.2%), ranged from 60% to 70%; and the FBRRs of six PYLs—*BPH3* + *BPH17*-PYL (79.3%), *BPH17* + *BPH21*-PYL (70.9%), *BPH20* + *BPH32*-PYL (71.6%), *BPH21* + *BPH25*-PYL (70.3%), *BPH25* + *BPH17-ptb*-PYL (78.9%) and *BPH32* + *BPH17-ptb*-PYL (74.4%), ranged from 70% to 80%. Additionally, DSs and FBRRs were positively correlated (Pearson's C = 0.89; $p < 0.001$).

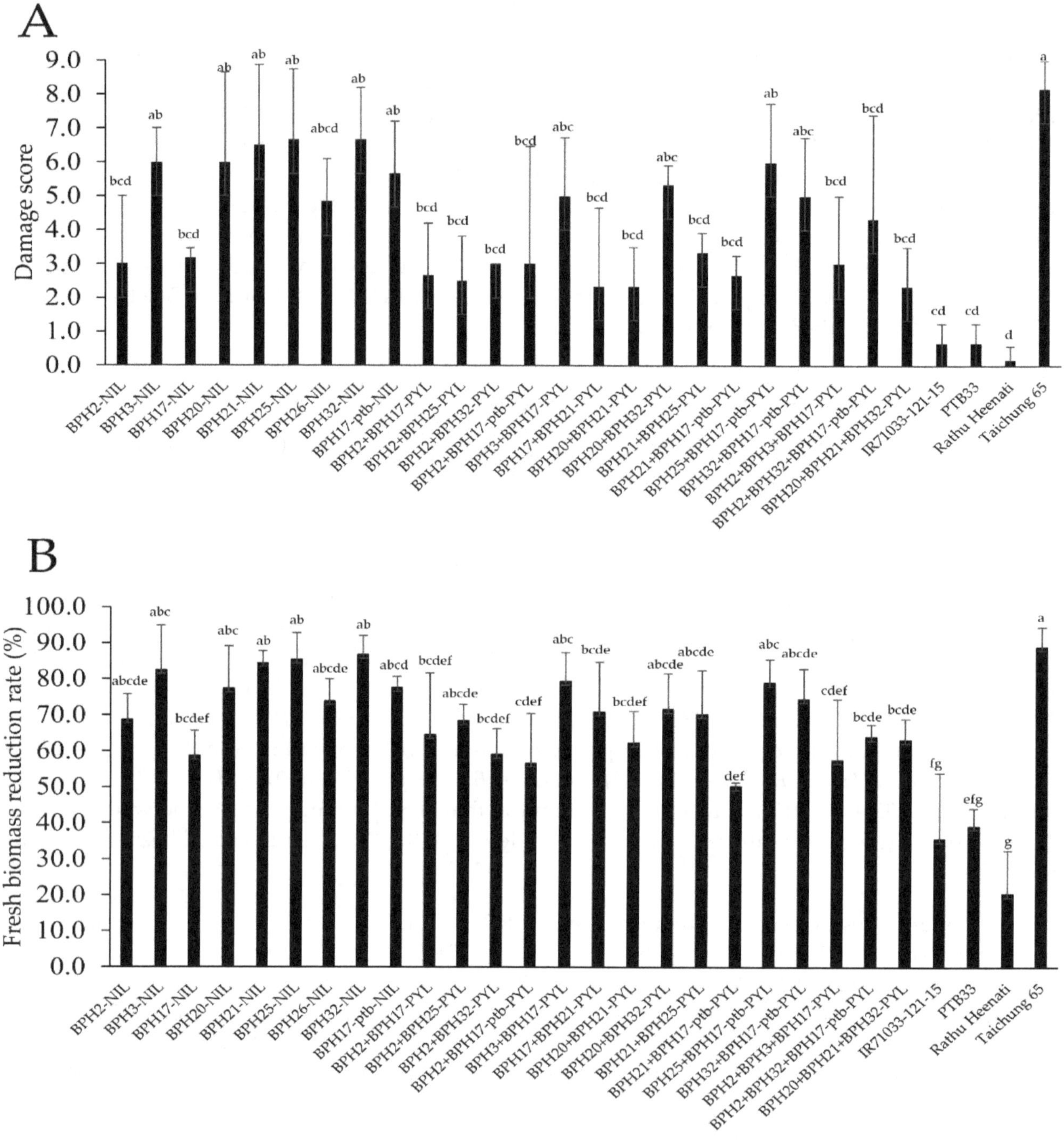

Figure 2. Damage scores (**A**) and fresh biomass reduction rates (**B**) of near-isogenic lines and pyramided lines infested with the Hadano-1966 *Nilaparvata lugens* population using the modified seedbox screening test at the seedling stage. The lower damage scores and fresh biomass reduction rates indicate higher resistance levels.

2.4. Comparison of Adult Mortality (ADM) of the Hadano-66 Population to the NILs and PYLs

Levels of adult mortality (ADM) of the donor parents IR71033-121-15, PTB33 and Rathu Heenati (100%) were significantly higher than that of T65 (17.6%) (Table 4). Among the NILs, *BPH2*-NIL and *BPH17*-NIL had the highest ADM rates, 68.9% and 59.0%, respectively. The ADM rates of other NILs were not significantly different from that of T65. The PYLs carrying the *BPH2*—*BPH2* + *BPH17*-PYL (75.0%), *BPH2* + *BPH25*-PYL (87.5%), *BPH2* + *BPH32*-PYL (84.0%) and *BPH2* + *BPH17-ptb*-PYL (84.0%), showed the highest ADM rates among PYLs with two genes and were higher than any of the corresponding NILs. The ADM rates of seven PYLs—*BPH3* + *BPH17*-PYL, *BPH17* + *BPH21*-PYL, *BPH20* + *BPH32*-PYL, *BPH21* + *BPH25*-PYL, *BPH21* + *BPH17-ptb*-PYL, *BPH25* + *BPH17-ptb*-PYL and *BPH32* + *BPH17-ptb*-PYL, ranged from 50.0% to 68.0%, while the ADM of *BPH20* + *BPH21*-PYL was 33.3%. The ADM rates of PYLs for three genes—*BPH2* + *BPH3* + *BPH17*-PYL (96.0%), *BPH2* + *BPH32* + *BPH17-ptb*-PYL (95.8%) and *BPH20* + *BPH21* + *BPH32*-PYL (92.0%), were higher than those of the corresponding NILs and PYLs for two genes, and were similar to the ADM rates of the donor parents (100%). Furthermore, the ADM rates were negatively correlated with the DSs (Pearson's C = −0.79; $p < 0.001$) and FBRRs (Pearson's C = −0.76; $p < 0.001$).

Table 4. The adult mortality of *Nilaparvata lugens* on near-isogenic lines and pyramided lines carrying brown planthopper resistance genes.

Entry	Adult Mortality (%)	
	Hadano-66	Koshi-2013
BPH2-NIL	68.9 ± 28.5 abcd	4.0 ± 8.9 b
BPH3-NIL	30.0 ± 38.0 def	0.0 ± 0.0 b
BPH17-NIL	59.0 ± 25.1 abcde	20.0 ± 14.1 b
BPH20-NIL	24.0 ± 22.7 def	4.0 ± 8.9 b
BPH21-NIL	36.0 ± 37.5 cdef	12.0 ± 17.9 b
BPH25-NIL	16.0 ± 15.8 f	16.0 ± 16.7 b
BPH26-NIL	50.0 ± 41.4 abcdef	4.0 ± 8.9 b
BPH32-NIL	14.0 ± 16.5 f	12.0 ± 17.9 b
BPH17-ptb-NIL	22.0 ± 19.9 ef	20.0 ± 20.0 b
BPH2+BPH17-PYL	75.0 ± 19.5 abcd	32.0 ± 17.9 b
BPH2+BPH25-PYL	87.5 ± 17.9 ab	12.0 ± 11.0 b
BPH2+BPH32-PYL	84.0 ± 16.7 abc	16.0 ± 16.7 b
BPH2+BPH17-ptb-PYL	84.0 ± 16.7 abc	16.0 ± 16.7 b
BPH3+BPH17-PYL	50.0 ± 32.5 abcdef	24.0 ± 16.7 b
BPH17+BPH21-PYL	58.3 ± 27.5 abcdef	16.0 ± 21.9 b
BPH20+BPH21-PYL	33.3 ± 22.2 cdef	24.0 ± 16.7 b
BPH20+BPH32-PYL	62.5 ± 38.5 abcde	36.0 ± 21.9 b
BPH21+BPH25-PYL	64.0 ± 26.1 abcde	24.0 ± 16.7 b
BPH21+BPH17-ptb-PYL	54.2 ± 35.0 abcdef	8.0 ± 17.9 b
BPH25+BPH17-ptb-PYL	68.0 ± 26.8 abcde	16.0 ± 16.7 b
BPH32+BPH17-ptb-PYL	62.5 ± 32.9 abcde	20.0 ± 14.1 b
BPH2+BPH3+BPH17-PYL	96.0 ± 8.9 ab	36.0 ± 38.5 b
BPH2+BPH32+BPH17-ptb-PYL	95.8 ± 8.9 ab	20.8 ± 20.1 b
BPH20+BPH21+BPH32-PYL	92.0 ± 11.0 ab	28.0 ± 26.8 b
IR71033-121-15	100.0 ± 0.0 a	44.0 ± 16.7 ab
PTB33	100.0 ± 0.0 a	36.0 ± 21.9 b
Rathu Heenati	100.0 ± 0.0 a	84.0 ± 35.8 a
Taichung 65	17.6 ± 16.7 f	5.0 ± 10.0 b

Parameter values (means ± standard deviations) followed by the same letter are not significantly different between each *Nilaparvata lugens* population ($p < 0.05$, Tukey–Kramer multiple comparison tests).

2.5. Comparison of ADM Rates for the Koshi-2013 Population on the NILs and PYLs

T65 was susceptible to the Koshi-2013 population with an ADM of 5.0%. Rathu Heenati had the highest ADM among entries (84.0%), which was significantly higher than T65. The ADM rates of the other donor parents, IR71033-121-15 (44.0%) and PTB33 (36.0%), were lower than that of Rathu Heenati. The ADM rates of the NILs were less than or equal to 20.0%. Among the PYLs, the ADM rates of *BPH2* + *BPH17*-PYL (32.0%), *BPH20* + *BPH32*-PYL (36.0%) and *BPH2* + *BPH3* + *BPH17*-PYL (36.0%) were highest. The ADM rates of the six PYLs, *BPH3* + *BPH17*-PYL, *BPH20* + *BPH21*-PYL, *BPH21* + *BPH25*-PYL, *BPH32* + *BPH17-ptb*-PYL, *BPH2* + *BPH32* + *BPH17-ptb*-PYL and *BPH20* + *BPH21* + *BPH32*-PYL ranged from 20% to 28%. The ADM rates of the other PYLs, *BPH2* + *BPH32*-PYL, *BPH2* +

BPH17-ptb-PYL, *BPH17* + *BPH21*-PYL, *BPH25* + *BPH17-ptb*-PYL, *BPH2* + *BPH25*-PYL and *BPH21* + *BPH17-ptb*-PYL ranged from 8.0% to 16.0%.

2.6. Agronomic Characteristics of the NILs and PYLs

Six agronomic traits—days to heading (DTH), panicle length (PL), culm length (CL), flag leaf length (LL), flag leaf width (LW) and panicle number per plant (PN) of the NILs and PYLs are presented in Table 5. The DTHs and PNs of the NILs and PYLs were not significantly different from those of T65. The PLs, CLs, LLs and LWs were similar for NILs and T65, except that the *BPH2*-NIL had longer culms, *BPH25*-NIL had shorter panicles and *BPH3*-NIL had wider flag leaves. The PLs, CLs, LLs and LWs were not significantly different between the PYLs and T65, except that *BPH2* + *BPH17*-PYL, *BPH2* + *BPH25*-PYL, *BPH2* + *BPH32*-PYL, *BPH2* + *BPH17-ptb*-PYL, *BPH21* + *BPH25*-PYL and *BPH2* + *BPH3* + *BPH17*-PYL had longer culms; *BPH21* + *BPH25*-PYL had longer flag leaves; *BPH2* + *BPH32*-PYL, *BPH17* + *BPH21*-PYL and *BPH20* + *BPH32*-PYL had narrower flag leaves; and *BPH3* + *BPH17*-PYL had wider flag leaves.

Table 5. Agronomic traits of near-isogenic lines and pyramided lines for brown planthopper resistance genes.

Entry	Average of Agronomic Trait (AVE ± SD)					
	DTH (day)	CL (cm)	PL (cm)	LL (cm)	LW (cm)	PN
BPH2-NIL	104.0 ± 2.0	120.3 ± 3.2 *	23.8 ± 0.7	36.8 ± 4.5	1.1 ± 0.0	16.6 ± 3.2
BPH3-NIL	100.8 ± 1.3	104.3 ± 4.8	18.9 ± 1.0	30.2 ± 3.3	1.3 ± 0.0 *	16.2 ± 1.5
BPH17-NIL	98.6 ± 1.3	103.2 ± 7.5	23.4 ± 2.2	33.8 ± 8.0	1.2 ± 0.1	15.2 ± 3.1
BPH20-NIL	100.8 ± 1.6	94.2 ± 5.6	20.5 ± 2.0	29.7 ± 4.5	1.1 ± 0.0	14.8 ± 4.1
BPH21-NIL	100.8 ± 1.5	104.2 ± 2.9	20.9 ± 1.2	32.6 ± 3.0	1.0 ± 0.1	13.8 ± 2.6
BPH25-NIL	100.4 ± 0.9	100.0 ± 1.9	18.1 ± 1.9 **	26.7 ± 2.8	1.1 ± 0.1	17.6 ± 1.1
BPH26-NIL	98.4 ± 0.9	102.1 ± 0.9	22.0 ± 1.6	27.3 ± 3.2	1.1 ± 0.0	13.4 ± 1.8
BPH32-NIL	100.2 ± 1.8	97.4 ± 1.5	20.9 ± 0.9	29.3 ± 1.0	1.2 ± 0.0	15.2 ± 1.6
BPH17-ptb-NIL	99.0 ± 0.0	97.4 ± 2.3	20.7 ± 1.0	29.6 ± 1.9	1.1 ± 0.0	14.6 ± 1.7
BPH2+BPH17-PYL	104.6 ± 1.3	115.1 ± 7.9 *	19.8 ± 1.6	27.3 ± 2.6	1.2 ± 0.0	13.8 ± 2.0
BPH2+BPH25-PYL	103.0 ± 1.4	116.2 ± 1.7 *	21.0 ± 0.9	30.2 ± 4.7	1.1 ± 0.1	14.2 ± 1.6
BPH2+BPH32-PYL	104.8 ± 1.3	111.4 ± 8.0 *	19.6 ± 1.6	27.8 ± 5.5	1.0 ± 0.0 **	17.8 ± 3.7
BPH2+BPH17-ptb-PYL	100.8 ± 0.8	110.7 ± 2.3 **	23.0 ± 1.1	26.9 ± 5.0	1.2 ± 0.1	15.2 ± 2.7
BPH3+BPH17-PYL	102.4 ± 1.3	103.3 ± 0.5	18.8 ± 0.9	29.7 ± 3.3	1.5 ± 0.1 *	17.4 ± 2.8
BPH17+BPH21-PYL	98.0 ± 0.0	91.5 ± 1.8	21.9 ± 1.7	30.4 ± 3.7	1.0 ± 0.1 *	14.6 ± 1.5
BPH20+BPH21-PYL	102.8 ± 1.1	77.9 ± 3.3	23.2 ± 2.4	34.7 ± 6.1	1.2 ± 0.0	18.2 ± 2.3
BPH20+BPH32-PYL	105.2 ± 0.4	88.3 ± 1.8	19.3 ± 1.7	34.3 ± 5.8	0.9 ± 0.1 *	16.0 ± 5.2
BPH21+BPH25-PYL	102.8 ± 1.3	110.5 ± 4.9 **	23.0 ± 1.9	41.7 ± 3.5 *	1.1 ± 0.1	18.0 ± 3.6
BPH21+BPH17-ptb-PYL	102.8 ± 2.5	107.5 ± 4.6	22.4 ± 1.6	33.6 ± 4.5	1.1 ± 0.1	16.0 ± 2.6
BPH25+BPH17-ptb-PYL	99.6 ± 1.9	99.6 ± 2.4	19.9 ± 2.0	28.9 ± 3.7	1.2 ± 0.1	15.8 ± 4.8
BPH32+BPH17-ptb-PYL	99.0 ± 0.0	107.6 ± 3.5	20.4 ± 2.2	26.7 ± 3.4	1.1 ± 0.0	18.0 ± 1.6
BPH2+BPH3+BPH17-PYL	103.4 ± 2.9	119.0 ± 2.6 *	22.4 ± 0.7	31.4 ± 2.2	1.1 ± 0.1	16.0 ± 3.4
BPH2+BPH32+BPH17-ptb-PYL	100.6 ± 1.1	102.4 ± 4.4	19.1 ± 0.3	26.2 ± 3.2	1.0 ± 0.1	16.6 ± 5.2
BPH20+BPH21+BPH32-PYL	101.2 ± 0.4	91.7 ± 3.1	19.3 ± 1.4	26.6 ± 4.5	1.0 ± 0.1	17.8 ± 2.8
Taichung 65	99.2 ± 0.4	91.4 ± 2.7	21.2 ± 0.7	29.7 ± 4.2	1.1 ± 0.1	14.6 ± 1.5

DTH: days to heading, CL: culm length, PL: panicle length, LL: flag leaf length, LW: flag leaf width, PN: panicle number per plant. * $p < 0.01$; ** $p < 0.05$ (Dunnett's multiple comparison tests against Taichung 65).

3. Discussion

The seven NILs we developed carried BPH resistance genes on the short arm of chromosome 4 (*BPH17*-NIL, *BPH20*-NIL and *BPH17-ptb*-NIL), on the short arm of chromosome 6 (*BPH3*-NIL and *BPH32*-NIL) and on the long arm of chromosome 12 (*BPH2*-NIL and *BPH21*-NIL). One of the resistance genes on chromosome 12, *BPH2*, was originally identified from ASD7 which was used as a donor parent for many modern resistant varieties (e.g., IR36, IR42 and so on) [31,67]. *BPH2* from ASD7 is identical to *BPH26* in DNA sequence and resistance level [10]. *BPH2* from ASD7 was resistant against the Hatano-66 population (synonym of Hadano-66) but susceptible to Nishigoshi-05, a BPH population collected in Koshi, Kumamoto Prefecture in 2005 [51]. PTB33 was reported to carry one dominant and one recessive gene [40] that were confirmed to be *BPH2* and *BPH3* using conventional genetic analysis [41]. However, there was no report of the exact location of *BPH2* from PTB33. In our study, *BPH2*-NIL had similar resistance patterns to *BPH2* in ASD7: *BPH2*-NIL was highly resistant (ADM of 68.9%) against the Hadano-66 population but less effective (ADM of 4.0%) against the recently collected population, Koshi-2013. Moreover, *BPH2*-NIL (PTB33) and *BPH26*-NIL had similar resistance levels against both Hadano-66 and Koshi-2013, suggesting that PTB33, ADR52 and ASD7 might harbor the same resistance gene. Further sequence analysis for *BPH2* from PTB33 is necessary to understand its genetic basis. Another gene on chromosome 12, *BPH21*, was originally identified from IR71033-121-15, an introgression line derived from *O. minuta* and estimated to be located between two markers, S12094A and B122, on the long arm of chromosome 12 [17]. Recently, *BPH21* has been reported to be allelic to *BPH26* [6] and *BPH18* [68] based on amino acid sequences. Both *BPH18* and *BPH26* were isolated and located at 22.9 Mbp on chromosome 12 [9,10]. Therefore, we estimated that the location of *BPH21* was around 22.9 Mbp on chromosome 12, and the region carrying *BPH21* from IR71033-121-15 was selected using RM1246 (19.2 Mbp) and RM28493 (23.3 Mbp) in this study.

The *BPH17* locus on chromosome 4S from Rathu Heenati has been reported by Sun et al. (2005) [16]. *BPH17* was mapped between two markers, RHD9 (6.2 Mbp) and RHC10 (7.0 Mbp), on chromosome 4S and isolated by Liu et al. (2014) [8]. The amino acid sequence and chromosomal location of *BPH17* from Rathu Heenati were the same as those of *BPH17-ptb* from PTB33 [8]. In this study, resistance of *BPH17*-NIL and *BPH17-ptb*-NIL against the Hadano-66 population differed; however, both NILs had similar effects on the Koshi-2013 population. The different resistant levels might be because the loci were derived from different accessions or varieties of rice. Therefore, the amino acid sequences of PTB33 and Rathu Heenati used in this study on the *BPH17* locus should be determined for future research. Additionally, *BPH20* was detected between two markers, B42 (8.7 Mbp) and B44 (8.9 Mbp) on chromosome 4 [17]. Two NILs for *BPH17* and *BPH20* on the genetic background of 9311 varieties developed by Xiao et al (2016) [51] showed different resistance levels against a BPH population from China [68]. In our study, the resistance levels of *BPH17* and *BPH20* were different in both MSST and antibiosis tests against the Hadano-66 population and against the Koshi-2013 population, which corresponds well with previous research by Xiao et al. (2016) [51]. Therefore, the genes on chromosome 4S of IR71033-121-15, PTB33 and Rathu Heenati might be different. To confirm this, further sequence analyses are needed for the three loci *BPH17*, *BPH17-ptb* and *BPH20*.

Among six genes/QTLs that have been identified on the short arm of chromosome 6 of *O. sativa* and its wild relatives [3,12], *BPH3* and *BPH32* have been widely introduced to elite rice cultivars to improve BPH resistance and were related to durable and broad-spectrum resistance in PTB33 and Rathu Heenati [67,69]. In previous research, *BPH3* was mapped onto chromosome 6 between two markers, RM19291 (1.2 Mbp) and RM8072 (1.4 Mbp) [69]. *BPH32* from PTB33 was identified at the same location as *BPH3* from Rathu Heenati, but the amino acid sequence of *BPH3* was not identical to that of *BPH32* [12]. In our study, the resistance levels of the *BPH3*-NIL were slightly different from those of the *BPH32*-NIL, suggesting that *BPH3* might be different from *BPH32*. A comparison of amino acid sequences between *BPH3* and *BPH32* would be necessary to confirm whether these resistance genes are different.

Among the developed NILs, the *BPH3*-NIL, *BPH17*-NIL, *BPH17-ptb*-NIL and *BPH32*-NIL had around 97.0% of their chromosomal segments from the recurrent parent. This proportion coincides with the theoretical ratio for substituting chromosomal segments from recurrent parents by backcrossing four times. The other NILs had fewer chromosomal segments from T65 than the theoretical rate. The substituted chromosomal segments from the donor parents might be related to undesirable traits such as the suppression of the associated BPH resistance gene. Additionally, due to the low density of available polymorphic SSR markers between T65 and donor DNA around the target genes, the BPH resistance genes on the NILs were selected using two flanking markers that were relatively far apart. Furthermore, *BPH17*, *BPH20* and *BPH17-ptb* on the NILs were selected by flanking markers with longer intervals because of the low density of polymorphic markers between donor parents and T65 around the chromosome 4S region. The intervals between each of the flanking marker pairs for *BPH17* and *BPH17-ptb* were 3.7 Mbp, and that for *BPH20* was 5.7 Mbp. Similarly, the interval for each of the two flanking markers for *BPH2* and *BPH21* on chromosome 12 was 4.1 Mbp because the exact locations of genes had not been identified before we started to develop the NILs by MAS. Therefore, many of the NILs had relatively long chromosome segments derived from the donor parents and there is a possibility that the remaining chromosomal segments from donor DNA around the target genes included linkage drag associated with susceptibility. In further research, ensuring that flanking markers are tightly linked to target genes will avoid linkage drag from donors through MAS and backcrossing.

An improvement of rice resistance levels against BPH is necessary since many genes have become less effective against BPH across Asia [45]. In this study, we developed 15 PYLs carrying two or three genes for BPH resistance. The PYLs tended to increase resistance against the two BPH populations, Hadano-66 and Koshi-2013. Among the 15 PYLs, 12 and nine PYLs had higher ADM rates than corresponding NILs against Hadano-66 and Koshi-2013, respectively; ten PYLs had lower FBRRs compared to corresponding NILs in the MSST against the Hadano-66 population. For example, *BPH2* + *BPH32*-PYL (84.0%) and *BPH2* + *BPH32* + *BPH17-ptb*-PYL (95.8%) had higher resistance levels than those of the *BPH2*-NIL (68.9%), *BPH32*-PYL (14.0%) and *BPH17-ptb*-PYL (22.0%) in antibiosis tests against the Hadano-66 population. The ADM rates of *BPH2* + *BPH17*-PYL (32.0%) and *BPH2* + *BPH3* + *BPH17*-PYL (36.0%) were higher than those of *BPH2*-NIL (4.0%), *BPH3*-NIL (0%) and *BPH17*-NIL (20.0%) against the Koshi-2013 population. Additionally, the FBRR of *BPH2* + *BPH17* (42.3%) was lower than for *BPH2*-NIL (67.6%) and *BPH17*-NIL (58.8%). However, the effectiveness of the PYLs was not consistently higher than that of the corresponding NILs. The effect of PYLs was influenced by specific interactions between gene loci, the specific BPH populations and the screening methods. For example, the resistance levels of *BPH3* + *BPH17*-PYL (50.0%) and *BPH17* + *BPH21*-PYL (58.3%) were not higher than that of *BPH17*-NIL (59.0%) in antibiosis tests (or ADM rates) against the Hadano-66 population. *BPH2* + *BPH25*-PYL (87.5%) showed higher ADM against the Hadano-66 population in comparison to *BPH2*-NIL (68.9%) and *BPH25*-NIL (16.0%), while the ADM rate of *BPH2* + *BPH25*-PYL (12.0%) was lower than that of *BPH25*-NIL (16.0%) against the Koshi-2013 population. A similar tendency has been reported for gene combinations between *BPH1* and *BPH2* [70]; *BPH18* and *BPH32*; *BPH20* and *BPH32*; and *BPH2, BPH18* and *BPH32* [57]. That the resistance levels of most of the PYLs were not significantly higher than those of the corresponding NILs might be related to the relatively small number of replications used in the bioassays (five replications for the antibiosis and three replications for the MSST).

In a previous study, virulence of a BPH population collected during 2005 in Japan had increased compared with the virulence of a population collected in 1966 [51]. Through antibiosis tests, we evaluated BPH resistance against the populations collected in 1966 (Hadano-66) and in 2013 (Koshi-2013). Both represented BPH arriving as migrants to Japan. The Hadano-66 population was virulent to T65 (with no resistance gene) but avirulent to all plants with resistance genes, including Mudgo (*BPH1*), ASD7 (*BPH2*), Rathu Heenai (*BPH3* and *BPH17*), Babawee (*BPH4*), Chin Saba (*BPH8*), Balamawee (*BPH9*) and two NILs, *BPH25*-NIL and *BPH26*-NIL [51,71]. In the present study, most of the NILs, all of the PYLs and the donor parents were still effective against the Hadano-66 population. In contrast, all of the NILs and most of the PYLs were susceptible to the Koshi-2013 population, suggesting that BPH recently arriving to Japan from China has greater virulence than was evident about 50 years ago (i.e., 1966). Among the PYLs, two PYLs, *BPH20* + *BPH32*-PYL and *BPH2* + *BPH3* + *BPH17*-PYL, had relatively high resistance, suggesting that PYLs with combinations of these genes are likely to provide good resistance against the current BPH populations that arrive to Japan (Koshi-2013). Finding new sources of resistance genes will be necessary to further improve resistance against contemporary BPH populations as they gain virulence.

In comparison to the corresponding NILs and PYLs, the resistance levels of PTB33, Rathu Heenati and IR71033-121-15 were higher. This suggests that PTB33, Rathu Heenati and IR71033-121-15 might also contain other BPH resistance gene(s). The other genetic factor(s) for BPH resistance can be revealed by analyzing the segregating populations derived from crosses between the developed PYLs and their donor parents in future studies. Additionally, Rathu Heenati had *QBPH4.1* (5.8-7.8 Mbp) and *QBPH4.2* (15.2-17.2 Mbp) on chromosome 4S rather than *BPH3* and *BPH17* [13]. Therefore, the NILs and PYLs carrying *QBPH4.1* and *QBPH4.2* should be developed and evaluated in further analyses. On the other hand, the lower resistance levels of the NILs and PYLs might be related to the relatively high ratio of substituted chromosomal segments from donors in the NILs (from 3.8 to 55.0 Mbp) and PYLs. There is a possibility that the retained donor chromosomal segments in the genetic background of the NILs and PYLs might be linked to the suppression of BPH resistance. To gain further knowledge of BPH resistance controlled by multiple genes, it will be essential to reduce the donor parent chromosomal segments on the NILs and PYLs by further backcrossing and MAS.

4. Materials and Methods

4.1. Plant Materials

To develop NILs with BPH resistance genes, a *japonica* rice variety, T65, that is susceptible to BPH, was used as a recurrent parent, and three rice varieties resistant to BPH were donor parents. The donor lines were IR71033-121-15, PTB33 and Rathu Heenati. IR71033-121-15 contains two BPH resistance genes, *BPH20* and *BPH21*, from the wild rice species *O. minuta* (Accession number: IRGC101141) [17]. PTB33 (Accession number: IRGC19325) that originated from India contains *BPH2*, *BPH17-ptb* and *BPH32*. Rathu Heenati (Acc. no. IRGC 11730), that originated from Sri Lanka, carries *BPH3* and *BPH17* [16,39]. T65 was crossed with these donor parents and F_1 plants were backcrossed four times with T65 to generate BC_4F_1 plants (Figure 3). At each generation of backcrossing, plants carrying BPH resistance genes from the donor parents were selected by MAS using flanking SSR markers of the target BPH resistance genes (Table 1). The selected BC_4F_1 plants were self-pollinated to produce BC_4F_3, BC_4F_4 and BC_4F_5 plants with BPH resistance genes. Finally, seven NILs with either *BPH2*, *BPH3*, *BPH17*, *BPH20*, *BPH21*, *BPH32* or *BPH17-ptb* were developed. The NILs were used to survey the genetic background and evaluate BPH resistance levels as well as agronomic traits. Two additional NILs, *BPH25*-NIL and *BPH26*-NIL were used in the development of the PYLs [54].

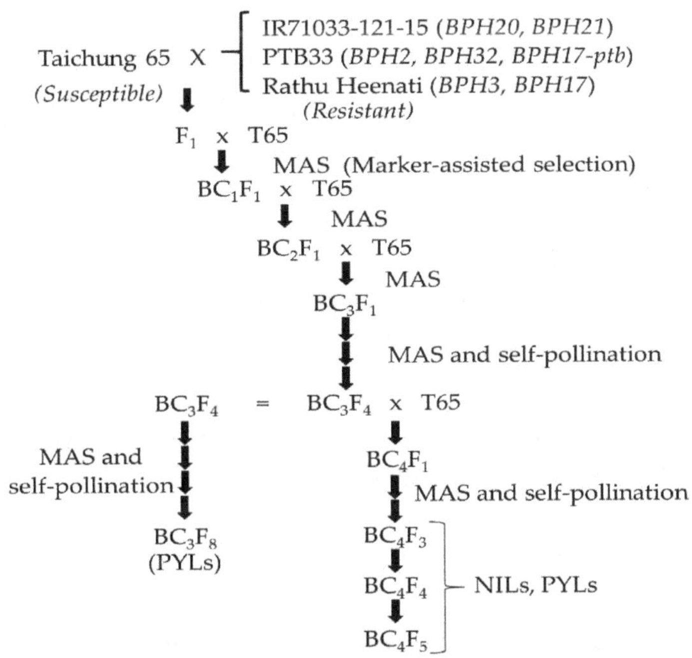

Figure 3. Breeding scheme for the development of near-isogenic lines and pyramided lines containing brown planthopper resistance genes from donor parents, IR71033-121-15, PTB33 and Rathu Heenati.

4.2. The Development of PYLs with BPH Resistance Genes

All the PYLs for two or three BPH resistance genes were developed using the NILs descended from the BC_4F_1 generation, except *BPH20* + *BPH21*-PYL and *BPH32* + *BPH17-ptb*-PYL that were descended from the BC_3F_1 generation. The F_1 plants derived from crosses between NILs were self-pollinated to produce F_2 plants. From 96 F_2 plants, plants that were homozygous for two or three BPH resistance genes were selected by MAS. Several plants from 96 F_3 plants with similar agronomic traits to T65 were selected as final PYLs. The following PYLs carrying two or three BPH resistance genes were evaluated for BPH resistance and agronomic traits: *BPH2* + *BPH17*-PYL, *BPH2* + *BPH25*-PYL, *BPH2* + *BPH32*-PYL, *BPH2* + *BPH17-ptb*-PYL, *BPH3* + *BPH17*-PYL, *BPH17* + *BPH21*-PYL, *BPH20* + *BPH21*-PYL, *BPH20* + *BPH32*-PYL, *BPH21* + *BPH25*-PYL, *BPH21* + *BPH17-ptb*-PYL, *BPH25* + *BPH17-ptb*-PYL, *BPH32* + *BPH17-ptb*-PYL, *BPH2* + *BPH3* + *BPH17*-PYL, *BPH2* + *BPH32* + *BPH17-ptb*-PYL and *BPH20* + *BPH21* + *BPH32*-PYL.

4.3. The MAS for BPH Resistance Genes

To conduct MAS, approximately 2 cm of leaves from two–week old seedlings were collected and dried in a freeze drier for 48 h, and total DNA was extracted using the potassium acetate method [72]. The genotypes of SSR markers on plants in each generation were determined by polymerase chain reaction (PCR) and electrophoresis. The PCR amplification mix (8 μL) contained 3 μL of 1X GoTaq® Green Master Mix (pH 8.5), 0.25 μM of primer and 4 μL of 20 times-diluted DNA. Each PCR amplification included one cycle at 96 °C for 5 min, 35 cycles at 96 °C for 30 s, 55 °C for 30 s and 72 °C for 30 s, followed by one extension cycle at 25 °C for 1 min. PCR products were analyzed by electrophoresis at 200 V using 4% agarose gel with 0.5 μg/mL ethidium bromide in 0.5X TBE buffer for 1 h and photographed under ultraviolet light. During MAS for resistance genes on chromosome 4S, the plants with *BPH17* and *BPH17-ptb*, were selected using two markers, RM8213 and MS10, and the plants with *BPH20* were selected using MS10 and RM5900 (Table 1). The plants with *BPH3* and *BPH32* on the short arm of chromosome 6 were selected using two flanking markers, RM508 and RM588. The plants carrying *BPH2* and *BPH21* located on the long arm of chromosome 12 were screened using RM1246 and RM28493. The plants with *BPH25* were selected using S00310 and MSSR1, and the plants with *BPH26* were selected using RM309, RM28438, InD14, RM28466, RM28481 and MSSR2.

4.4. The Genetic Background Survey of the NILs

In the genetic background survey of the NILs, the bulk DNA from five plants was used. A total of 384 SSR markers distributed on 12 rice chromosomes were used during polymorphism tests with T65 and the donor parents [62]. Among the 384 SSR markers, 254 SSR markers with polymorphisms between IR71033-121-15 and T65 were utilized to identify substituted chromosomal segments from IR71033-121-15 on *BPH20*-NIL and *BPH21*-NIL. Additionally, 244 of 384 SSR markers with polymorphisms between PTB33 and T65 were used to detect substituted chromosomal segments from PTB33 on *BPH2*-NIL, *BPH32*-NIL and *BPH17-ptb*-NIL. To identify substituted chromosomal segments from Rathu Heenati on *BPH3*-NIL and *BPH17*-NIL, 204 of 384 SSR markers with polymorphisms between Rathu Heenati and T65 were used. The whole genome compositions of the developed NILs were graphically displayed following the concept of the graphical genotype proposed by Young and Tanksley (1989) using GGT software version 2.0 [73].

4.5. The BPH Populations and the Characterization of BPH Resistance

Two BPH populations from Japan (Hadano-66 and Koshi-2013) were used to evaluate the NILs and PYLs for their resistance. Hadano-66 was collected in Hadano City, Kanagawa Prefecture, Japan in 1966 [51], and Koshi-2013 was collected in Koshi City, Kumamoto Prefecture, Japan in 2013. Both BPH strains were maintained on the susceptible *japonica* rice variety, Reiho, at 25 °C with 16 h/8 h of light/dark at Kyushu Okinawa Agricultural Research Center of the National Agriculture and Food Research Organization in Japan.

To evaluate resistance, an adaptation of the modified seedbox screening test (MSST) [45,74] was applied at 25 °C using the Hadano-66 strain. To conduct the test, 30 seeds of each of the NILs, PYLs and parent lines were sown to single rows in a plastic tray (23.0 × 30.0 × 2.5 cm) with 2.5 cm between successive rows of seedlings. Two sets of trays—one tray infested by BPH and the other without infestation (the control tray), were used to measure the effects of BPH on plant biomass. One row of Rathu Heenati was added as a resistant control, while three rows of T65 were sown at the center and the two edges as a susceptible control. At seven days after sowing (DAS), the plants in the trays were thinned to 20 plants per row. One tray was infested by the second and third instar nymphs at a density of around 20 BPHs per plant. The experiment was replicated three times. When all the plants of T65 were completely desiccated due to BPH feeding, the DSs of all lines were graded following the standard evaluation system for rice of the International Rice Research Institute [75]. The plants from each row in the two trays were cut above the soil surface and weighed. The fresh biomass reduction rate (FBRR) was calculated using the following formula:

$$\begin{array}{c}\text{Fresh biomass reduction rate}\\ (\text{FBRR})(\%)\end{array} = \left[1 - \frac{\text{Infested plant weight (g)}}{\text{Non-infested plant weight (g)}}\right] \times 100. \quad (1)$$

4.6. Antibiosis Tests

Antibiosis tests were conducted at 25 °C following the method described by Myint et al. (2009) [51]. Five plants of each NIL, PYL and parent line were individually sown in 200 mL plastic cups. At four weeks after sowing, the plants were trimmed to 15 cm height and covered with a plastic cage with insect screen windows for ventilation. Each cage was infested with five thin-abdomen brachypterous female BPHs. At five days after infestation, the ADM was recorded (i.e., the number of dead females).

4.7. Characterization of NILs and PYLs for Agronomic Traits

The NILs and PYLs were grown in a paddy field at Saga University (Saga, Japan) in 2018 and characterized for their agronomic traits compared to those of T65. Seedlings were transplanted at 28 DAS as one plant per hill, with 20 cm between hills and 25 cm between rows. Each entry was planted as at least three rows (12 plants per a row). Six agronomic traits: DTH, CL, PL, LL, LW and PN were measured for five plants in the same row. DTH was the days from sowing until 50% of panicles flowered. CL was measured from the soil surface to the panicle neck. PL is the length from tip to panicle neck of the longest panicle. The flag leaf width and length were measured from the largest and longest flag leaf of each sampled plant. Panicle number is the number of reproductive panicles of each plant at maturity.

4.8. Statistical Analysis

Mean values of BPH resistance (DS, FBRR and ADM) for the NILs and PYLs and agronomic traits were compared using one-way ANOVA. Dunnett's test and Tukey Kramer's test were conducted for multiple comparisons of BPH resistance and agronomic traits, respectively, using R software version 3.5.2.

Author Contributions: P.S.V., D.S.B., F.G.H., H.Y. and D.F. designed the research. S.S.-M. and M.M. provided insects for conducting the research. C.D.N., H.V., D.Z. and D.F. performed the research. C.D.N., H.V. and D.F. developed the plant materials. C.D.N. and D.F. wrote the paper.

Acknowledgments: We thank staff of the Insect Pest Management Research Group, Kyushu Okinawa Agricultural Research Center, for rearing and preparing insect populations, and Atsushi Yoshimura, Yoshiyuki Yamagata and the staff belonging to the Plant Breeding Laboratory at Kyushu University for growing and maintaining the plant materials. Furthermore, we thank Kshirod K. Jena for aiding in the development of materials and Elmer Sanchez for developing materials and taking care of the plants. We also wish to thank the Government of Vietnam for the doctoral fellowship granted to C.D.N.

References

1. Bottrell, D.G.; Schoenly, K.G. Resurrecting the ghost of green revolutions past: The brown planthopper as a recurring threat to high-yielding rice production in tropical Asia. *J. Asia Pac. Entomol.* **2012**, *15*, 122–140. [CrossRef]
2. Wei, J.; Jia, D.; Mao, Q.; Zhang, X.; Chen, Q.; Wu, W.; Chen, H.; Wei, T. Complex interactions between insect-borne rice viruses and their vectors. *Curr. Opin. Virol.* **2018**, *33*, 18–23. [CrossRef] [PubMed]
3. Fujita, D.; Kohli, A.; Horgan, F.G. Rice resistance to planthoppers and leafhoppers. *CRC Crit. Rev. Plant Sci.* **2013**, *32*, 162–191. [CrossRef]
4. Horgan, F.G. Integrating gene deployment and crop management for improved rice resistance to Asian planthoppers. *Crop Prot.* **2018**, *110*, 21–33. [CrossRef]
5. Horgan, F.G. Insect herbivores of rice: Their natural regulation and ecologically based management. In *Rice Production Worldwide*; Chauhan, B.S., Jabran, K., Mahajan, G., Eds.; Springer: Cham, Switzerland, 2017; pp. 279–302.
6. Zhao, Y.; Huang, J.; Wang, Z.; Jing, S.; Wang, Y.; Ouyang, Y.; Cai, B.; Xin, X.-F.; Liu, X.; Zhang, C.; et al. Allelic diversity in an NLR gene *BPH9* enables rice to combat planthopper variation. *Proc. Natl. Acad. Sci. USA* **2016**, *113*, 12850–12855. [CrossRef] [PubMed]
7. Du, B.; Zhang, W.; Liu, B.; Hu, J.; Wei, Z.; Shi, Z.; He, R.; Zhu, L.; Chen, R.; Han, B.; et al. Identification and characterization of *Bph14*, a gene conferring resistance to brown planthopper in rice. *Proc. Natl. Acad. Sci. USA* **2009**, *106*, 22163–22168. [CrossRef] [PubMed]
8. Liu, Y.; Wu, H.; Chen, H.; Liu, Y.; He, J.; Kang, H.; Sun, Z.; Pan, G.; Wang, Q.; Hu, J.; et al. A gene cluster encoding lectin receptor kinases confers broad-spectrum and durable insect resistance in rice. *Nat. Biotechnol.* **2014**, *33*, 301–305. [CrossRef] [PubMed]
9. Ji, H.; Kim, S.R.; Kim, Y.H.; Suh, J.P.; Park, H.M.; Sreenivasulu, N.; Misra, G.; Kim, S.M.; Hechanova, S.L.; Kim, H.; et al. Map-based cloning and characterization of the *BPH18* gene from wild rice conferring resistance to brown planthopper (BPH) insect pest. *Sci. Rep.* **2016**, *6*, 34376. [CrossRef] [PubMed]
10. Tamura, Y.; Hattori, M.; Yoshioka, H.; Yoshioka, M.; Takahashi, A.; Wu, J.; Sentoku, N.; Yasui, H. Map-based cloning and characterization of a brown planthopper resistance gene *BPH26* from *Oryza sativa* L. ssp. *indica* cultivar ADR52. *Sci. Rep.* **2014**, *4*, 5872. [PubMed]
11. Wang, Y.; Cao, L.; Zhang, Y.; Cao, C.; Liu, F.; Huang, F.; Qiu, Y.; Li, R.; Luo, X. Map-based cloning and characterization of *BPH29*, a B3 domain-containing recessive gene conferring brown planthopper resistance in rice. *J. Exp. Bot.* **2015**, *66*, 6035–6045. [CrossRef] [PubMed]
12. Ren, J.; Gao, F.; Wu, X.; Lu, X.; Zeng, L.; Lv, J.; Su, X.; Luo, H.; Ren, G. *Bph32*, a novel gene encoding an unknown SCR domain-containing protein, confers resistance against the brown planthopper in rice. *Sci. Rep.* **2016**, *6*, 37645. [CrossRef] [PubMed]
13. Kamolsukyeunyong, W.; Ruengphayak, S.; Chumwong, P.; Kusumawati, L.; Chaichoompu, E.; Jamboonsri, W.; Saensuk, C.; Phoonsiri, K.; Toojinda, T.; Vanavichit, A. Identification of spontaneous mutation for broad-spectrum brown planthopper resistance in a large, long-term fast neutron mutagenized rice population. *Rice* **2019**, *12*, 16. [PubMed]
14. Qiu, Y.; Guo, J.; Jing, S.; Zhu, L.; He, G. Development and characterization of *japonica* rice lines carrying the brown planthopper-resistance genes *BPH12* and *BPH6*. *Theor. Appl. Genet.* **2012**, *124*, 485–494. [CrossRef] [PubMed]
15. Yang, H.; You, A.; Yang, Z.; Zhang, F.; He, R.; Zhu, L.; He, G. High-resolution genetic mapping at the *Bph15* locus for brown planthopper resistance in rice (*Oryza sativa* L.). *Theor. Appl. Genet.* **2004**, *110*, 182–191. [CrossRef] [PubMed]
16. Sun, L.; Su, C.; Wang, C.; Zhai, H.; Wan, J. Mapping of a major resistance gene to the brown planthopper in the rice cultivar Rathu Heenati. *Breed. Sci.* **2005**, *55*, 391–396. [CrossRef]

17. Rahman, M.L.; Jiang, W.; Chu, S.H.; Qiao, Y.; Ham, T.H.; Woo, M.O.; Lee, J.; Khanam, M.S.; Chin, J.H.; Jeung, J.U.; et al. High-resolution mapping of two rice brown planthopper resistance genes, *Bph20(t)* and *Bph21(t)*, originating from *Oryza minuta*. *Theor. Appl. Genet.* **2009**, *119*, 1237–1246. [CrossRef] [PubMed]
18. Hu, J.; Xiao, C.; Cheng, M.; Gao, G.; Zhang, Q.; He, Y. Fine mapping and pyramiding of brown planthopper resistance genes *QBph3* and *Qbph4* in an introgression line from wild rice *O. officinalis*. *Mol. Breed.* **2015**, *35*, 3. [CrossRef]
19. Hu, J.; Xiao, C.; Cheng, M.; Gao, G.; Zhang, Q.; He, Y. A new finely mapped *Oryza australiensis*-derived QTL in rice confers resistance to brown planthopper. *Gene* **2015**, *561*, 132–137. [CrossRef] [PubMed]
20. Mohanty, S.K.; Panda, R.S.; Mohapatra, S.L.; Nanda, A.; Behera, L.; Jena, M.; Sahu, R.K.; Sahu, S.C.; Mohapatra, T. Identification of novel quantitative trait loci associated with brown planthopper resistance in the rice landrace Salkathi. *Euphytica* **2017**, *213*, 38. [CrossRef]
21. Qiu, Y.; Guo, J.; Jing, S.; Zhu, L.; He, G. High-resolution mapping of the brown planthopper resistance gene *Bph6* in rice and characterizing its resistance in the 9311 and Nipponbare near isogenic backgrounds. *Theor. Appl. Genet.* **2010**, *121*, 1601–1611. [CrossRef] [PubMed]
22. Hirabayashi, H.; Kaji, R.; Angeles, E.R.; Ogawa, T.; Brar, D.S.; Khush, G.S. RFLP analysis of a new gene for resistance to brown planthopper derived from *O. officinalis* on rice chromosome 4. *Breed. Res.* **1999**, *48* (Suppl. 1), 82. (In Japanese)
23. Li, R.; Li, L.; Wei, S.; Wei, Y.; Chen, Y.; Bai, D.; Yang, L.; Huang, F.; Lu, W.; Zhang, X.; et al. The evaluation and utilization of new genes for brown planthopper resistance in common wild rice (*Oryza rufipogon* Griff.). *Mol. Entomol.* **2010**, *1*, 1.
24. Huang, D.; Qiu, Y.; Zhang, Y.; Huang, F.; Meng, J.; Wei, S.; Li, R.; Chen, B. Fine mapping and characterization of *BPH27*, a brown planthopper resistance gene from wild rice (*Oryza rufipogon* Griff.). *Theor. Appl. Genet.* **2013**, *126*, 219–229. [CrossRef] [PubMed]
25. He, J.; Liu, Y.; Liu, Y.; Jiang, L.; Wu, H.; Kang, H.; Liu, S.; Chen, L.; Liu, X.; Cheng, X.; et al. High-resolution mapping of brown planthopper (BPH) resistance gene *Bph27(t)* in rice (*Oryza sativa* L.). *Mol. Breed.* **2013**, *31*, 549–557. [CrossRef]
26. Lakshminarayana, A.; Khush, G.S. New genes for resistance to the brown planthopper in rice. *Crop Sci.* **1977**, *17*, 96–100. [CrossRef]
27. Jairin, J.; Phengrat, K.; Teangdeerith, S.; Vanavichit, A.; Toojinda, T. Mapping of a broad-spectrum brown planthopper resistance gene, *Bph3*, on rice chromosome 6. *Mol. Breed.* **2007**, *19*, 35–44. [CrossRef]
28. Kawaguchi, M.; Murata, K.; Ishii, T.; Takumi, S.; Mori, N.; Nakamura, C. Assignment of a brown planthopper (*Nilaparvata lugens* Stål) resistance gene *bph4* to the rice chromosome 6. *Breed. Sci.* **2001**, *51*, 13–18. [CrossRef]
29. Myint, K.K.M.; Fujita, D.; Matsumura, M.; Sonoda, T.; Yoshimura, A.; Yasui, H. Mapping and pyramiding of two major genes for resistance to the brown planthopper (*Nilaparvata lugens* [Stål]) in the rice cultivar ADR52. *Theor. Appl. Genet.* **2012**, *124*, 495–504. [CrossRef] [PubMed]
30. Sun, L.; Liu, Y.; Jiang, L.; Su, C.; Wang, C.; Zhai, H.; Wan, J. Identification of quantitative trait loci associated with resistance to brown planthopper in the *indica* rice cultivar Col.5 Thailand. *Hereditas* **2007**, *144*, 48–52. [CrossRef] [PubMed]
31. Athwal, D.S.; Pathak, M.D.; Bacalangco, E.H.; Pura, C.D. Genetics of resistance to brown planthoppers and green leafhoppers in *Oryza sativa* L. *Crop Sci.* **1971**, *11*, 747–750. [CrossRef]
32. Jeon, Y.H.; Ahn, S.N.; Choi, H.C.; Hahn, T.R.; Moon, H.P. Identification of a RAPD marker linked to a brown planthopper resistance gene in rice. *Euphytica* **1999**, *107*, 23–28. [CrossRef]
33. Sun, L.H.; Wang, C.M.; Su, C.C.; Liu, Y.Q.; Zhai, H.Q.; Wan, J.M. Mapping and marker-assisted selection of a brown planthopper resistance gene *bph2* in rice (*Oryza sativa* L.). *Acta Genet. Sin.* **2006**, *33*, 717–723. [CrossRef]
34. Kabir, M.A.; Khush, G.S. Genetic analysis of resistance to brown planthopper in rice (*Oryza sativa* L.). *Plant Breed.* **1988**, *100*, 54–58.
35. Qiu, Y.; Guo, J.; Jing, S.; Zhu, L.; He, G. Fine mapping of the rice brown planthopper resistance gene *BPH7* and characterization of its resistance in the 93-11 background. *Euphytica* **2014**, *198*, 369–379. [CrossRef]
36. Murata, K.; Fujiwara, M.; Murai, H.; Takumi, S.; Mori, N.; Nakamura, C. *Bph9*, a dominant brown planthopper resistance gene, is located on the long arm of rice chromosome 12. *Rice Genet. Newslett.* **2000**, *17*, 84–86.

37. Ishii, T.; Brar, D.S.; Multani, D.S.; Khush, G.S. Molecular tagging of genes for brown planthopper resistance and earliness introgressed from *Oryza australiensis* into cultivated rice, *O. sativa*. *Genome* **1994**, *37*, 217–221. [CrossRef] [PubMed]
38. Jena, K.K.; Jeung, J.U.; Lee, J.H.; Choi, H.C.; Brar, D.S. High-resolution mapping of a new brown planthopper (BPH) resistance gene, *Bph18(t)*, and marker-assisted selection for BPH resistance in rice (*Oryza sativa* L.). *Theor. Appl. Genet.* **2006**, *112*, 288–297. [CrossRef] [PubMed]
39. Jairin, J.; Teangdeerith, S.; Leelagud, P.; Phengrat, K.; Vanavichit, A.; Toojinda, T. Detection of brown planthopper resistance genes from different rice mapping populations in the same genomic location. *ScienceAsia* **2007**, *33*, 347–352. [CrossRef]
40. Sidhu, G.S.; Khush, G.S. Genetic analysis of brown planthopper resistance in twenty varieties of rice, *Oryza sativa* L. *Theor. Appl. Genet.* **1978**, *53*, 199–203. [CrossRef] [PubMed]
41. Angeles, E.R.; Khush, G.S.; Heinrichs, E.A. Inheritance of resistance to planthoppers and leafhoppers in rice. In Proceedings of the International Rice Genetics Symposium, Manila, Philippines, 27–31 May 1985; International Rice Research Institute: Manila, Philippines, 1986; pp. 537–549.
42. Saxena, R.C.; Barrion, A.A. Biotypes of the brown planthopper *Nilaparvata lugens* (Stål) and strategies in deployment of host plant resistance. *Insect Sci. Appl.* **1985**, *6*, 271–289. [CrossRef]
43. Ketipearachchi, Y.; Kaneda, C.; Nakamura, C. Adaptation of the brown planthopper (BPH), *Nilaparvata lugens* (Stål) (Homoptera: Delphacidae), to BPH resistant rice cultivars carrying *bph8* or *Bph9*. *Appl. Entomol. Zool.* **1998**, *33*, 497–505. [CrossRef]
44. Pathak, M.D.; Khush, G.S. Studies of varietal resistance in rice to the brown planthopper at the International Rice Research Institute. In *Brown Planthopper—Threat to Rice Production in Asia*; International Rice Research Institute: Los Baños, Philippines, 1979; pp. 285–301.
45. Horgan, F.G.; Ramal, A.F.; Bentur, J.S.; Kumar, R.; Bhanu, K.V.; Sarao, P.S.; Iswanto, E.H.; Van Chien, H.; Phyu, M.H.; Bernal, C.C.; et al. Virulence of brown planthopper (*Nilaparvata lugens*) populations from South and South East Asia against resistant rice varieties. *Crop Prot.* **2015**, *78*, 222–231. [CrossRef]
46. Claridge, M.F.; Hollander, J.D. Virulence to rice cultivars and selection for virulence in populations of the brown planthopper *Nilaparvata lugens*. *Entomol. Exp. Appl.* **1982**, *32*, 213–221. [CrossRef]
47. Alam, S.N.; Cohen, M.B. Detection and analysis of QTLs for resistance to the brown planthopper, *Nilaparvata lugens*, in a doubled-haploid rice population. *Theor. Appl. Genet.* **1998**, *97*, 1370–1379. [CrossRef]
48. Peñalver Cruz, A.; Arida, A.; Heong, K.L.; Horgan, F.G. Aspects of brown planthopper adaptation to resistant rice varieties with the *Bph3* gene. *Entomol. Exp. Appl.* **2011**, *141*, 245–257. [CrossRef]
49. Ferrater, J.B.; Naredo, A.I.; Almazan, M.L.P.; de Jong, P.W.; Dicke, M.; Horgan, F.G. Varied responses by yeast-like symbionts during virulence adaptation in a monophagous phloem-feeding insect. *Arthropod-Plant Interact.* **2015**, *9*, 215–224. [CrossRef]
50. Horgan, F.G.; Ferrater, J.B. Benefits and potential trade-offs associated with yeast-like symbionts during virulence adaptation in a phloem-feeding planthopper. *Entomol. Exp. Appl.* **2017**, *163*, 112–125. [CrossRef]
51. Myint, K.K.M.; Yasui, H.; Takagi, M.; Matsumura, M. Virulence of long-term laboratory populations of the brown planthopper, *Nilaparvata lugens* (Stål), and whitebacked planthopper, *Sogatella furcifera* (Horváth) (Homoptera: Delphacidae), on rice differential varieties. *Appl. Entomol. Zool.* **2009**, *44*, 149–153. [CrossRef]
52. Qiu, Y.; Guo, J.; Jing, S.; Tang, M.; Zhu, L.; He, G. Identification of antibiosis and tolerance in rice varieties carrying brown planthopper resistance genes. *Entomol. Exp. Appl.* **2011**, *141*, 224–231. [CrossRef]
53. Ali, M.P.; Alghamdi, S.; Begum, M.; Anwar Uddin, A.B.M.; Alam, M.Z.; Huang, D. Screening of rice genotypes for resistance to the brown planthopper, *Nilaparvata lugens* Stål. *Cereal Res. Commun.* **2012**, *40*, 502–508. [CrossRef]
54. Yara, A.; Phi, C.N.; Matsumura, M.; Yoshimura, A.; Yasui, H. Development of near-isogenic lines for *BPH25(t)* and *BPH26(t)*, which confer resistance to the brown planthopper, *Nilaparvata lugens* (Stål.) in indica rice 'ADR52'. *Breed. Sci.* **2010**, *60*, 639–647. [CrossRef]
55. Liu, Y.; Chen, L.; Liu, Y.; Dai, H.; He, J.; Kang, H.; Pan, G.; Huang, J.; Qiu, Z.; Wang, Q.; et al. Marker assisted pyramiding of two brown planthopper resistance genes, *Bph3* and *Bph27 (t)*, into elite rice Cultivars. *Rice* **2016**, *9*, 27. [CrossRef] [PubMed]
56. Xiao, Y.; Li, J.; Yu, J.; Meng, Q.; Deng, X.; Yi, Z.; Xiao, G. Improvement of bacterial blight and brown planthopper resistance in an elite restorer line Huazhan of *Oryza*. *F. Crop. Res.* **2016**, *186*, 47–57. [CrossRef]

57. Jena, K.K.; Hechanova, S.L.; Verdeprado, H.; Prahalada, G.D.; Kim, S.R. Development of 25 near-isogenic lines (NILs) with ten BPH resistance genes in rice (*Oryza sativa* L.): Production, resistance spectrum, and molecular analysis. *Theor. Appl. Genet.* **2017**, *130*, 2345–2360. [CrossRef] [PubMed]
58. Wang, H.; Shi, S.; Guo, Q.; Nie, L.; Du, B.; Chen, R.; Zhu, L.; He, G. High-resolution mapping of a gene conferring strong antibiosis to brown planthopper and developing resistant near-isogenic lines in 9311 background. *Mol. Breed.* **2018**, *38*, 107. [CrossRef]
59. Horgan, F.G.; Almazan, M.L.P.; Vu, Q.; Ramal, A.F.; Bernal, C.C.; Yasui, H.; Fujita, D. Unanticipated benefits and potential ecological costs associated with pyramiding leafhopper resistance loci in rice. *Crop Prot.* **2019**, *115*, 47–58. [CrossRef] [PubMed]
60. Hu, J.; Li, X.; Wu, C.; Yang, C.; Hua, H.; Gao, G.; Xiao, J.; He, Y. Pyramiding and evaluation of the brown planthopper resistance genes *Bph14* and *Bph15* in hybrid rice. *Mol. Breed.* **2012**, *29*, 61–69. [CrossRef]
61. Fujita, D.; Myint, K.K.M.; Matsumura, M.; Yasui, H. The genetics of host-plant resistance to rice planthopper and leafhopper. In *Planthoppers: New Threats to the Sustainability of Intensive Rice Production Systems in Asia*; Heong, K.L., Hard, B., Eds.; International Rice Research Institute: Los Baños, Philippines, 2009; pp. 389–399.
62. McCouch, S.R.; Teytelman, L.; Xu, Y.; Lobos, K.; Clare, K.; Walton, M.; Fu, B.; Maghirang, R.; Li, Z.; Xing, Y.; et al. Development and mapping of 2240 new SSR markers for rice (*Oryza sativa* L.). *DNA Res.* **2002**, *9*, 199–207. [CrossRef] [PubMed]
63. International Rice Genome Sequencing Project (IRGSP). The map-based sequence of the rice genome. *Nature* **2005**, *436*, 793–800.
64. Temnykh, S.; DeClerck, G.; Lukashova, A.; Lipovich, L.; Cartinhour, S.; McCouch, S. Computational and experimental analysis of microsatellites in rice (*Oryza sativa* L.): Frequency, length variation, transposon associations, and genetic marker potential. *Genome Res.* **2001**, *11*, 1441–1452. [CrossRef] [PubMed]
65. Temnykh, S.; Park, W.D.; Ayres, N.; Cartinhour, S.; Hauck, N.; Lipovich, L.; Cho, Y.G.; Ishii, T.; McCouch, S.R. Mapping and genome organization of microsatellite sequences in rice (*Oryza sativa* L.). *Theor. Appl. Genet.* **2000**, *100*, 697–712. [CrossRef]
66. The Rice Annotation Project (RAP). Available online: https://rapdb.dna.affrc.go.jp/index.html (accessed on 8 November 2019).
67. Khush, G.S.; Virk, P.S. *IR Varieties and Their Impact*; International Rice Research Institute: Los Baños, Philippines, 2005.
68. Xiao, C.; Hu, J.; Ao, Y.T.; Cheng, M.X.; Gao, G.J.; Zhang, Q.L.; He, G.C.; He, Y.Q. Development and evaluation of near-isogenic lines for brown planthopper resistance in rice cv. 9311. *Sci. Rep.* **2016**, *6*, 38159. [CrossRef] [PubMed]
69. Jairin, J.; Teangdeerith, S.; Leelagud, P.; Phengrat, K.; Vanavichit, A.; Toojinda, T. Physical mapping of *Bph3*, a brown planthopper resistance locus in rice. *Maejo Int. J. Sci. Technol.* **2007**, *1*, 166–177.
70. Sharma, P.N.; Torii, A.; Takumi, S.; Mori, N.; Nakamura, C. Marker-assisted pyramiding of brown planthopper (*Nilaparvata lugens* Stål) resistance genes *Bph1* and *Bph2* on rice chromosome 12. *Hereditas* **2004**, *140*, 61–69. [CrossRef] [PubMed]
71. Myint, K.K.M.; Matsumura, M.; Takagi, M.; Yasui, H. Demographic parameters of long-term laboratory strains of the brown planthopper, *Nilaparvata lugens* Stål, (Homoptera: Delphacidae) on resistance genes, *bph20*(t) and *Bph21(t)* in rice. *J. Fac. Agric. Kyushu Univ.* **2009**, *54*, 159–164.
72. Dellaporta, S.L.; Wood, J.; Hicks, J.B. A plant DNA minipreparation: Version II. *Plant Mol. Biol. Rep.* **1983**, *1*, 19–21. [CrossRef]
73. Young, N.D.; Tanksley, S.D. Restriction fragment length polymorphism maps and the concept of graphical genotypes. *Theor. Appl. Genet.* **1989**, *77*, 95–101. [CrossRef] [PubMed]
74. Naeemullah, M.; Sharma, P.N.; Tufail, M.; Mori, N.; Matsumura, M.; Takeda, M.; Nakamura, C. Characterization of brown planthopper strains based on their differential responses to introgressed resistance genes and on mitochondrial DNA polymorphism. *Appl. Entomol. Zool.* **2009**, *44*, 475–483. [CrossRef]
75. *Standard Evaluation System for Rice (SES)*, 5th ed.; International Rice Research Institute: Manila, Philippines, 2014.

10

Screening of Diverse Ethiopian Durum Wheat Accessions for Aluminum Tolerance

Edossa Fikiru Wayima [1], Ayalew Ligaba-Osena [2], Kifle Dagne [3], Kassahun Tesfaye [3], Eunice Magoma Machuka [4], Samuel Kilonzo Mutiga [4] and Emmanuel Delhaize [5,*]

1 Department of Plant Sciences, College of Agriculture and Natural Resources, Madda Walabu University, Bale-Robe P.O. Box 247, Ethiopia
2 Laboratory of Molecular Biology and Biotechnology, The University of North Carolina at Greensboro, 321 McIver St., 308 Sullivan Science Building, Greensboro, NC 27412, USA
3 Cellular and Molecular Biology Department, College of Natural Sciences, Addis Ababa University, Addis Ababa P.O. Box 1176, Ethiopia
4 Biosciences Eastern and Central Africa-International Livestock Research Institute (BecA-ILRI) Hub, Nairobi P.O. Box 30709-00100, Kenya
5 CSIRO Agriculture and Food, GPO Box 1700, Canberra 2601, Australia
* Correspondence: manny.delhaize@csiro.au

Abstract: Acid soils and associated Al^{3+} toxicity are prevalent in Ethiopia where normally Al^{3+}-sensitive durum wheat (*Triticum turgidum* ssp *durum* Desf.) is an important crop. To identify a source of Al^{3+} tolerance, we screened diverse Ethiopian durum germplasm. As a center of diversity for durum wheat coupled with the strong selection pressure imposed by extensive acid soils, it was conceivable that Al^{3+} tolerance had evolved in Ethiopian germplasm. We used a rapid method on seedlings to rate Al^{3+} tolerance according to the length of seminal roots. From 595 accessions screened using the rapid method, we identified 21 tolerant, 180 intermediate, and 394 sensitive accessions. When assessed in the field the accessions had tolerance rankings consistent with the rapid screen. However, a molecular marker specific for the D-genome showed that all accessions rated as Al^{3+}-tolerant or of intermediate tolerance were hexaploid wheat (*Triticum aestivum* L.) that had contaminated the durum grain stocks. The absence of Al^{3+} tolerance in durum has implications for how Al^{3+} tolerance evolved in bread wheat. There remains a need for a source of Al^{3+}-tolerance genes for durum wheat and previous work that introgressed genes from bread wheat into durum wheat is discussed as a potential source for enhancing the Al^{3+} tolerance of durum germplasm.

Keywords: aluminum; resistance; Ethiopia; durum; hydroponics; soil acidity; *Triticum turgidum* ssp. *durum* Desf.

---

1. Introduction

Durum (*Triticum turgidum* ssp *durum* Desf.) wheat and bread wheat (*Triticum aestivum* L.) provide the Ethiopian population with a large proportion of its caloric intake [1,2]. Despite a general increase in production and productivity of wheat during the last two decades (1998–2016), Ethiopia imports a substantial amount of both durum and bread wheat. Durum production as a proportion of the total wheat produced in Ethiopia has declined over the past few decades from about 80% in the 1980s to an estimated current proportion of only about 20%. Despite a reduction in the proportion of grain produced in Ethiopia, durum constitutes 50% to 80% of the wheat grain imported in any given year [1], indicating that demand for durum remains strong. One of the abiotic stresses that may be contributing to the decline in durum production is the prevalence of acid soils in Ethiopia. Durum wheat is very sensitive of the toxic Al^{3+} found in acid soil and is the most sensitive of the small-grained crops to Al^{3+}

toxicity [3]. By contrast, bread wheat has Al^{3+}-tolerant alleles of *TaALMT1* [4] and *TaMATE1B* [5] along with other yet to be cloned Al^{3+}-tolerance genes [6].

Worldwide, acidic soils are one of the most important limitations to agricultural production [7]. In Ethiopia, acid soils account for about 34% of agricultural land area that range from slight to strongly acidic soils [8]. Acid soils generally occupy the western part of the country extending from southwest to northwest, although strongly acidic soils occur mainly in the western part of the country including the lowlands. Acid soils are particularly prevalent in the highlands of Ethiopia [9] and the application of lime commonly results in improved yields of various crops [10,11]. The productivity of wheat over acidic areas of Ethiopia is low as compared to parts of the country where soil acidity is absent. For instance, in areas with strongly acidic soils, which occur widely in the western and southwestern parts of the country, the productivity of wheat is as low as 0.8–2.0 t/ha. These areas include West Wallaga, Illu-Ababora, Jimma, Gamo Gofa, Asosa, and Metekel zones. On the other hand, a relatively higher grain yield, ranging from 2.5 to 3.1 t/ha, is obtained in regions with near neutral soil pH including West Arsi, Arsi, Siltie, East Shewa, and Bale zones (Supplementary Materials file, Table S1).

Aluminum is solubilized in acid soils into the toxic Al^{3+} which in sensitive plants typically affects the viability of the root apex. Al^{3+} inhibits root growth resulting in reduced uptake of water and nutrients ultimately hindering plant growth and development [12]. In addition to its toxic effects on plant tissues, Al^{3+} affects nutrient availability within the soil. All of these effects significantly reduce crop yield. While management, primarily by application of lime, is important for neutralizing acid soils, this can be costly and it can take years to correct acidity at depth. A complementary strategy for improving crop production on acid soils is the use of Al^{3+}-tolerant germplasm developed through breeding or genetic modification along with liming practices. In several crop species, variation in Al^{3+} tolerance has been identified and selective breeding programs can be implemented to increase production on acid soils [13–15]. However as noted above, durum wheat is one of the most Al^{3+} sensitive of the small-grained crops and shows little variation in Al^{3+} tolerance [16]. Bread wheat is hexaploid and possesses the A-, B-, and D-genomes whereas durum wheat is tetraploid possessing only the A- and B-genomes. Although, as noted below, a major gene for Al^{3+} tolerance is found on the D-genome of bread wheat, there are other genes for Al^{3+} tolerance in bread wheat found on the A- and B-genomes [6]. Early reports that identified Al^{3+}-tolerant durum genotypes with a level of tolerance similar to that of bread wheat [17] can now be attributed to misidentification or contamination of durum grain stocks with bread wheat [18], highlighting the importance of verifying the genetic identity of germplasm. Bread wheat, in contrast to durum, shows a large variation in Al^{3+} tolerance and much of this is conditioned by alleles of the *TaALMT1* gene located on chromosome 4D [19] and *TaMATE1B* located on chromosome 4B. *TaALMT1* encodes an Al^{3+}-activated membrane channel permeable to malate with the malate exuded by root apices binding toxic Al^{3+} to protect the developing roots. Another Al^{3+} tolerance gene in bread wheat is *TaMATE1B* that encodes a citrate transporter in the plasma membrane and in some bread wheat genotypes confers a lower level of tolerance than *TaALMT1* [20] whereas in other genotypes *TaMATE1B* appears to be the predominant Al^{3+} tolerance gene [21]. Both *TaALMT1* and *TaMATE1B* have been introgressed from bread wheat into durum wheat. This required use of the pairing homeologous (*ph1c*) mutant to enable *TaALMT1* to be transferred from chromosome 4D in bread wheat to chromosome 4B in durum wheat [20]. In contrast to bread wheat, for durum seedlings grown on acid soil, *TaMATE1B* appears to provide a greater level of Al^{3+} tolerance than *TaALMT1* for reasons that are not understood.

Ethiopia has been considered as a center of diversity for durum germplasm and landraces grown by farmers are a potential source of agronomically important genes [22,23]. A 90 K single nucleotide polymorphism (SNP) chip analysis supports the notion that durum wheat in Ethiopia is particularly diverse [24]. Furthermore, others undertaking similar phylogenetic analyses based on SNP markers have suggested that Ethiopia is a second center of origin for durum wheat [25]. We speculated that because of the diversity of the durum germplasm coupled with a strong selection pressure imposed by extensive areas of acid soils in Ethiopia useful levels of Al^{3+} tolerance could have evolved in Ethiopian durum landraces. Cultivated durum genotypes sourced from various countries were found to be all Al^{3+} sensitive [16] whereas others [26] identified relatively tolerant lines in a similar population. The Ethiopian landraces are likely to be more diverse in their genetic makeup than the populations previously screened for tolerance and provide an opportunity to identify unique genes that confer tolerance of acid soils and Al^{3+} toxicity in particular. As an initial strategy it would be preferable to identify Al^{3+} tolerance in germplasm already adapted to Ethiopian conditions instead of introgressing genes from other sources. In this study we screened a diverse set of Ethiopian durum germplasm for Al^{3+} tolerance using a rapid hydroponic screen and then assessed selected lines more thoroughly in the hydroponic screen and finally in field trials on acid soil. Despite Ethiopia having conditions that would favor the evolution of Al^{3+} tolerance, the diverse durum germplasm was found to be Al^{3+} sensitive and highlighted the importance of verifying the identity of grain stocks. The absence of Al^{3+} tolerance genes in diverse durum germplasm suggests that genes encoding Al^{3+} tolerance found on the A- and B-genomes of bread wheat arose subsequent to the hybridization with the D-genome that produced hexaploid wheat.

2. Materials and Methods

2.1. Germplasm

A total of 595 durum wheat accessions obtained from the Ethiopian Biodiversity Institute (EBI: www.ibc.gov.et; Supplementary Materials file, Table S2), were screened in a series of non-replicated trials for their Al^{3+} tolerance using a rapid hydroponic screen. Based on their performance at this preliminary stage, accessions were classified as Al^{3+}-tolerant, -intermediate, or -sensitive (see below). Subsequently all the tolerant as well as selected accessions of the intermediate and sensitive classes, totaling 150 accessions, were evaluated in a replicated hydroponic experiment and a field trial. For the experiments that screened previously identified Al^{3+}-tolerant lines of durum the selected lines are described by Raman et al. [26]. A durum line (Langdon 4D (4B)) that has the 4B chromosome substituted by the 4D chromosome of bread wheat [27] was included as a positive control of a confirmed Al^{3+}-tolerant durum line [18].

2.2. Hydroponic Culture

A rapid hydroponic screening was undertaken using an apparatus comprised of a plastic basin (for holding nutrient solution), a plate that held seed of different wheat accessions separately submerged in nutrient solution and an aeration system (Supplementary Materials file, Figure S1). The method is based on one described previously [16]. Seedlings grown submerged in the aerated nutrient solution remained viable and could be transplanted to soil when required to bulk up grain. Grain harvested from single plants of selected seedlings from this "preliminary" screen was then used in a replicated hydroponic experiment using the same growth conditions. For the replicated experiment, the average performance of 10 seedlings for each accession was determined and the experiment repeated three times.

The nutrient solution comprised of 500 μM KNO_3, 500 μM $CaCl_2$, 500 μM NH_4NO_3, 150 μM $MgSO_4$, 10 μM KH_2PO_4, 2 μM $FeCl_3$, and 5 μM of $Al_2SO_4.18H_2O$ as described previously [28]. The pH of the solution was adjusted to about 4.3 with 1 M HCl. Dry grain was immersed in the solution and seedlings allowed to grow submerged for 5 days in the solution culture with the nutrient solution changed every day to maintain pH and Al^{3+} concentration relatively constant. After 5 days seedlings had typically developed three seminal roots and the total root length was recorded to the nearest 0.1 cm.

We used conventional hydroponics with the same nutrient solution composition as described above to compare the Al^{3+} tolerance of a set of durum and bread wheat genotypes in a range of $AlCl_3$ concentrations. Relative root length for the various genotypes was calculated after 3 days growth as root length in solution that contained Al^{3+} relative to root length in control solution that lacked Al^{3+}. Errors associated with relative root length were calculated as described previously [29].

2.3. Determining the Identity of the Al^{3+}-Tolerant Accessions

The morphological similarity between durum and bread wheat made it difficult to establish whether the identified Al^{3+}-tolerant lines were durum or bread wheat by using phenotypic traits alone. Therefore, a polymerase chain reaction (PCR) assay that targeted the Dgas44 sequence was used to determine whether the lines were durum or bread wheat [30]. Dgas44 is a D-genome specific repetitive sequence that can be used to distinguish hexaploid wheat from tetraploid wheats that lack the D-genome [31]. The sequence of forward and reverse primers of Dgas44 marker, respectively, were 5′-CTTCTGACGGGTCAGGGGCAC-3′ and 5′-CTGAATGCCCCTGCGGCTTAAG-3′.

Ten grain of bulked up samples used in the field trial along with one verified bread wheat cultivar (Enkoy) and one durum wheat accession (8317), distinguished by its reddish/pinkish seed color, were planted in pots. Enkoy variety was included as a positive control (i.e., possesses the D-genome since it is a bread wheat cultivar), while accession 8317 was used as a negative control since it is a known durum accession that lacks the D-genome. Young green leaves were collected separately from three individual plants for DNA extraction. Leaf samples were freeze-dried in liquid nitrogen and pulverized with a Geno/Grinder 2000 and genomic DNA was extracted with a ZymoResearch kit (Plant/Seed MiniPrep) following the manufacturer's protocol. The extracted genomic DNA was quantified with a NanoDrop 2000 UV-Vis Spectrophotometer (Thermo Scientific, Waltham, MA, USA) and PCR of samples undertaken as previously described [18].

2.4. Field Experiment

The field experiment was conducted at Bedi (38°36′3″ E, 9°5′59″ N), which is located in Watabicha Minjaro Kebele, Welmera District, West Shewa Zone, Oromia Regional State, Ethiopia (Supplementary Materials file, Figure S2). It is situated at approximately 35 km west of Addis Ababa and about 25 km away from Holeta town in the north-east. Crops are generally planted from around mid-June, though it varies with crop type and the time of onset of rains.

Samples of the reddish-brown soil to a depth of 20 cm were collected from the experimental plot with an auger in a regular pattern following a line transect. The soil samples were submitted the following day to JIJE LABOGLASS PLC (Addis Ababa, Ethiopia) for analysis of the major parameters using standard procedures as shown in Table 1. Characterized as a clay type, the pH of a water extract of the soils was found to be 4.92 indicating acidity and the likely presence of Al^{3+} toxicity. Note that a water extract will generally have a higher pH value than if a $CaCl_2$ solution is used for the extraction. For the limed treatments 451 kg of $CaCO_3$ was manually applied to the soil the year prior to the trial covering an area of 200 m^2. The lime was obtained from Guder Lime Factory (Guder, West Shewa Zone, Ethiopia) through collaboration with Oromia Agriculture and Rural Development Bureau. The plot was then ploughed immediately and re-ploughed after a week to thoroughly mix the lime with the soil to a depth of about 20 cm so that Al^{3+} would be detoxified in the soil solution.

A total of 150 accessions were selected based on their response to Al^{3+} toxicity in the preliminary hydroponic screen described above. These accessions were planted during the major rainy season on limed and un-limed plots in the field using a randomized complete block design. The two treatments of the experiment (limed and un-limed blocks) were each replicated twice, and the replicated blocks were spaced 1 m apart such that the long edge of the blocks were arranged adjacent to one another. All 150 accessions were planted side by side in each block resulting in a total area of 30 × 2 m for each block. Individual accessions were planted within the blocks randomly in single rows 2 m long with 20 cm spacing between rows. Contamination of un-limed plots with lime through erosion and splashing was avoided by considering slope and spacing factors in the experimental layout. Specifically, plots at a higher slope were assigned as un-limed plots and were separated from the limed plots by 2 m. N and P fertilizers were applied after 3–4 weeks of planting as urea and di-ammonium phosphate at blanket recommendation rate (100 kg/ha) since there was no pre-determined site-specific fertilizer application rate. Hand-weeding was done at early seedling stage (about one month) and before the booting stage. Mature plants were harvested to measure total biomass and grain yield.

Table 1. Chemical and physical characteristics of the soils of the study area.

Soil Parameter		Value
pH (water extract; 1:1.25)		4.92
Buffer pH (water extract 1:2)		5.57
Electrical conductivity (dS/m)		0.07
Organic matter (%)		4.01
Total N (%)		0.78
Available P (mg/kg)		8.87
Exchangeable acidity (meq/100 g)		2.23
CEC and exchangeable bases (cmol (+)/kg)	CEC	27.19
	Ca	5.27
	Mg	0.66
	K	0.91
Texture	Clay (%)	59
	Silt (%)	28
	Sand (%)	13
	Soil class	Clay

2.5. Statistical Analysis

Normality of distributions of the replicated hydroponic and field data were tested with SPSS version 24 (IBM SPSS statistics for Windows 2016, Version 24.0. Armonk, NY, USA: IBM Corp.) and RStudio (version 1.0.143: Integrated Development for R. RStudio, Inc., Boston, MA, USA. Retrieved from www.rstudio.com). The non-parametric (Kruskal–Wallis) and parametric ANOVA were computed either with RStudio or SPSS in order to determine whether there was significant difference between accessions for their performance in Al^{3+} treatments. The general linear model in which both accessions and replications were considered as fixed factors was used for analysis of variance. The median absolute deviation, a non-parametric statistic, was computed with RStudio to assess the level of variability attributed to the trait. Furthermore, post-hoc analysis was conducted using Tukey's honestly significant difference for parametric statistics, while a nonparametric post-hoc was done with Dunn's test, a Kruskal–Wallis test based post-hoc with "agricolae" package of RStudio. The coefficient of determination (r^2) was calculated to examine the relationship between various variables.

3. Results

3.1. Hydroponic Screen

Durum wheat accessions (595) were initially screened for Al^{3+} tolerance using a rapid hydroponic method of submerged seedlings. These accessions sometimes comprised of two or more genotypes since there was a large variation in performance between individual plants of the accession (Figure 1). Furthermore, there were visually observable differences within an accession such as variation in grain color. To take this heterogeneity into account an accession was scored based on its best-performing seedling. The use of average performance of plants in representing an accession would have resulted in rejection of many accessions because of a poor average performance such that a single plant within the accession with an acceptable level of Al^{3+} tolerance would be lost.

Accessions were classified into three phenotypic classes based on their total root length as tolerant (≥3.1 cm), intermediate (≤3.0 cm but ≥2.1 cm) and susceptible (≤2.0 cm). Moreover, accession 6956 was included in the tolerant class because of its exceptionally long primary seminal root. Using this criterion 21 accessions were classed as tolerant, 180 as intermediate, and 394 as susceptible to Al^{3+} toxicity (Supplementary Materials file, Table S3).

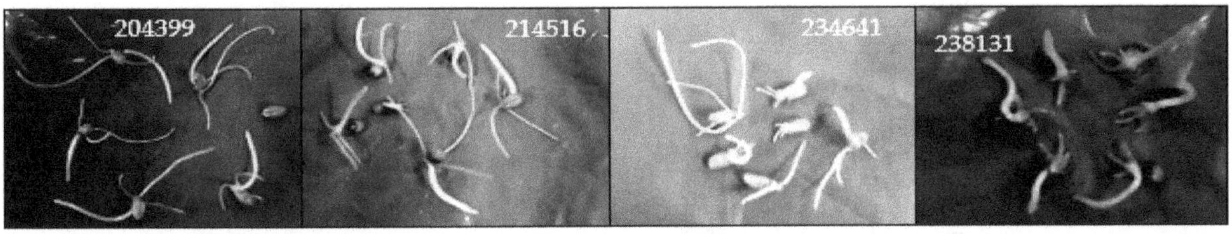

Figure 1. Examples of seedlings grown for 5 days in Al^{3+}-containing nutrient solution. Accessions 204399 and 214516 were scored as tolerant as was accession 234641 that was clearly segregating for tolerance. Accession 238131 was scored as sensitive. Individual seedlings identified within an accession to be tolerant were planted out to bulk up grain and the harvested grain used in a further hydroponic screen and a field trial.

A total of 150 accessions that included all 21 of the Al^{3+}-tolerant class, 79 of the best performers of the intermediate class, and a random selection of 50 of the sensitive class were evaluated in three replications to more precisely characterize their Al^{3+} tolerance. In this experiment accessions were represented in each experiment by the average performance of ≥10 seedlings using grain harvested from individually selected seedlings. The performance of accessions in the replicated experiment using the same hydroponic method as the preliminary screen was well correlated with the preliminary screen ($r^2 = 0.72$; Figure 2).

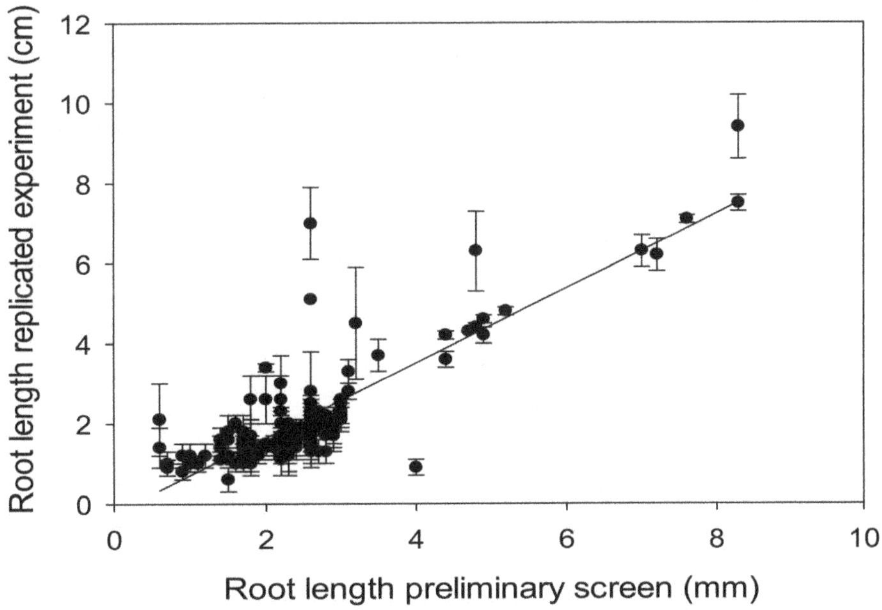

Figure 2. Relationship between total root length of the individual seedlings selected (total root length from the preliminary screen, $n = 1$ for each accession) that had the longest roots and the total root length of the resulting progeny of the selections showing the average of 10 or more seedlings (in three replications). Error bars indicate the standard error of the mean for the three replications.

3.2. Field Experiment

The set of 150 selections used in the replicated hydroponic screen was assessed in the field on an acid soil. As there was limited grain harvested from single plants of the selected seedlings, the accessions were assessed as single rows 2 m long on both an un-amended plot and an equivalent plot that had been amended with lime to neutralize the acidity. The grain yield and biomass means for the limed plots of all accessions combined was greater than the means of the acid plot indicating that soil acidity was clearly present at this site. Grain yield on the acid site was reduced on average to only 18% of the limed site whereas biomass was reduced to 28% of the limed site (Figure 3).

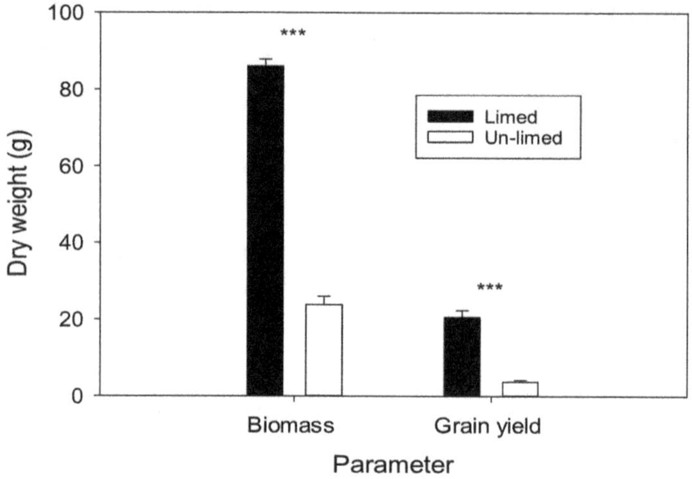

Figure 3. Liming improves biomass and grain yield of Ethiopian accessions grown on an acid soil. Ethiopian accessions (150) were grown in single rows 2 m long on an acid soil and the same soil that had lime incorporated to a depth of 20 cm. Final biomass and grain yield were determined and data of all accessions were combined with error bars indicating the standard error of the mean (n = 150). Student's t-test indicated significant differences between treatments (***; $p < 0.001$).

The performance of the Al^{3+}-tolerant lines on the un-limed acid plot was remarkable and clearly distinct from that of the sensitive accessions. Figure 4 shows the relative (un-limed/limed) biomass and grain yields of the accessions selected for the field trial against the root length as determined in the preliminary screen.

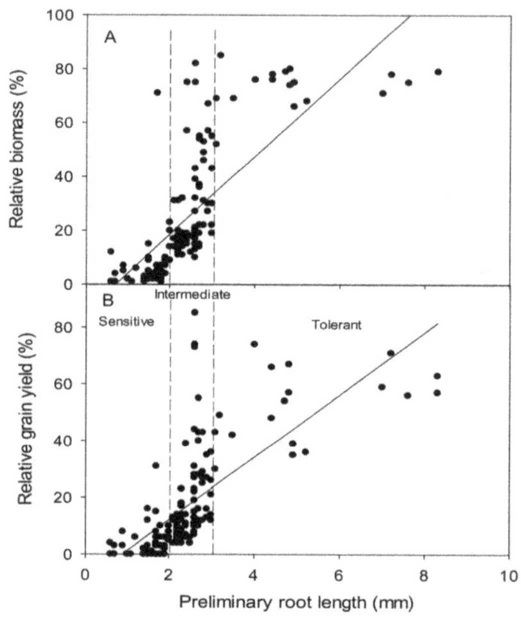

Figure 4. Root length as determined with the rapid screen identifies Al^{3+}-tolerant accessions based on both mature biomass (**A**) and grain yield (**B**). The preliminary root length is a measure of the total root length of the most tolerant seedling in the pool of seedlings assessed for each accession. The dashed

vertical lines denote the cut-off points for root lengths of seedlings initially classed as; tolerant: >3.0 cm; intermediate: 2.0–3.0 cm; sensitive: <2.0 cm. The single seedling was grown to seed and the resulting progeny grown in two replicates as 2 m rows on un-limed and limed plots. The mean of the replicates was used to calculate the relative biomass and relative grain yields where values for the un-limed plots were divided by the limed plots and multiplied by 100 to express data as a percentage. The solid line shows a linear regression with $r^2 = 0.56$ for (A) and $r^2 = 0.52$ for (B).

Expressing data in relative terms takes into account inherent differences in plant vigor when Al^{3+} is absent but a similar relationship was found when using unmodified data of only the acid plot (Supplementary Materials file, Figure S3). Biomass and final grain yield of genotypes grown in the acid trial were strongly correlated (Figure 5A). When biomass and grain yield data were combined within each grouping of lines selected from the preliminary screen (sensitive, intermediate, and tolerant), they showed rankings consistent with their grouping based on the screen (Figure 5B).

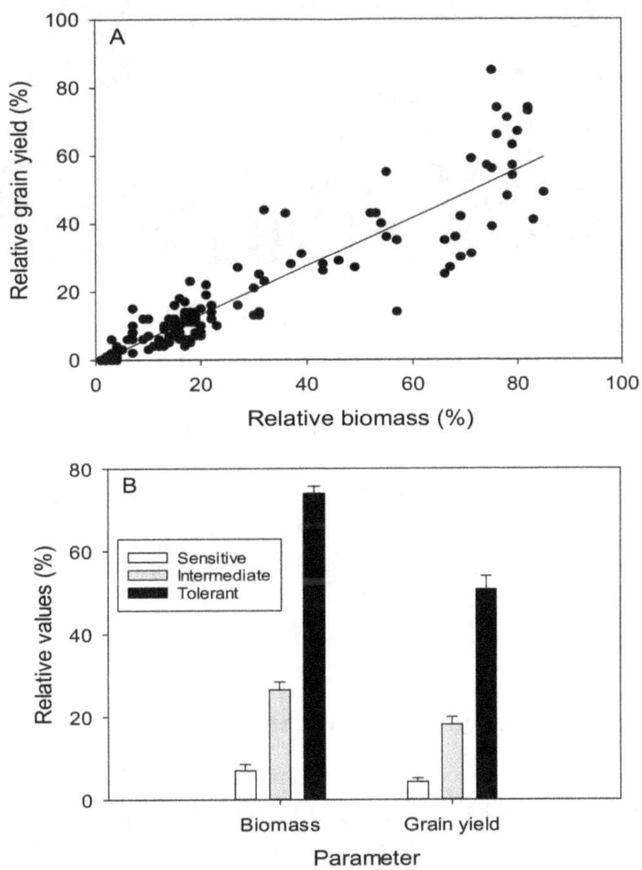

Figure 5. Relative biomass and relative grain yield are highly correlated and the classification of accessions into groups with the preliminary screen of Al^{3+}-tolerant, -intermediate, and -sensitive is consistent with final biomass and grain yield. Biomass and grain yield of accessions grown on the un-limed plot are expressed as a percent of the limed plot using the procedure described in the legend of Figure 4. (**A**) Relationship between relative biomass and relative grain yield for all accessions grown in the field ($r^2 = 0.84$). (**B**) Relative biomass and relative grain yield of all accessions allocated to the three classes were combined with error bars denoting the standard error and a one-way ANOVA of the data showed significant differences between all groups for each of biomass and grain yield (ANOVA on ranks $p < 0.001$ between classes; $n = 47$ for Sensitive, $n = 83$ for Intermediate, and $n = 20$ for Tolerant classes).

Since previous studies have found that stocks of durum wheat lines can be contaminated with bread wheat, we sought to verify that both Al^{3+}-tolerant and -intermediate selections were indeed durum wheat. It can sometimes be difficult to distinguish the species based solely on the phenotypes so we used the Dgas molecular marker that is unique to the D-genome to distinguish the species. Using Dgas we found that all the Al^{3+}-tolerant and -intermediate accessions were actually bread wheat (Figure 6), with only the most sensitive genotypes being durum wheat.

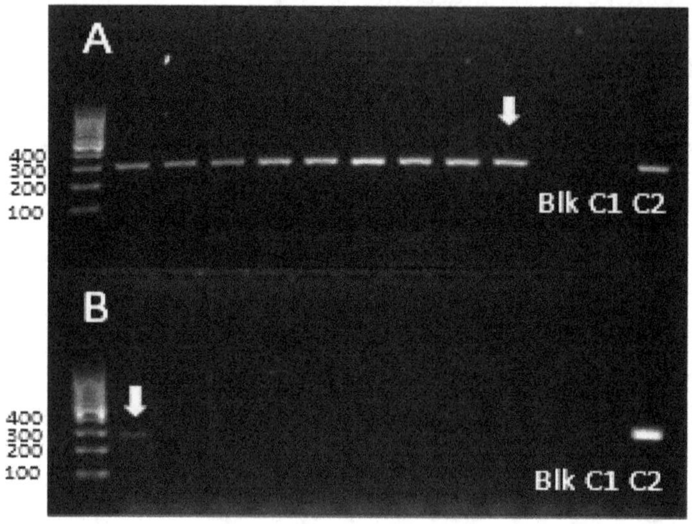

Figure 6. Examples of accessions screened by PCR for the Dgas sequence. Panel (**A**) is of Al^{3+}-tolerant selections while panel (**B**) are all Al^{3+}-sensitive selections except for the first lane after the DNA ladder. Most samples are shown as duplicates in lanes side by side except for the last lane of (A) (arrow) whose duplicate is the first lane of (B) (arrow). The lane on the left is a 100 bp DNA ladder with sizes of the smallest four markers shown on the right. A band at about 300 bp is indicative of the presence of the D-genome. Blk is a no DNA sample while C1 is a known durum wheat (accession 8317) and C2 is a known bread wheat variety (Enkoy). All lines (38 lines) that yielded more than 3.5 g per row in the field trial (Supplementary Materials file, Figure S3B) were analyzed and all possessed the 300 bp band indicating that they were hexaploid wheat.

3.3. *Other Potential Sources of Al^{3+}-Tolerant Durum Wheat*

As discussed in the Introduction, the Al^{3+} tolerance of durum has been enhanced by introgression of genes from bread wheat although not into Ethiopian germplasm [20]. As an alternate source of genes, a previous report identified three out of 420 tetraploid genotypes screened that were comparatively Al^{3+} tolerant although they did not approach the tolerance shown by a tolerant bread wheat used as a check line [26]. The lines had been confirmed as tetraploid with a molecular marker specific for the D-genome, so these genotypes were a potential source of genes that could be used in direct crosses to Ethiopian lines. To establish if these three lines had a useful level of Al^{3+} tolerance, we assessed their performance in hydroponic culture against bread wheat lines that varied in tolerance as well as a 4D (4B) substitution line of durum. However, all three lines were rated as sensitive with the best performer having a similar level of tolerance as Al^{3+}-sensitive bread wheat (Figure 7). By contrast, a durum line where the 4B chromosome was substituted by the 4D chromosome of hexaploid wheat had a level of Al^{3+} tolerance comparable to the tolerant hexaploids.

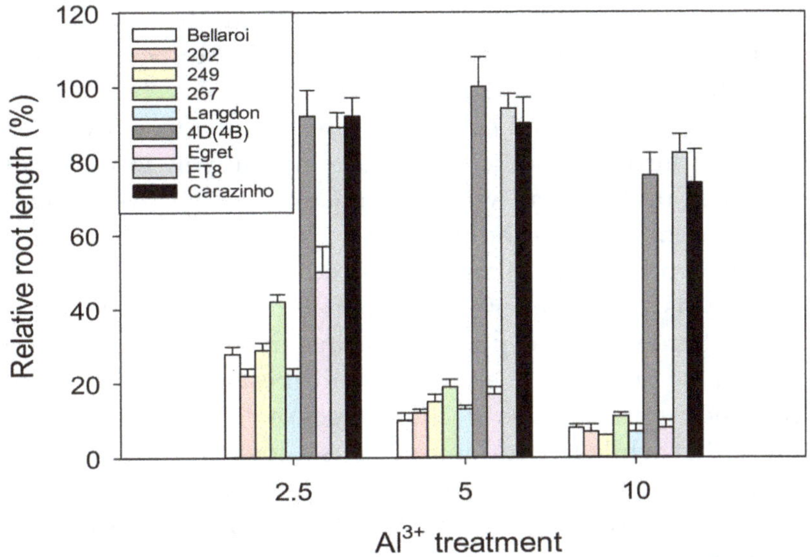

Figure 7. Relative Al^{3+} tolerance of durum wheat genotypes previously identified as being relatively Al^{3+} tolerant in comparison to a durum substitution line with chromosome 4D of bread wheat and bread wheat lines possessing Al^{3+}-tolerant and -sensitive alleles of the *TaALMT1* gene. Bellaroi is an Australian durum wheat cultivar; lines 202, 249, and 267 are accessions previously classified as being Al^{3+}-tolerant durum genotypes [26]; Langdon is the parental durum cultivar used to generate the 4D substitution line; 4D (4B) is a durum line in the Langdon background where chromosome 4B has been substituted with chromosome 4D; Egret is an Al^{3+}-sensitive bread wheat cultivar; ET8 and Carazinho are Al^{3+}-tolerant bread wheat lines. Root length is expressed as percent relative to a zero control and error bars indicate cumulative errors of three to five seedlings.

4. Discussion

4.1. A Rapid Screen Proves Robust and Correlates with Grain Yield in Field Trials when Ranking Wheat Germplasm for Al^{3+} Tolerance

In this work we screened Ethiopian durum accessions for Al^{3+} tolerance first using a high throughput method with hydroponics, then with a replicated hydroponic screen and finally in the field on acid soil. Ethiopian durum accessions that include landraces are reported to show great diversity for many traits and molecular analyses have verified the unique nature of this germplasm [24]. However, despite this diversity our work shows that there does not appear to be useful levels of Al^{3+} tolerance in Ethiopian durum wheat and that seedlings identified to be Al^{3+} tolerant were in fact contaminating bread wheat. It should be noted that many of the accessions classed as tolerant only had a few Al^{3+}-tolerant seedlings indicating that the durum grain stocks had been contaminated with bread wheat grain. This was verified by the use of a molecular marker that confirmed Al^{3+}-tolerant seedlings to be bread wheat. We conclude that the rapid screen was sufficiently robust as a preliminary screen to identify Al^{3+}-tolerant bread wheat contaminants. The accessions ranked as having intermediate tolerance in the preliminary screen could still have conferred a useful level of Al^{3+} tolerance for durum if they ranked similarly in the field. However, in the field the best performing accessions ranked as having intermediate Al^{3+} tolerance from the preliminary screen were all identified as bread wheat (Figure 6 and Supplementary Materials Figure S3B). Nevertheless, there exist Al^{3+}-sensitive genotypes of bread wheat so the rapid screen does not replace the use of a molecular marker in identifying bread wheat lines. Verifying the identity of durum wheat grain stocks is of particular importance for germplasm banks but should also be confirmed by researchers. Bread wheat genotypes in the past have been mistakenly identified as durum wheat and the most relevant to this study was the incorrect identification of Al^{3+}-tolerant durum lines that were subsequently found to be bread wheat [17,18].

In some cases, bread wheat has morphological characteristics in the field that are similar to durum wheat emphasizing the need for molecular analysis to establish the species identity [32].

The absence of Al^{3+} tolerance in unmodified *Triticum turgidum* ssp *durum* Desf. is consistent with previous studies that screened durum cultivars with presumably a lower level of diversity than the Ethiopian germplasm and also showed no or only a comparatively low level of Al^{3+} tolerance [16]. The simple and rapid screen undertaken with minimal equipment over only 5 days growth with seedlings submerged in nutrient solution showed a remarkable consistency in classifying the relative tolerance of germplasm grown in the field and measured for mature biomass and even grain yield (Figure 5B). Although there was a good general agreement of the hydroponic screens with performance of accessions in the field, there were exceptions. For instance, several accessions rated as having intermediate tolerance with the preliminary screen were the best or amongst the best for grain yield in the field (Figure 4B). This finding emphasizes the importance of verifying selections on acid soil whether in the field or in pots and that while hydroponic screens allow for rapid assessment of germplasm, the relative tolerance of genotypes can differ when grown in soil. For instance, a durum line carrying the *TaMATE1B* gene introgressed from hexaploid wheat (see below) shows marginal Al^{3+}-tolerance in hydroponics that is considerably less than a line carrying *TaALMT1*, whereas the situation is reversed when the lines are grown on acid soil, with *TaMATE1B* lines outperforming *TaALMT1* lines [20].

4.2. The Evolution of Al^{3+} Tolerance in Bread Wheat Occurred Subsequent to the Hybridization of the D-Genome

Here we show that durum lines previously reported to be relatively Al^{3+}-tolerant [26] had a level of tolerance similar to sensitive wheat (Figure 7), confirming the absence of useful Al^{3+} tolerance in durum wheat germplasm (excluding lines where genes from hexaploid wheat have been introgressed as discussed below). This observation is puzzling given reports of the presence of Al^{3+} tolerance genes on the A- and B-genomes of bread wheat. Although *TaALMT1*, the major gene for Al^{3+} tolerance in bread wheat, is located on chromosome 4D, there are several loci on the A- and B-genomes reported to confer Al^{3+} tolerance [6]. For example, *TaMATE1B* the only Al^{3+} tolerance gene other than *TaALMT1* that has been cloned from bread wheat, is located on chromosome 4B [33]. We surmise that the absence of Al^{3+} tolerance genes in diverse durum germplasm suggests that the multiple Al^{3+} tolerance genes found in bread wheat arose subsequent to the hybridization of the D-genome with the A- and B-genomes some 10,000 years ago. Transposable elements have been shown to enhance the level of expression of genes encoding transport proteins such as those of the MATE family [34] that confer Al^{3+} tolerance. The event of polyploidization commonly results in genome instability including the activation of transposable elements [35]. It is conceivable that activation of transposons when the hexaploid was formed has been key in enabling bread wheat to evolve Al^{3+} tolerance.

4.3. Strategies to Enhance the Al^{3+} Tolerance of Ethiopian Durum Wheat

To date the only durum germplasm verified to be Al^{3+} tolerant are those that were developed by introgression of Al^{3+} tolerance genes from bread wheat into durum wheat [20]. In those cases, the *ph1c* mutant was used to introgress the *TaALMT1* gene located on chromosome 4D of bread wheat and steps taken to avoid the hybrid necrosis that can occur when crossing bread to durum wheat to introgress the *TaMATE1B* gene on chromosome 4B. This germplasm is not considered to be genetically modified since it was developed by so-called natural means and can therefore be used in the field without restrictions. The germplasm is a source of Al^{3+} tolerance genes in a durum background that could be used to introgress one or both genes derived from bread wheat into selected Ethiopian durum germplasm. Many of the farms that grow durum wheat in Ethiopia are small holdings and it has been found that landraces otherwise known as "farmer's varieties" perform better than "improved" germplasm under many situations [36]. With this in mind it may be useful to cross the Al^{3+}-tolerant germplasm described above into selected landraces adapted to local regions. Using local landraces as recurrent parents in multiple backcrosses while tracking the presence of Al^{3+} tolerance genes using molecular markers should maintain the germplasm with various valuable landrace traits that could otherwise be lost [37]. The *TaMATE1B* gene seems a preferred candidate at this stage since it appears to be more effective than *TaALMT1* in a durum background and there exists a co-dominant marker that can be used to track the tolerant allele [20]. A recent publication has shown that the *TaMATE1B* gene introgressed into a durum genetic background confers a marked ability of roots from mature plants to withstand Al^{3+} toxicity when grown in an acid soil [38]. A simple backcrossing program where a single gene is introgressed into landraces could be useful in establishing whether or not *TaMATE1B* can improve durum grain production of landraces on acid soils of Ethiopia. We speculate that introgressing both *TaALMT1* with *TaMATE1B* may provide the greatest level of Al^{3+} tolerance to durum wheat as has been found for some bread wheat genotypes.

5. Conclusions

Here we show that despite the diversity of Ethiopian durum germplasm and large regions of acid soils, conditions conducive for the evolution of Al^{3+} tolerance, a useful level of Al^{3+} tolerance was not detected. A high throughput screen identified Al^{3+}-tolerant seedlings within accessions but all of them were shown to be contaminating bread wheat. This finding highlights the importance of correct identification of germplasm and confirmation of species identity with molecular markers particularly for closely-related species. The finding that Ethiopian durum germplasm is Al^{3+} sensitive is consistent with Al^{3+} tolerance having evolved in bread wheat subsequent to the hybridization of the D-genome with the A- and B-genomes. A field trial on acid soil showed biomass and grain yields that correlated with classifications based on a high throughput screen, confirming the utility of the rapid screen for preliminary assessment of germplasm.

Supplementary Materials:
Figure S1: Photographs of the components of the hydroponic system. (A) The plastic cups to which dry grain was added. The holes allowed aeration of grain while the cups themselves separated the accessions from one another. The slots in the plastic stands at the end of a series of cups enabled the long air stone to be inserted and held firm. (B) Top view showing air stones and pumps along with the plastic basin used to hold the nutrient solution. (C) Top view of assembled equipment showing plastic cups with a partitioning plate inserted within each cup so that two accessions could be placed into each cup and separated from one another. Nutrient solution was added to the basin to fill and cover the cups that held the grain. The holes in the cups allowed sufficient aeration for growth of seedlings. Nutrient solution was changed every day to maintain pH at about 4.3 and Al_2SO_4 at 5 μM. Figure S2: Location of the study area in Ethiopia. Figure S3: Relationships between Al^{3+} tolerance determined with a rapid screen and (A) biomass or (B) grain yield for durum accessions grown on an acid plot in the field. Table S1: Wheat production across zones and special districts of Ethiopia showing yields averaged over 5 years. Table S2: Passport data of durum accessions used in hydroponics screening for Al^{3+} tolerance. Table S3: Preliminary screen of Al^{3+} tolerance of 594 Ethiopian durum wheat accessions grown in Al^{3+}-containing nutrient solution where total root length after 5 days growth was used to classify genotypes.

Author Contributions: Conceptualization, E.F.W.; methodology, E.F.W. and E.D.; formal analysis, E.F.W. and E.D.; investigation, E.F.W., E.D., A.L.O.; resources, E.F.W., A.L.O., K.T., E.M.M.; data curation, E.F.W.; writing—original draft preparation, E.F.W. and E.D.; writing—review and editing, A.L.O., and K.D.; supervision, E.D., A.L.O., K.D., K.T. and S.K.M.; project administration, K.T. and E.M.M.; funding acquisition, E.F.W., A.L.O. and K.T.

References

1. Minot, N.; Warner, J.; Lemma, S.; Kasa, L.; Gashaw, A.; Rashid, S. *The Wheat Supply Chain in Ethiopia: Patterns, Trends, and Policy Options*; International Food Policy Research Institute (IFPRI): Washington, DC, USA, 2015.
2. Sall, A.T.; Chiari, T.; Legesse, W.; Seid-Ahmed, K.; Ortiz, R.; van Ginkel, M.; Bassi, M.F. Durum wheat (*Triticum durum* Desf.): Origin, cultivation and potential expansion in sub-Saharan Africa. *Agronomy* **2019**, *9*, 263. [CrossRef]

3. Bona, L.; Wright, R.J.; Baligar, V.C. A rapid method for screening cereals for acid soil tolerance. *Cereal Res. Commun.* **1991**, *19*, 465–468.
4. Raman, H.; Ryan, P.R.; Raman, R.; Stodart, B.J.; Zhang, K.; Martin, P.; Wood, R.; Sasaki, T.; Yamamoto, Y.; Mackay, M.; et al. Analysis of *TaALMT1* traces the transmission of aluminum resistance in cultivated common wheat (*Triticum aestivum* L.). *Theor. Appl. Genet.* **2008**, *116*, 343–354. [CrossRef] [PubMed]
5. Aguilera, J.G.; Minozzo, J.A.D.; Barichello, D.; Fogaca, C.M.; da Silva, J.P.; Consoli, L.; Pereira, J.F. Alleles of organic acid transporter genes are highly correlated with wheat resistance to acidic soil in field conditions. *Theor. Appl. Genet.* **2016**, *129*, 1317–1331. [CrossRef] [PubMed]
6. Raman, H.; Stodart, B.; Ryan, P.R.; Delhaize, E.; Emebiri, L.; Raman, R.; Coombes, N.; Milgate, A. Genome-wide association analyses of common wheat (*Triticum aestivum* L.) germplasm identifies multiple loci for aluminium resistance. *Genome* **2010**, *53*, 957–966. [CrossRef] [PubMed]
7. Von Uexküll, H.R.; Mutert, E. Global extent, development and economic impact of acid soils. *Plant Soil* **1995**, *171*, 1–15. [CrossRef]
8. Wayima, E.F. Classification of Ethiopian soils with pH. *J. Soil Sci. Environ. Manag.* **2019**, in press.
9. Kidanemariam, A.; Gebrekidan, H.; Mamo, T.; Kibret, K. Impact of altitude and land use type on some physical and chemical properties of acidic soils in Tsegede highlands, northern Ethiopia. *Open J. Soil Sci.* **2012**, *2*, 223–233. [CrossRef]
10. Kidanemariam, A.; Gebrekidan, H.; Mamo, T.; Tesfaye, K. Wheat crop response to liming materials and N and P fertilizers in acidic soils of Tsegede highlands, northern Ethiopia. *Agric. For. Fish.* **2013**, *2*, 126–135. [CrossRef]
11. Mosissa, F. Progress of soil acidity management research in Ethiopia. *Greener J. Soil Sci. Plant Nutr.* **2018**, *5*, 9–22. [CrossRef]
12. Kochian, L.V.; Pineros, M.A.; Liu, J.P.; Magalhaes, J.V. Plant adaptation to acid soils: The molecular basis for crop aluminum resistance. *Ann. Rev. Plant Biol.* **2015**, *66*, 571–598. [CrossRef] [PubMed]
13. Magalhaes, J.V.; Garvin, D.F.; Wang, Y.H.; Sorrells, M.E.; Klein, P.E.; Schaffert, R.E.; Li, L.; Kochian, L.V. Comparative mapping of a major aluminum tolerance gene in sorghum and other species in the Poaceae. *Genetics* **2004**, *167*, 1905–1914. [CrossRef] [PubMed]
14. Bian, M.; Waters, I.; Broughton, S.; Zhang, X.Q.; Zhou, M.X.; Lance, R.; Sun, D.F.; Li, C.D. Development of gene-specific markers for acid soil/aluminium tolerance in barley (*Hordeum vulgare* L.). *Mol. Breed.* **2013**, *32*, 155–164. [CrossRef]
15. Ma, J.F.; Nagao, S.; Sato, K.; Ito, H.; Furukawa, J.; Takeda, K. Molecular mapping of a gene responsible for Al-activated secretion of citrate in barley. *J. Exp. Bot.* **2004**, *55*, 1335–1341. [CrossRef] [PubMed]
16. Ryan, P.R.; Raman, H.; Gupta, S.; Sasaki, T.; Yamamoto, Y.; Delhaize, E. The multiple origins of aluminium resistance in hexaploid wheat include *Aegilops tauschii* and more recent *cis* mutations to *TaALMT1*. *Plant J.* **2010**, *64*, 446–455. [CrossRef] [PubMed]
17. Foy, C.D. Tolerance of durum wheat lines to an acid, aluminum-toxic subsoil. *J. Plant Nutr.* **1996**, *19*, 1381–1394. [CrossRef]
18. Han, C.; Ryan, P.R.; Yan, Z.; Delhaize, E. Introgression of a 4D chromosomal fragment into durum wheat confers aluminium tolerance. *Ann. Bot.* **2014**, *114*, 135–144. [CrossRef]
19. Delhaize, E.; Ma, J.F.; Ryan, P.R. Transcriptional regulation of aluminium tolerance genes. *Trends Plant Sci.* **2012**, *17*, 341–348. [CrossRef]
20. Han, C.; Zhang, P.; Ryan, P.R.; Rathjen, T.M.; Yan, Z.H.; Delhaize, E. Introgression of genes from bread wheat enhances the aluminium tolerance of durum wheat. *Theor. Appl. Genet.* **2016**, *129*, 729–739. [CrossRef]
21. Garcia-Oliveira, A.L.; Martins-Lopes, P.; Tolra, R.; Poschenrieder, C.; Guedes-Pinto, H.; Benito, C. Differential physiological responses of Portuguese bread wheat (*Triticum aestivum* L.) genotypes under aluminium stress. *Diversity* **2016**, *8*, 26. [CrossRef]
22. Mengistu, D.K.; Kidane, Y.G.; Fadda, C.; Pe, M.E. Genetic diversity in Ethiopian durum wheat (*Triticum turgidum* var durum) inferred from phenotypic variations. *Plant Genet. Res. Charact. Util.* **2018**, *16*, 39–49. [CrossRef]
23. Mengistu, D.K.; Kiros, A.Y.; Pe, M.E. Phenotypic diversity in Ethiopian durum wheat (*Triticum turgidum* var. durum) landraces. *Crop J.* **2015**, *3*, 190–199. [CrossRef]
24. Mengistu, D.K.; Kidane, Y.G.; Catellani, M.; Frascaroli, E.; Fadda, C.; Pe, M.E.; Dell'Acqua, M. High-density

molecular characterization and association mapping in Ethiopian durum wheat landraces reveals high diversity and potential for wheat breeding. *Plant Biotechnol. J.* **2016**, *14*, 1800–1812. [CrossRef] [PubMed]
25. Kabbaj, H.; Sall, A.T.; Al-Abdallat, A.; Geleta, M.; Amri, A.; Filali-Maltouf, A.; Belkadi, B.; Ortiz, R.; Bassi, F.M. Genetic diversity within a global panel of durum wheat (*Triticum durum*) landraces and modern germplasm reveals the history of alleles exchange. *Front. Plant Sci.* **2017**, *8*, 1277. [CrossRef] [PubMed]
26. Raman, H.; Hare, R.; Graham, K.; Coombes, N.; Raman, R. Characterisation of durum germplasm for aluminium resistance using nutrient solution culture. In *International Wheat Genetics Symposium*; Appels, R., Lagudah, R.E.E., Langridge, P., Mackay, M., McIntyre, L., Sharp, P., Eds.; Sydney University Press: Brisbane, Australia, 2008; pp. 1–3.
27. Joppa, L.R.; Williams, N.D. Langdon durum disomic substitution lines and aneuploid analysis in tetraploid wheat. *Genome* **1988**, *30*, 222–228. [CrossRef]
28. Delhaize, E.; Ryan, P.R.; Hebb, D.M.; Yamamoto, Y.; Sasaki, T.; Matsumoto, H. Engineering high-level aluminum tolerance in barley with the *ALMT1* gene. *Proc. Natl. Acad. Sci. USA* **2004**, *101*, 15249–15254. [CrossRef] [PubMed]
29. Zhou, G.F.; Delhaize, E.; Zhou, M.X.; Ryan, P.R. The barley *MATE* gene, *HvAACT1*, increases citrate efflux and Al^{+3} tolerance when expressed in wheat and barley. *Ann. Bot.* **2013**, *112*, 603–612. [CrossRef] [PubMed]
30. Bryan, G.J.; Dixon, A.; Gale, M.D.; Wiseman, G. A PCR-based method for the detection of hexaploid bread wheat adulteration of durum wheat and pasta. *J. Cereal Sci.* **1998**, *28*, 135–145. [CrossRef]
31. McNeil, D.; Lagudah, E.S.; Hohmann, U.; Appels, R. Amplification of DNA-sequences in wheat and its relatives—the Dgas44 and R350 families of repetitive sequences. *Genome* **1994**, *37*, 320–327. [CrossRef]
32. Zeven, A.C.; Waninge, J. The presence of 3 groups of Scalavatis and other hexaploid bread wheat plants contaminating durum-wheat fields in Cyprus. *Euphytica* **1989**, *43*, 117–124. [CrossRef]
33. Tovkach, A.; Ryan, P.R.; Richardson, A.E.; Lewis, D.C.; Rathjen, T.M.; Ramesh, S.; Tyerman, S.D.; Delhaize, E. Transposon-mediated alteration of *TaMATE1B* expression in wheat confers constitutive citrate efflux from root apices. *Plant Physiol.* **2013**, *161*, 880–892. [CrossRef]
34. Pereira, J.F.; Ryan, P.R. The role of transposable elements in the evolution of aluminium resistance in plants. *J. Exp. Bot.* **2019**, *70*, 41–54. [CrossRef] [PubMed]
35. Madlung, A.; Tyagi, A.P.; Watson, B.; Jiang, H.; Kagochi, T.; Doerge, R.W.; Martienssen, R.; Comai, L. Genomic changes in synthetic Arabidopsis polyploids. *Plant J.* **2005**, *41*, 221–230. [CrossRef] [PubMed]
36. Kidane, Y.G.; Mancini, C.; Mengistu, D.K.; Frascaroli, E.; Fadda, C.; Pe, M.E.; Dell'Acqua, M. Genome wide association study to identify the genetic base of smallholder farmer preferences of durum wheat traits. *Front. Plant Sci.* **2017**, *8*, 1230. [CrossRef] [PubMed]
37. Tsegaye, B.; Berg, T. Genetic erosion of Ethiopian tetraploid wheat landraces in Eastern Shewa, Central Ethiopia. *Genet. Resour. Crop Evol.* **2007**, *54*, 715–726. [CrossRef]
38. Pooniya, V.; Palta, J.A.; Chen, Y.; Delhaize, E.; Siddique, K.H. Impact of the *TaMATE1B* gene on above and below-ground growth of durum wheat grown on an acid and Al^{3+}-toxic soil. *Plant Soil* **2019**, 1–12. [CrossRef]

11

Identification of Anther Length QTL and Construction of Chromosome Segment Substitution Lines of *Oryza longistaminata*

Takayuki Ogami, Hideshi Yasui, Atsushi Yoshimura and Yoshiyuki Yamagata *

Plant Breeding Laboratory, Faculty of Agriculture, Kyushu University. 744, Motooka, Nishi-ku, Fukuoka 819-0395, Japan; ohgammy9.30@gmail.com (T.O.); hyasui@agr.kyushu-u.ac.jp (H.Y.); ayoshi@agr.kyushu-u.ac.jp (A.Y.)

* Correspondence: yoshiyuk@agr.kyushu-u.ac.jp.

Abstract: Life histories and breeding systems strongly affect the genetic diversity of seed plants, but the genetic architectures that promote outcrossing in *Oryza longistaminata*, a perennial wild species in Africa, are not understood. We conducted a genetic analysis of the anther length of *O. longistaminata* accession W1508 using advanced backcross quantitative trait locus (QTL) analysis and chromosomal segment substitution lines (CSSLs) in the genetic background of *O. sativa* Taichung 65 (T65), with simple sequence repeat markers. QTL analysis of the BC_3F_1 population (n = 100) revealed that four main QTL regions on chromosomes 3, 5, and 6 were associated to anther length. We selected a minimum set of BC_3F_2 plants for the development of CSSLs to cover as much of the W1508 genome as possible. The additional minor QTLs were suggested in the regional QTL analysis, using 21 to 24 plants in each of the selected BC_3F_2 population. The main QTLs found on chromosomes 3, 5, and 6 were validated and designated *qATL3*, *qATL5*, *qATL6.1*, and *qATL6.2*, as novel QTLs identified in *O. longistaminata* in the mapping populations of 94, 88, 70, and 95 BC_3F_4 plants. *qATL3*, *qATL5*, and *qATL6.1* likely contributed to anther length by cell elongation, whereas *qATL6.2* likely contributed by cell multiplication. The QTLs were confirmed again in an evaluation of the W1508ILs. In several chromosome segment substitution lines without the four validated QTLs, the anthers were also longer than those of T65, suggesting that other QTLs also increase anther length in W1508. The cloning and diversity analyses of genes conferring anther length QTLs promotes utilization of the genetic resources of wild species, and the understanding of haplotype evolution on the differentiation of annuality and perenniality in the genus *Oryza*.

Keywords: anther length; cell elongation; genetic architecture; outcrossing; perennial species; rice

---

1. Introduction

Life histories and breeding systems strongly affect the genetic diversity of seed plants. Annuals tend to allocate their resources to sexual reproduction to produce as many flowers as possible for the one-time dispersal of seeds. By contrast, perennial species tend to primarily allocate resources to vegetative growth because of the need to occupy physical space, and to use local water and nutrient resources over an extended life span depending on the ecological circumstances [1].

Perennial species tend to show higher heterozygosity when compared with the annual species or domesticated species [2]. Heterozygosity has been found to correlate with fitness-related traits, such as survival probability, reproductive success, and disease resistance [3–7]. Populations of perennials maintain high genetic diversity by producing few but large floral organs, and by promoting relatively high outcrossing rates with mechanisms to prevent self-pollination—such as self-incompatibility or monoecious flowers [8,9]. An understanding of the genetic architecture of a favorable trait that promotes outcrossing will help to answer a question: What set of polymorphisms or genes contribute to outcrossing characteristics? However, the genetic basis of breeding system-associated traits is not fully understood.

There are eight species of the genus *Oryza* with the AA genome: Two Asian wild species (*O. rufipogon* Griff. and *O. nivara* Sharma et Shastry), two wild species in Africa (*O. longistaminata* A. Chev. & Roehr. and *O. barthii* A. Chev.), one wild species in South America (*O. glumaepatula* Steud.), one wild species in Australia (*O. meridionalis* Ng.), and two cultivated species [10]. Of the two cultivated rice species, *O. sativa* L. (Asian rice) is domesticated from the Asian perennial wild species *O. rufipogon*, and *O. glaberrima* Steud. (African rice) is domesticated from the African annual wild species *O. barthii*. The Asian cultivated and wild species form a species complex in which the production of hybrid progeny is possible. The differentiation of annual (*O. nivara*) and perennial (*O. rufipogon*) species reflects adaptation to their ecological niches [11,12]. In Africa, the annual species *O. barthii* and the perennial species *O. longistaminata* occupy different ecological habitats. *Oryza longistaminata* is characterized by a rhizome [13], and has a particularly large anther [14], as indicated by the species name '*longistaminata*' (long stamen). Thus, the anther is one of the key traits of this species. Additionally, it has high heterozygosity [15] that is likely due to its self-incompatibility and high outcrossing rate, achieved by the large reproductive organs [16]. Cultivated rice is a self-pollinated species with an outcrossing rate of less than 4% [17]. The outcrossing rate of perennial wild species tends to be higher than that of cultivated species. That of *O. longistaminata* and *O. rufipogon* reaches 100% in certain combinations of hybridization, although the rate for wild species typically ranges from 3.2% to 50% [18,19]. To fully understand perenniality and their outcrossing characteristics, we need to investigate the genetic architectures of their reproductive systems.

Chromosomal segment substitution lines (CSSLs) facilitate the genetic analysis and characterization of quantitative traits of donor varieties, or species in the genetic background of a recurrent parent. CSSLs are powerful genetic tools for the identification of quantitative trait loci (QTLs) [20–23] in trials across different years and environments [24–26]. Studies of anther QTLs have been conducted with F_2 populations and recombinant inbred line (RIL) populations derived from the interspecific crosses between *O. rufipogon* and *O. sativa* [27–29], or intraspecific crosses between the *indica* and *japonica* subspecies of *O. sativa* [30]. However, QTLs that confer anther length in *O. longistaminata* have not been examined and validated using near-isogenic lines (NIL) in the genetic background of *O. sativa*.

In this study, we elucidated the genetic basis of anther length in *O. longistaminata*, by using advanced backcross QTL analysis and CSSLs in the genetic background of *O. sativa* L. cv. Taichung 65 (T65). QTLs conferring anther length identified in BC_3F_1 and BC_3F_2 populations were validated in the BC_3F_4 population, in which single QTL regions were segregating. We validated the QTLs again by constructing CSSLs of *O. longistaminata*. NILs were evaluated to characterize the histological cause of increased anther length. CSSLs chromosome segments derived from *O. longistaminata* acc. W1508 were named as W1508ILs, using the term 'introgression lines' (ILs) to refer to CSSLs based on interspecific hybridization.

2. Results

2.1. Genetic Variation in Anther Length in the Backcrossed Population

We measured the anther length of *O. barthii*, *O. glumaepatula*, and *O. longistaminata* (Figure 1) to evaluate the genetic variation in anther length. The anther length of *O. longistaminata* were longer than those of all *O. barthii* and *O. glumaepatula* accessions, except for *O. glumaepatula* W1183. To identify the QTLs derived from *O. longistaminata* in a uniform genetic background, we developed CSSLs carrying W1508 (*O. longistaminata*) chromosomal segments in the genetic background of *O. sativa* T65.

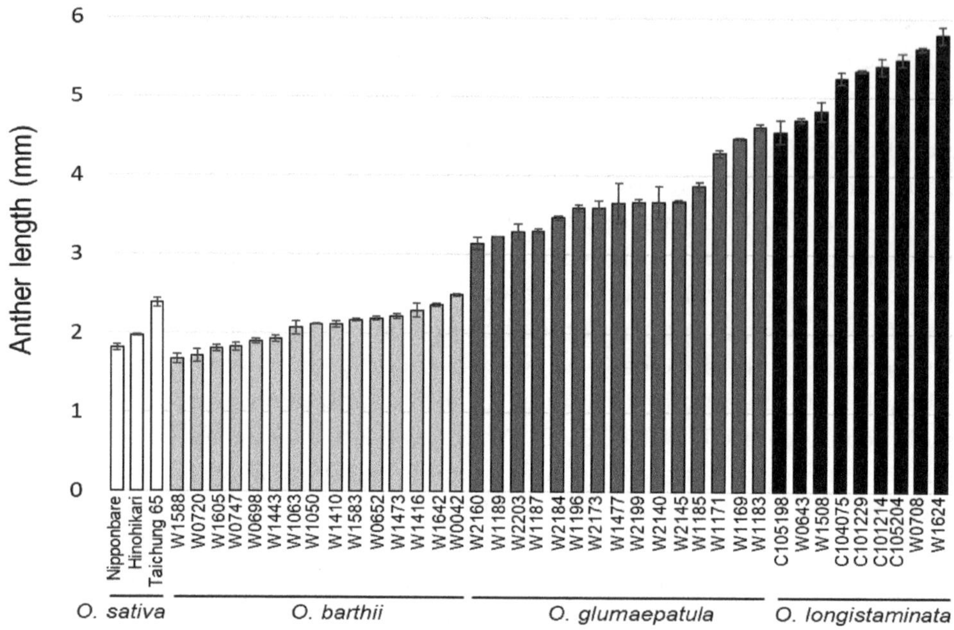

Figure 1. Distribution of anther length (mm) in cultivars of *O. sativa* and accessions of *O. barthii*, *O. glumaepatula*, and *O. longistaminata*. Mean ± SE, n = 3.

We developed 372 BC_3F_1 plants by recurrent backcrossing to F_1, BC_1, and BC_2 plants to develop the CSSLs of W1508 (Figure 2). We genotyped 100 BC_3F_1 plants, and selected 26 BC_3F_1 plants that covered as much of the W1508 genome as possible, to minimize the number of candidate CSSLs (Figure S1). Simple interval mapping (n = 100) suggested that *O. longistamina* alleles at one QTL on chromosome (Chr.) 3, one QTL on Chr. 5, and two QTLs on Chr. 6 increased the anther length (Table S1). The other peaks were below the experiment-wise threshold levels of the logarithm of odds (LOD) at $LOD_{\alpha = 0.05} = 2.65$ and $LOD_{\alpha = 0.01} = 3.40$. Multiple QTL analysis using forward/backward model selection also suggested that these four QTLs additively increase anther length without epistasis, which explains more than half (55.5%) of the phenotypic variation in the BC_3F_1 population (Figure S2, Table S1).

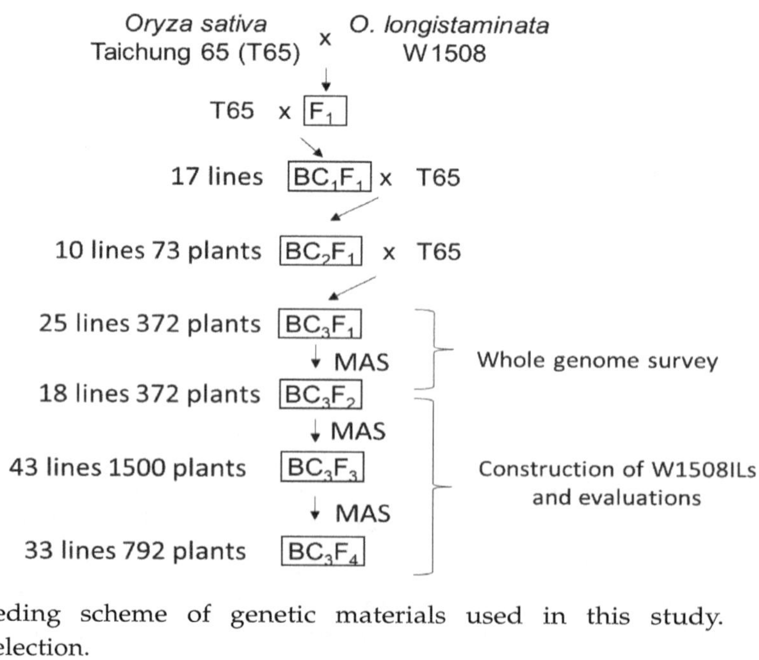

Figure 2. Breeding scheme of genetic materials used in this study. MAS represents marker-assisted selection.

To construct CSSLs, and to perform regional QTL analysis using 21 to 24 plants of the BC_3F_2 populations derived from the selected 26 BC_3F_1 plants covering the whole genome, we defined a heterozygous "target region" as a genomic region in BC_3F_1 plants for the fixation of the W1508-derived segments on the homozygous condition in the progeny (Figure S1). The anther length of T65 was 2.33 mm and that of W1508 was 4.33 mm. The mean anther length of the 26 BC_3F_2 lines ranged from 2.23 to 2.85 mm. Nine BC_3F_2 lines had a significantly higher mean anther length than T65, but none had a significantly lower mean anther length (Figure 3). Our working hypothesis was that the W1508 segments in the 9 BC_3F_2 lines with a higher mean anther length (boxed in Figure S1) most likely had QTLs associated with longer anther length. Simple interval mapping in the 26 BC_3F_2 populations at the target regions detected 13 minor QTLs on Chrs. 1, 2, 3, 4, 5, 6, 9, 10, and 11 (Table S2).

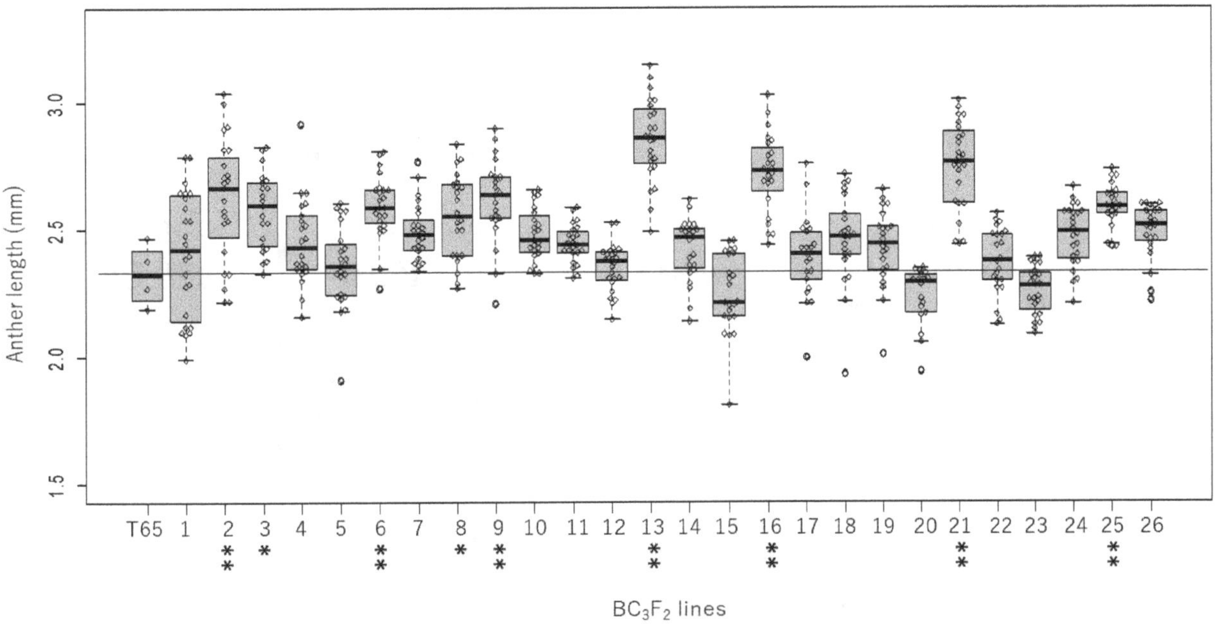

Figure 3. Box plots of anther length among the 26 BC_3F_2 populations. A thin horizontal line shows the average anther length in the Taichung 65 (T65) genetic background. Thick horizontal bars represent average values. Ranges between upper and lower quantiles are indicated by grey boxes. Maximum and minimum values excluding outlier are indicated by upper and lower whiskers, respectively. * and ** represent significant differences at P = 0.05 and P = 0.01, respectively.

2.2. Validation of QTL

For the identification of major QTLs conferring anther length, we conducted QTL analysis on BC_3F_2 13, 16, and 21 families that had the highest anther length means of 2.88 mm, 2.82 mm, and 2.68 mm, respectively (Figure 4). In the BC_3F_2 21 population (n = 44), there was segregation on chromosomes 3, 6, and 11 (Figure 4a). The QTL positively regulating anther length located between *RM3436* and *RM5959* on chromosome 3 was named *qATL3.3*. It was validated using the more advanced BC_3F_4 6 (n = 96) generation at 151.5 cM on chromosome 3, with the LOD value 18.27. *qATL3.3* explained 58.4% of the phenotypic variance in the BC_3F_4 population, with 0.17 mm of additive effect and 0.03 mm of dominant effect (Table S2).

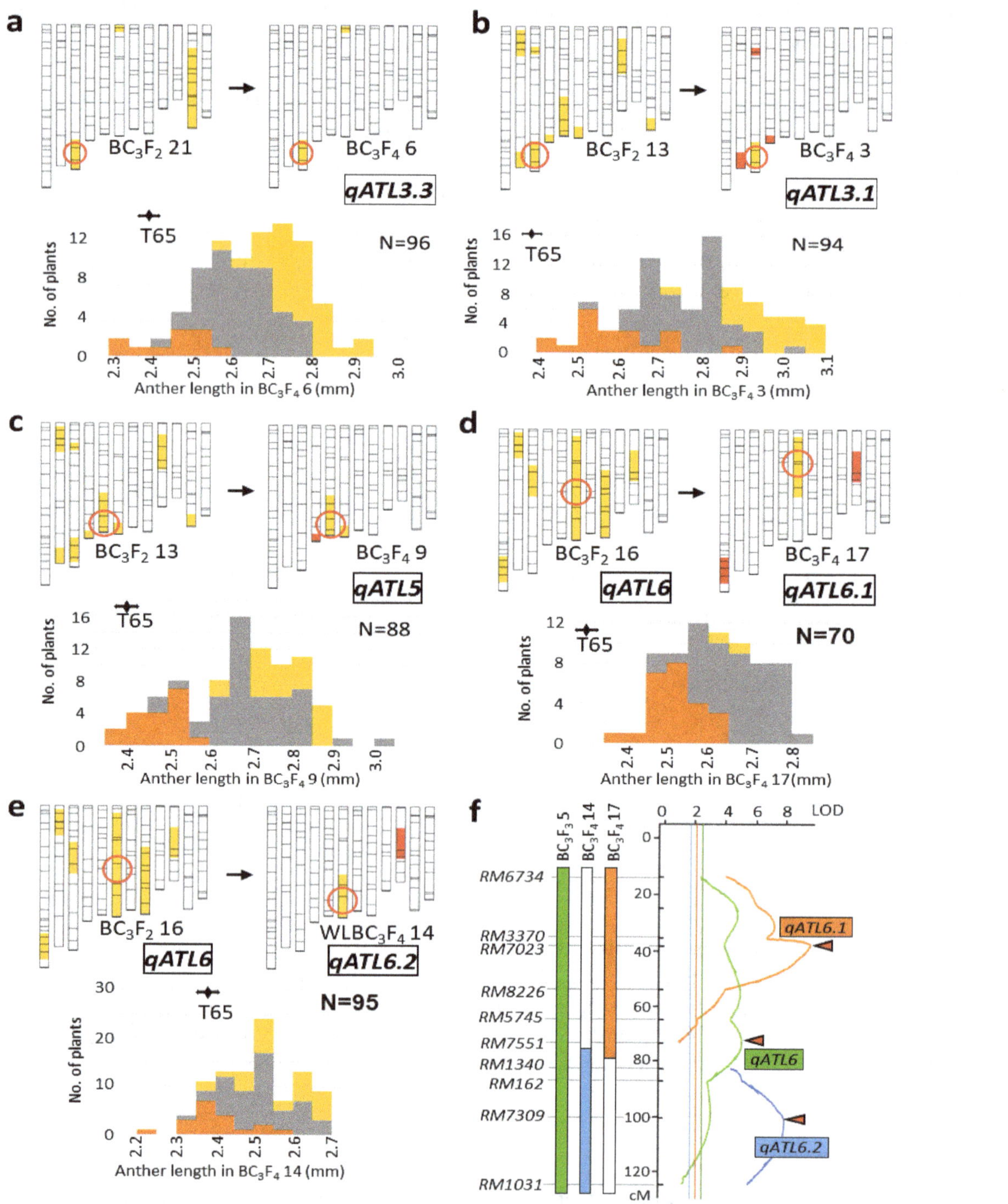

Figure 4. Validation of the anther length quantitative trait locus (QTLs) located on chromosomes 3, 5, and 6 in BC_3F_4 populations segregating at only one QTL region. (**a**)–(**e**) Graphical genotypes of the parents of BC_3F_3 and BC_3F_4 populations, and frequency distributions of anther length in BC_3F_4 populations for (**a**) *qATL3.3*, (**b**) *qATL3.1*, (**c**) *qATL5*, (**d**) *qATL6.1*, and (**e**) *qATL6.2*. Graphical genotypes: Yellow, heterozygous; red, homozygous for the W1508 allele. Frequency distributions shown in colors by genotypes at *RM5959* (**a**), *RM3525* (**b**), *RM1054* (**c**), *RM7023* (**d**), and *RM7309* (**e**): Orange, homozygous for T65; yellow, homozygous for W1508; gray, heterozygous; (**f**) logarithm of odds (LOD) curves in simple interval mapping of anther length for detection of *qATL6* (green) in BC_3F_3 5, *qATL6.1* (orange) in BC_3F_4 17, and *qATL6.2* (blue) in BC_3F_4 14.

Among the 71 plants in the BC_3F_2 13 population, the range of anther length was 2.26–3.32 mm, with a population mean of 2.82 mm. Simple interval mapping revealed two QTL: *qATL3.1* at 137.6 cM on Chr. 3, and *qATL5* at 122 cM on Chr. 5. Segregation distortion at *qATL5* was detected at the closest marker, *RM1054*, with a ratio of 22 T65 homozygotes: 43 T65 heterozygotes: 6 W1508 homozygotes. *qATL3.1* and *qATL5* were each independently validated in the more advanced BC_3F_4 3 and nine populations, respectively, derived from the BC_3F_2 13 population (Figure 4b,c). *qATL3.1* was detected again at 142.5 cM on Chr. 3, with an LOD score of 15.70, an additive effect of 0.19 mm, and a dominance effect of −0.002 mm, explaining 47.6% of the phenotypic variance (Table S2). *qATL3.3* and *qATL3.1* seemed to be identical QTLs, because substitution of the *O. longistaminata* chromosome segment ranged between the SSR markers *RM3525* and *RM5959*. We unified the QTL names *qATL3.3* and *qATL3.1* as the single QTL *qATL3.1*. In BC_3F_4 9, *qATL5* was detected again at 121.4 cM on the long arm of Chr. 5, with an LOD score of 17.28, an additive effect of 0.16 mm, and a dominance effect of 0.08 mm, explaining 59.6% of the phenotypic variation.

The BC_3F_2 16 population (n = 86) segregated at genomic regions on Chrs. 1, 2, 3, 6, 8, and 10, with anther lengths ranging from 2.21 to 3.10 mm and a population mean of 2.68 mm (Figure 4d,e). In the QTL analysis, the QTLs derived from *O. longistaminata* on Chr. 6 (*qATL6*) increased anther length. Because the genomic region of *qATL6* ranged widely from 13.8 to 114.8 cM, with LOD scores greater than the 5% significance level, we assumed that multiple QTLs on Chr. 6 were linked (Figure 4f). We examined two progeny lines of the BC_3F_2 16 population, BC_3F_4 17 and 14, that had the separated *O. longistaminata* chromosome segments on the short and long arms, respectively. In BC_3F_4 17, QTL *qATL6.1* was detected at 38.3 cM on Chr. 6S (Figure 4d,f), which, with an additive effect of 0.06 mm and a dominance effect of 0.08 mm, explained 48.8% of the phenotypic variation in anther length. *RM7023* segregation distortion of the nearest-neighbor marker, *RM7023*, gaved at a ratio of 25 T65 homozygotes: 47 T65 heterozygotes: 1 W1508 homozygotes. In BC_3F_4 14, a QTL designated as *qATL6.2* was detected on the long arm of Chr. 6 (Figure 4e,f) at *RM7309*, with an LOD score of 7.91, an additive effect of 0.08 mm, and a dominance effect of 0.04 mm.

2.3. Characterization of Anther Length QTL Using Near-Isogenic Lines

To understand the cytological characteristics of the anther length in near-isogenic lines (NILs) for *qATL3.1*, *qATL5*, *qATL6.1*, and *qATL6.2*, we investigated the longitudinal lengths of epidermal cells (cell length, CL) in the mid-anther region. The anther length of the NILs for *qATL3.1*, *qATL5*, *qATL6.1*, and *qATL6.2* were significantly longer than those of T65, according to Dunnett's multiple comparison (Figure 5a). The CL of the *qATL3.1* NIL was significantly longer than that of T65, but that of the *qATL5* and *qATL6.1* NILs was not different (Figure 5b). The average CL in the *qATL6.2* NIL was significantly smaller than that in T65. These results suggest that the positive effect of *qATL3.1* on anther length was due primarily to cell elongation in the longitudinal direction, whereas that of the other QTLs, particularly *qATL6.2*, might have been due to cell number. Next, we investigated the anther length and pollen grain numbers (PGN) of T65, W1508, and the QTL NILs. PGNs increased in the order of 1910.0 in T65, 2091.1 in the *qATL6B* NIL, 2598.1 in the *qATL6.1* NIL, 2690.1 in the *qATL5* NIL, 3058.8 in the *qATL3.1* NIL, and 6862.2 in W1508.

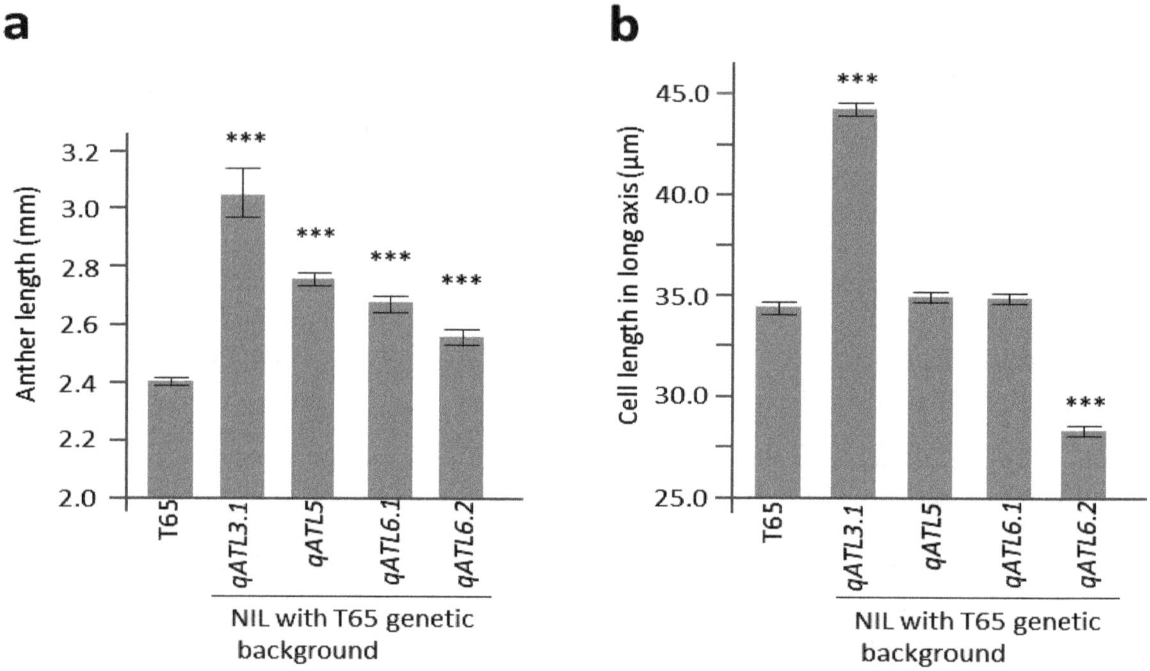

Figure 5. Characterization of QTLs using nearly isogenic lines (NIL) carrying W1508 chromosomal segments around QTL regions detected in this study. (**a**) Anther length (mm) in the NILs. Mean ± SE, n = 3; (**b**) cell length (µm) in the long axis of anthers in the NILs. * and ** represent significant differences in Dunnett's multiple comparisons to T65 control at 5% and 1% levels, respectively. The numbers of observed cells were 1460, 997, 953, 879, and 1033 in *qATL3.1* NIL, *qATL5* NIL, *qATL6.1* NIL, *qATL6.2* NIL, and T65 respectively.

2.4. Evaluation on W1508Ils

The target chromosome segments of W1508 selected in BC_3F_1 were fixed to homozygotes for the W1508 genes in BC_3F_2, BC_3F_3, BC_3F_4, and BC_3F_5, to construct the W1508ILs (Figure 6). We selected 32 lines covering the whole genomic region of W1508 using 125 SSR markers. Introgression of W1508 chromosomes did not occur at *RM3148* on Chr. 1; *RM5635* on Chr. 4; *RM2381* on Chr. 7; *RM8219* on Chr. 9; *RM7557*, *RM26090*, and *RM5918* on Chr. 11; or *RM3455*, *RM7195*, *RM1261*, and *RM309* on Chr. 12. The mean anther length in four or five plants increased significantly in W1508ILs 1, 2, 3, 7, 8, 9, 15, and 29, and decreased significantly in W1508IL 20 (Table S3). W1508ILs 8 and 9 carried *qATL3*, and W1508IL 15 carried *qATL5*. However, the mean anther length in W1508IL 17 (carrying *qATL6.1*) or 18 (carrying *qATL6.2*) did not increase significantly in five plant samples, suggesting that four or five replicates were not sufficient to detect the QTLs, despite being major QTL, for increased anther length. The minor QTLs detected in the BC_3F_2 populations were likely associated with the mean anther length in W1508ILs 1 and 2 at *RM3604*, and in W1508IL 7 at *RM5755*.

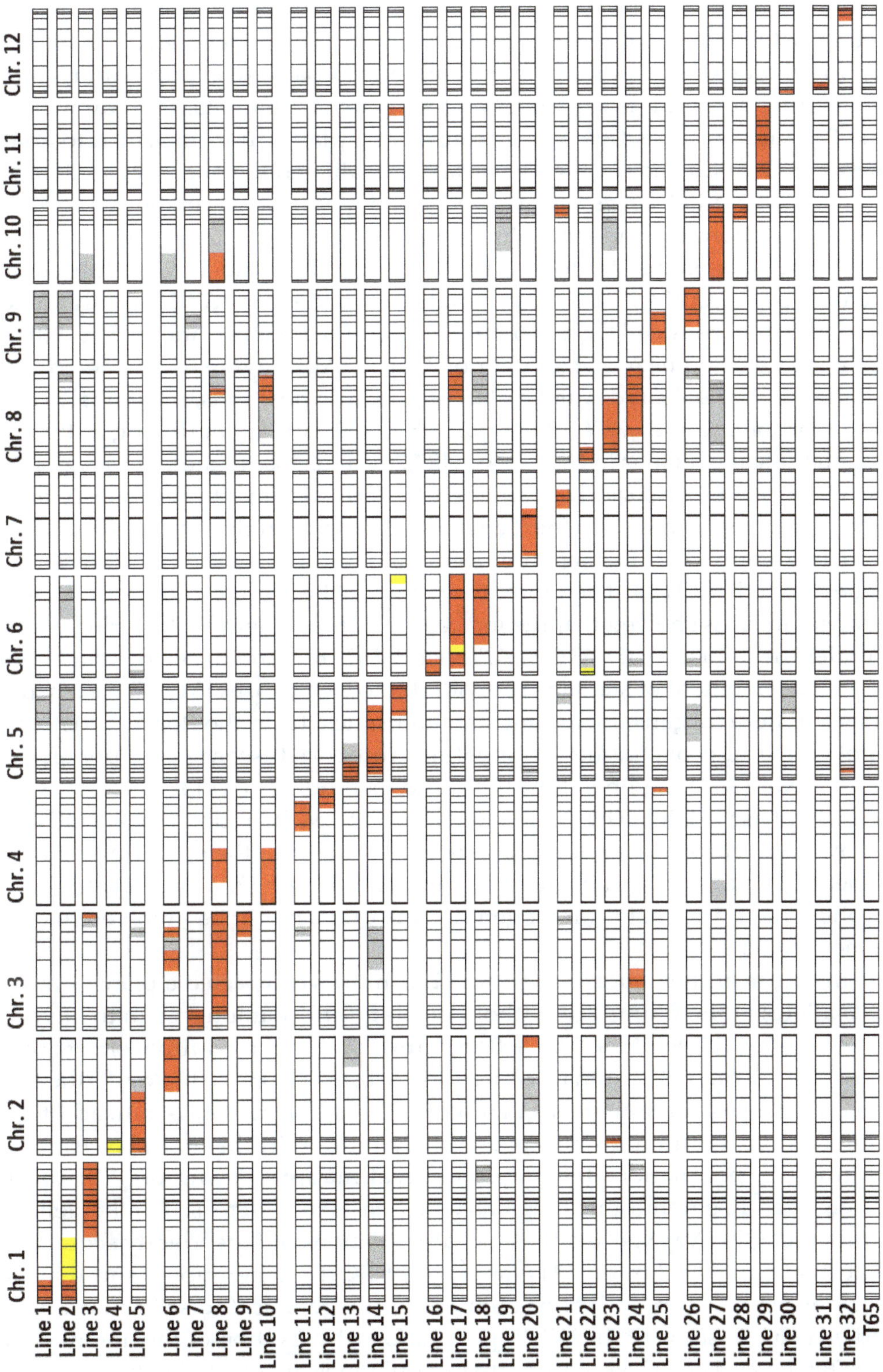

Figure 6. Construction of chromosome segment substitution lines of *O. longistaminata* accession W1508 (W1508ILs) in the genetic background of *O. sativa* L. cv. Taichung 65. Red and white boxes represent homozygous genotypes for W1508 and Taichung 65 alleles, respectively. Yellow boxes represent heterozygous genotypes. Missing genotypes at markers showing heterozygous genotypes at BC_3F_1 generation are indicated by grey.

3. Discussion

Polygenes or QTLs control most agronomic traits. For genetic analysis of quantitative traits, segregating populations such as F_2, RIL, and advanced backcross populations allow complicated genetic architectures to be disassembled to a few genetic elements, thereby increasing the power of detection. Because our objective was to construct a platform from which to analyze the genetic difference between *O. sativa* and *O. longistaminata*, we evaluated the phenotypic character anther length, which is a key character in *O. longistaminata*, and genotyped BC_3 populations and W1508ILs. We used genome-wide QTL exploration and model fitting procedures, and determined that phenotypic variation in anther length was explained by four QTLs on Chrs. 3, 5, and 6. Our subsequent single-factor segregation analyses of QTLs in BC_3F_4 populations validated *qATL3.1* (with an additive effect of 0.17 mm), *qATL3.3* (0.19 mm), *qATL5* (0.16 mm), *qATL6.1* (0.06 mm), and *qATL6.2* (0.08 mm). *qATL3.1* and *qATL3.3* were detected near one another on Chr. 3. Although we considered these two QTLs to be identical, their LOD peaks indicated different genomic positions, most likely owing to the slight segregation distortion around *qATL3.3* (df = 2, P = 0.006). QTLs regulating anther length were also reported around the same regions in *O. rufipogon* [27,28], and close to the *qATL5* region in F_2 populations derived from *O. sativa* × *O. rufipogon* [21], and in RILs derived from *O. sativa* subsp. *indica* × *O. rufipogon* W1944 [28,29]. QTLs that increase anther length have also been detected on Chr. 6 [27,28]. However, these previously identified QTLs have not been validated in *O. sativa* backcross populations in the absence of effects of other QTLs. Therefore, they may not be identical to the QTLs found here.

QTL analysis in NILs showed that the increase in anther length at flowering was likely due to cell elongation under the influence of *qATL3*, *qATL5*, and *qATL6.1*, and to cell division under the influence of *qATL6B*. The development of stamens is divided into two phases [31]. The early phase features the morphogenesis of floral organs, including stamen tissues, from the floral meristem. The late phase comprises pollen maturation, filament elongation, anther dehiscence, and pollen release, which are modulated by auxin (AUX), gibberellin (GA), and jasmonate (JA) [32–35]. AUX and GA can also regulate the early phase [34,36]. The expression of a MYB transcription factor, *HvGAMYB*, in barley, is upregulated by GA in early anther development in the nuclei of the epidermis, endothecium, middle layer, and tapetum, leading to sterility and shorter anther length [37]. Because AUX, GA, and JA affect stamen morphogenesis and male gametogenesis, the QTLs might be involved in epidermal cells via plant hormone signaling.

Because the mean anther length in T65 was 2.27 mm, we expected a plant homozygous for the *O. longistaminata* alleles at the four QTLs without epistatic effect among them to have an anther length of 3.21 mm (2.27 + 2 (0.17 + 0.16 + 0.06 + 0.08) mm), due to the additive effects of *qATL3.1*, *qATL5*, *qATL6.1*, and *qATL6.2*, or, of 3.25 mm (2.27 + 2 (0.19 + 0.16 + 0.06 + 0.08) mm) due to *qATL3.3*, *qATL5*, *qATL6.1*, and *qATL6.2*. As the anther length of W1508 was longer at 4.33 mm, additional, unidentified QTLs and interactions with minor effects are likely hidden in the *O. longistaminata* genome. QTL exploration in BC_3F_2 populations by simple interval mapping using only 21 to 24 plants in each population suggested 10 chromosomal regions with effects that increased anther length, and three that decreased anther length. We expected that the target regions independently segregated with *qATL3.1*, *qATL5*, *qATL6.1*, and *qATL6.2* in the QTL analysis of BC_3F_2 generations, and that the QTL analysis nested in each of the BC_3F_2 populations eliminated the genetic "noise" caused by *qATL3.1*, *qATL5*, *qATL6.1*, and *qATL6.2* in some of the BC_3F_2 populations, by fixation of the homozygous condition for the T65 allele (Table S2). The BC_3F_2 8, 10, and 11 populations, in which the BC_3F_1 parents did not retain the major QTL regions, might segregate QTLs additively with anther length increases of 0.15, 0.10, and 0.10 mm, respectively, at the SSR markers *RM5755*, *RM3317*, and *RM5635*. The BC_3F_2 25 population did not have the *O. longistaminata* segment at the major QTL regions, but the mean anther length of the population was significantly higher than that of T65 (Figure 3). Two populations—BC_3F_2 6 and 17—also did not have the major QTL region, and so most likely had another, unidentified QTL which could not be detected in the BC_3F_1 population. We also inferred chromosomal regions associated with increasing anther length at the major QTLs on Chrs. 3, 5, and 6, but the LOD peak positions were shifted, most likely owing to the few recombinations around QTL regions in populations with only 21 to 24 individuals. This result demonstrates that analysis in BC_3F_2 populations covering the whole genome of *O. longistaminata* increased the detection power, and that it was a practical way to perform the QTL scan where whole-genome genotyping is not available, despite reduced accuracy of the inferred map positions. This practical approach will be useful in the design of future experiments. The simple sum of the additive effects of the W1508 alleles at all QTLs reached 2.14 mm when homozygous (Table S2), which almost explained the difference between W1508 and T65 (4.33 mm − 2.27 mm = 2.06 mm). This suggested that multiple QTLs functioned additively, or by interacting with each other to form anthers as *O. longistaminata*.

The genetic analysis of anther length, which likely reflected on the lifestyle of the perennial wild rice species *O. longistaminata*, revealed that at least four major QTLs, *qATL3.1*, *qATL5*, *qATL6.1*, and *qATL6.2* increased anther length. Regional QTL analysis and CSSL analysis suggested that additional minor QTLs were also associated with regulation of anther length.

4. Materials and Methods

4.1. Plant Materials

We used *O. longistaminata* accession W1508, which was originally collected in Madagascar. *Oryza sativa* L. ssp. *japonica* Taichung 65 (T65) and W1508 were crossed, to develop F_1 plants with T65 cytoplasm. The National Institute of Genetics, Mishima, Japan, kindly provided *O. longistaminata* ratton and the seeds of *O. barthii* and *O. glumaepatula* accessions, and F_1 plants derived from a cross between T65 and W1508. T65 was used as the male parent in recurrent backcrosses to develop BC_3F_1 plants (Figure 2). The BC_3F_2 population was grown in 2015 and 2016.

4.2. Measurement of Anther Length

Panicles were collected at the heading stage and fixed and preserved in 70% ethanol. All anthers from 1 spikelet, just before anthesis, were removed and placed on a glass slide and then photographed under a microscope (Axioplan, Carl Zeiss, Jena, Germany) with a digital camera (DMX-1200, Nikon, Tokyo, Japan). The lengths of 4 to 6 anthers were measured in ImageJ software version 1.8 [38]. The average anther length was used as the phenotypic value of an individual plant.

4.3. Genotyping

Adult leaves were freeze-dried (FDU-1200, Eyela, Tokyo, Japan) and ground with a Multi-bead shocker (Yasui Kikai, Osaka, Japan). Genomic DNA was extracted by the potassium acetate method [39] from ground samples. The plants were genotyped for 124 simple sequence repeat (SSR) markers evenly distributed across the rice genomes (Table S1). For each marker, a 15 μL reaction mixture consisted of 50 mM KCl, 10 mM Tris (pH 9.0), 1.5 mM $MgCl_2$, 200 μM dNTPs, 0.2 μM primers, 0.75 U of GoTaq DNA polymerase (Promega, Fitchburg, WI, USA), and ~10 ng of genomic DNA template. PCR was performed on a GeneAmp PCR System 9700 (Applied Biosystems, Foster City, CA, USA). The thermal profile was an initial denaturation at 95 °C for 5 min: 35 cycles of 95 °C for 30 s, 55 °C for 30 s, and 72 °C for 30 s; and then a final elongation step at 72 °C for 7 min. The amplified products were electrophoresed in 4% agarose gel in 0.5× TBE buffer.

4.4. QTL Analysis

Simple interval mapping by an R/qtl library was used to detect QTLs conferring anther length [40]. LOD score thresholds with a significance level of 5% were empirically estimated by 1000 permutation tests. LOD peaks that exceeded the thresholds defined the QTLs. For multiple QTL mapping in the BC_3F_1 population, a forward/backward stepwise search was used for model selection, including epistasis of two loci by using the stepwiseqtl function in the R/qtl library, with a penalized LOD score criterion to balance model fitting and model complexity.

4.5. Measurement of Numbers of Pollen Grains

Six ethanol-fixed anthers from one spikelet per plant were crushed in a 1.5-mL tube containing 10 μL of 1% I_2–KI solution, and then suspended in 390 μL of distilled water. One μL of the suspension was applied to a glass slide, and pollen grains in 10 fields were photographed under a light microscope (Axioplan, Carl Zeiss, Jena, Germany). The pollen grains were counted in each field, and the numbers were multiplied by the dilution ratio to estimate the pollen grain number per anther.

4.6. Observation of Epidermal Cells of Anthers

Anthers were put in a drop of distilled water on a glass slide, and cut at both ends, and pollen was removed by pipetting back and forth. The epidermal cells of three anthers from one spikelet per plant were photographed at the center in the long axis of the anthers, through an optical microscope (Axioplan, Carl Zeiss, Jena, Germany) with a digital camera (DMX-1200, Nikon, Tokyo, Japan), and were counted in ImageJ. Three plants per line were evaluated as a replicate each, and the average value of epidermal cell length of nine anthers was calculated.

5. Conclusions

Genetic analysis of anther length, which is likely to relate to an outcrossing lifestyle in the perennial wild rice species *O. longistaminata*, revealed that at least four major QTLs, *qATL3*, *qATL5*, *qATL6.1*, and *qATL6.2* increase anther length. The regional QTL analysis and constructed CSSL series (designated 'W1508IL') suggested additional minor QTLs associated with the regulation of anther length. The cloning and diversity analysis of genes conferring anther length QTLs promotes utilization of the genetic resources of wild species, and the understanding of haplotype evolution on the differentiation of annuality and perenniality in the genus *Oryza*.

Supplementary Materials:
Figure S1: Graphical genotypes of BC_3F_1 for candidates for introgression lines of W1508 in T65 genetic background; Figure S2: Model selections of QTLs including main and interaction effects for anther length; Table S1: QTLs detected by simple interval mapping and multiple QTL mapping in BC_3F_1 population; Table S2: Summary of simple interval mapping for anther length in BC_3F_1, BC_3F_2, and BC_3F_4 populations derived from a cross between T65 and W1508; Table S3: Anther length of W1508ILs; Table S4: SSR markers for genotyping in this study.

Author Contributions: Funding acquisition, A.Y. and H.Y.; investigation, T.O. and Y.Y.; resources, A.Y. and H.Y.; supervision, H.Y.; validation, T.O.; writing, original draft, T.O.; writing, review and editing, Y.Y.

Acknowledgments: Wild rice species accessions used in this study were distributed from the National Institute of Genetics supported by the National Bioresource Project (NBRP), AMED, Japan, and were kindly shared from Maekawa, Institute of Plant Science and Resources, Okayama University, Japan.

References

1. Friedman, J.; Rubin, M.J. All in good time: Understanding annual and perennial strategies in plants. *Am. J. Bot.* **2015**, *102*, 497–499. [CrossRef] [PubMed]
2. Kuroda, Y.; Urajrong, H.; Sato, Y.I. Population genetic structure of wild rice (*Oryza rufipogon*) in mainland southeast Asia as revealed by microsatellite polymorphisms. *Topics* **2002**, *12*, 159–170. [CrossRef]
3. Annavi, G.; Newman, C.; Buesching, C.D.; Macdonald, D.W.; Burke, T.; Dugdale, H.L. Heterozygosity-fitness correlations in a wild mammal population: Accounting for parental and environmental effects. *Ecol. Evol.* **2014**, *4*, 2594–2609. [CrossRef] [PubMed]
4. Charpentier, M.J.E.; Williams, C.V.; Drea, C.M. Inbreeding depression in ring-tailed lemurs (*Lemur catta*): Genetic diversity predicts parasitism, immunocompetence, and survivorship. *Conserv. Genet.* **2008**, *9*, 1605–1615. [CrossRef]

5. Slate, J.; Kruuk, L.E.B.; Marshall, T.C.; Pemberton, J.M.; Clutton-Brock, T.H. Inbreeding depression influences lifetime breeding success in a wild population of red deer (*Cervus elaphus*). *Proc. R. Soc. Lond. B* **2000**, *267*, 1657–1662. [CrossRef] [PubMed]
6. Acevedo-Whitehouse, K.; Vicente, J.; Gortazar, C.; Hofle, U.; Fernandez-de-Mera, I.G.; Amos, W. Genetic resistance to bovine tuberculosis in the Iberian wild boar. *Mol. Ecol.* **2005**, *14*, 3209–3217. [CrossRef]
7. David, P. Heterozygosity-fitness correlations: New perspectives on old problems. *Heredity* **1998**, *80*, 531–537. [CrossRef] [PubMed]
8. Charlesworth, D. Evolution of plant breeding systems. *Curr. Biol.* **2006**, *16*, 726–735. [CrossRef]
9. Hamrick, J.L.; Godt, M.J.W. Effects of life history traits on genetic diversity in plant species. *Philos. Trans. R Soc. Biol. Sci.* **1977**, *351*, 1291–1298. [CrossRef]
10. Vaughan, D.A.; Lu, B.R.; Tomooka, N. The evolving story of rice evolution. *Plant Sci.* **2008**, *174*, 394–408. [CrossRef]
11. Banaticla-Hilario, M.C.N.; Berg, R.G.; Hamilton, N.R.S.; McNally, K.L. Local differentiation amidst extensive allele sharing in *Oryza nivara* and *O. rufipogon*. *Ecol. Evol.* **2013**, *3*, 3047–3062. [CrossRef]
12. Liu, R.; Zheng, X.M.; Zhou, L.; Zhou, H.F.; Ge, S. Population genetic structure of *Oryza rufipogon* and *Oryza nivara*: Implications for the origin of *O. nivara*. *Mol. Ecol.* **2015**, *24*, 5211–5228. [CrossRef] [PubMed]
13. Reuscher, S.; Furuta, T.; Bessho-Uehara, K.; Cosi, M.; Jena, K.K.; Toyoda, A.; Fujiyama, A.; Kurata, N.; Ashikari, M. Assembling the genome of the African wild rice *Oryza longistaminata* by exploiting synteny in closely related *Oryza* species. *Commun. Biol.* **2018**, *1*, 162. [CrossRef] [PubMed]
14. Hiroi, K.; Mamun, A.A.; Wada, T.; Takeoka, Y. A Study on the interspecific variation of spikelet structure in the genus *Oryza*. *Jpn. J. Crop Sci.* **1991**, *60*, 153–160. [CrossRef]
15. Zhang, Y.; Zhang, S.; Liu, H.; Fu, B.; Li, L.; Xie, M.; Song, Y.; Li, X.; Cai, J.; Wan, W.; et al. Genome and comparative transcriptomics of African wild rice *Oryza longistaminata* provide insights into molecular mechanism of rhizomatousness and self-incompatibility. *Mol. Plant* **2015**, *8*, 1683–1686. [CrossRef]
16. Chu, Y.E.; Morishima, H.; Oka, H.I. Partial self-incompatibility found in *Oryza perennis* subsp. *barthii*. *Gene Genet. Sys.* **1969**, *44*, 225–229. [CrossRef]
17. Sahadevan, P.C.; Namboodiri, K.M.N. Natural crossing in rice. *Proc. Indian Acad. Sci. Sect. B* **1963**, *58*, 176–185.
18. Oka, H.I.; Morishima, H. Variations in the breeding systems of wild rice, *Oryza perennis*. *Evolution* **1967**, *21*, 249–258. [CrossRef]
19. Sakai, K.I.; Narise, T. Studies on the breeding behavior of wild rice. *Annu. Rep. Natl. Inst. Genet. Jpn.* **1959**, *9*, 64–65.
20. Eshed, Y.; Zamir, D. A genomic library of *Lycopersicon pennellii* in *L. esculentum*: A tool for fine mapping genes. *Euphytica* **1994**, *79*, 175–179. [CrossRef]
21. Eshed, Y.; Zamir, D. An introgression line population of *Lycopersicon pennellii* in the cultivated tomato enables the identification and fine mapping of yield associated QTL. *Genetics* **1995**, *141*, 1147–1162. [PubMed]
22. Eshed, Y.; Zamir, D. Less-than-additive epistatic interactions of quantitative trait loci in tomato. *Genetics* **1996**, *143*, 1807–1817. [PubMed]
23. Chetelat, R.T.; Meglic, V. Molecular mapping of chromosome segments introgressed from *Solanum lycopersicoides* into cultivated tomato (*Lycopersicon esculentum*). *Theor. Appl. Genet.* **2000**, *100*, 232–241. [CrossRef]
24. Grandillo, S.; Ku, H.M.; Tanksley, S.D. Characterization of *fs8.1*, a major QTL influencing fruit shape in tomato. *Mol. Breed.* **1996**, *2*, 251–260. [CrossRef]
25. Yamamoto, T.; Kuboki, Y.; Lin, S.Y.; Sasaki, T.; Yano, M. Fine mapping of quantitative trait loci *Hd-1*, *Hd-2* and *Hd-3*, controlling heading date of rice, as single Mendelian factor. *Theor. Appl. Genet.* **1998**, *97*, 37–44. [CrossRef]
26. Monforte, A.J.; Tanksley, S.D. Fine mapping of a quantitative trait locus (QTL) from *Lycopersicon hirsutum* chromosome 1 affecting fruit characteristics and agronomic traits: Breaking linkage among QTLs affecting different traits and dissection of heterosis for yield. *Theor. Appl. Genet.* **2000**, *100*, 471–479. [CrossRef]
27. Xiong, L.Z.; Liu, K.D.; Dai, X.K.; Xu, C.G.; Zhang, Q.F. Identification of genetic factors controlling domestication related traits of rice using an F_2 population of a cross between *Oryza sativa* and *O. rufipogon*. *Theor. Appl. Genet.* **1999**, *98*, 243–251. [CrossRef]

28. Cai, W.; Morishima, H. (2002) QTL clusters reflect character associations in wild and cultivated rice. *Theor. Appl. Genet.* **2002**, *104*, 1217–1228. [CrossRef]
29. Uga, Y.; Fukuta, Y.; Ohsawa, R.; Fujimura, T. Variations of floral traits in Asian cultivated rice (*Oryza sativa* L.) and its wild relatives (*O rufipogon* Griff.). *Breed. Sci.* **2003**, *53*, 345–352. [CrossRef]
30. Uga, Y.; Siangliw, M.; Nagamine, T.; Ohsawa, R.; Fujimura, T.; Fukuta, Y. Comparative mapping of QTLs determining glume, pistil and stamen sizes in cultivated rice (*Oryza sativa* L.). *Plant Breed.* **2010**, *129*, 657–669. [CrossRef]
31. Goldberg, R.B.; Beals, T.P.; Sanders, P.M. Anther development: Basic principles and practical applications. *Plant Cell* **1993**, *5*, 1217–1229. [CrossRef] [PubMed]
32. McConn, M.; Browse, J. The critical requirement for linolenic acid is pollen development, not photosynthesis, in an Arabidopsis Mutant. *Plant Cell* **1996**, *8*, 403–416. [CrossRef] [PubMed]
33. Cecchetti, V.; Altamura, M.M.; Falasca, G.; Costantino, P.; Cardarelli, M. Auxin regulates *Arabidopsis* anther dehiscence, pollen maturation, and filament elongation. *Plant Cell* **2008**, *20*, 1760–1774. [CrossRef] [PubMed]
34. Cheng, Y.; Dai, X.; Zhao, Y. Auxin biosynthesis by the YUCCA flavin monooxygenases controls the formation of floral organs and vascular tissues in *Arabidopsis*. *Genes Dev.* **2006**, *20*, 1790–1799. [CrossRef] [PubMed]
35. Jewell, J.B.; Browse, J. Epidermal jasmonate perception is sufficient for all aspects of jasmonate-mediated male fertility in *Arabidopsis*. *Plant J.* **2016**, *85*, 634–647. [CrossRef] [PubMed]
36. Plackett, A.R.; Thomas, S.G.; Wilson, Z.A.; Hedden, P. Gibberellin control of stamen development: A fertile field. *Trends Plant Sci.* **2011**, *16*, 568–578. [CrossRef] [PubMed]
37. Murray, F.; Kalla, R.; Jacobsen, J.; Gubler, F. A role for HvGAMYB in anther development. *Plant J.* **2003**, *33*, 481–491. [CrossRef]
38. Schneider, C.A.; Rasband, W.S.; Eliceiri, K.W. NIH Image to ImageJ: 25 years of image analysis. *Nat. Methods* **2012**, *9*, 671–675. [CrossRef]
39. Dellaporta, S.L.; Wood, J.; Hicks, J.B. A plant DNA minipreparation: Version II. *Plant Mol. Biol. Rep.* **1983**, *1*, 19–21. [CrossRef]
40. Broman, K.W.; Wu, H.; Sen, S.; Churchill, G.A. R/qtl: QTL mapping in experimental crosses. *Bioinformatics* **2003**, *19*, 889–890. [CrossRef]

12

Durum Wheat Breeding in the Mediterranean Region

Ioannis N. Xynias [1,*,†], Ioannis Mylonas [2,†], Evangelos G. Korpetis [2], Elissavet Ninou [3], Aphrodite Tsaballa [2], Ilias D. Avdikos [4] and Athanasios G. Mavromatis [4]

1 School of Agricultural Sciences, University of Western Macedonia, Terma Kontopoulou, 53100 Florina, Greece
2 Institute of Plant Breeding and Genetic Resources, Hellenic Agricultural Organization–"Demeter", 57001 Thessaloniki, Greece; ioanmylonas@yahoo.com (I.M.); korpetis@yahoo.gr (E.G.K.); tsaballa80@gmail.com (A.T.)
3 Institute of Industrial and Fodder Crops, Hellenic Agricultural Organization Demeter–"Demeter", 41335 Larissa, Greece; lisaninou@gmail.com
4 Laboratory of Genetics and Plant Breeding, School of Agriculture, Aristotle University of Thessaloniki, 54124 Thessaloniki, Greece; avdikos.elias@gmail.com (I.D.A.); amavromat@agro.auth.gr (A.G.M.)
* Correspondence: ixynias@teiwm.gr or ioannis_xynias@hotmail.com
† These authors contributed equally to this work.

Abstract: This brief historical review focuses on durum wheat domestication and breeding in the Mediterranean region. Important milestones in durum wheat breeding programs across the countries of the Mediterranean basin before and after the Green Revolution are discussed. Additionally, the main achievements of the classical breeding methodology are presented using a comparison of old and new cultivars. Furthermore, current breeding goals and challenges are analyzed. An overview of classical breeding methods in combination with current molecular techniques and tools for cultivar development is presented. Important issues of seed quality are outlined, focusing on protein and characteristics that affect human health and are connected with the consumption of wheat end-products.

Keywords: Mediterranean basin; durum wheat; breeding; MAS; biochemical markers; quality

---

1. Introduction

Durum wheat (*Triticum turgidum* subsp. *durum* (Desf.) Husnot) is one of the most essential cereal species and is cultivated worldwide over almost 17 million ha, with a global production of 38.1 million tonnes in 2019 [1]. The largest producer is the European Union, with 9 million tonnes in 2018, followed by Canada, Turkey, United States, Algeria, Mexico, Kazakhstan, Syria, and India [2–6]. Durum wheat production and cultivation areas are concentrated in the Mediterranean. Moreover, the countries of the Mediterranean basin are the largest importers and the largest consumers of durum wheat products (flour, pasta, and semolina). Among European Union (EU) countries, Italy is considered the leader of durum wheat production, with an average production of 4.26 million tonnes in the last decade (1.28 million ha growing area), followed by France with 1.89 million tonnes (0.37 million ha), Greece with 1.07 million tonnes (0.37 million ha), and Spain with 0.98 million tonnes (0.38 million ha) (Table 1) [2]. Since durum wheat is mainly grown under rain-fed conditions in the Mediterranean basin, its productivity is profoundly affected by rainfall and biotic (pests and diseases) and abiotic (drought, sunlight, cold, and salinity) stresses.

Table 1. The world's leading durum wheat producing countries (2010–2019) [2–4].

Country	Average Production (Millions of Tonnes)
Canada	5.2
Italy	4.3
Turkey	3.7
USA	2.3
Kazakhstan	2.2
Syria	2.2
Algeria	2.2
France	1.9
Morocco	1.8
Greece	1.1
Spain	1.0
Tunisia	1.0

2. A historical Review of Durum Wheat Breeding

2.1. *Prehistory and Early History*

Wheat cultivation and human civilization have evolved together for at least 10,000 years since humans first attempted to produce food during the "Neolithic Revolution" [7]. The first step in the improvement of durum wheat involved the domestication of its wild progenitors [8] in the hilly area of southwest Asia at the Tigris and Euphrates basin (ancient Mesopotamia) and in the mountains of Iran, Turkey, Syria, and Jordan, in the area often referred to as the "Fertile Crescent" [9,10].

According to Shewry [7], an early and primitive form of plant breeding was carried out with the first selections from wild populations of *T. turgidum* subsp. *dicoccoides* (Körn. ex Asch. and Graebner) Thell. based primarily on yield, but also considering other genetic characteristics; mainly non-brittle rachis and free-threshing naked kernels. It has been proven that the first trait is controlled by two major genes, *brittle rachis 2 (Br-A2)* and *brittle rachis 3 (Br-A3),* on the short arms of chromosomes 3A and 3B, respectively [11], while the free kernels originated from a dominant mutation at the Q locus [7,12].

The cultivated emmer (*T. turgidum* subsp. *dicoccum* (Schrank ex Schübler) Thell.) was the first dominant wheat in Asia, Africa, and Europe in the early years of agriculture, in the seventh millennium BC. Emmer grains were found in the tombs of the Egyptian Pyramids and were widespread in ancient Babylon and Central Europe [13]. Seeds of emmer were discovered in archaeological finds dating back to between 7500 and 6700 BC [10]. It remained a dominant cereal until the first millennium BC when it was replaced by free-threshing durum wheat [14]. Its grains have been found in eight archaeological sites of the Iron Age, dating from the end of the 2nd millennium BC to the end of the 4th century BC in Northern Greece [15]. The first agricultural book on wheat was written using cuneiform on a clay tablet around 1700 BC and was discovered in Israel in 1950 [13].

The written history of wheat science began 2500 years ago when the Greek botanist Theophrastus (371–287 BC) wrote the study "Enquiry into Plants". Later, the famous Roman writers Columella, Varrone, Virgil, and Pliny the Elder described wheat as the primary agricultural food source in the Mediterranean region [16].

2.2. *Modern History*

The modern history of durum wheat breeding in the Mediterranean region began in the early 20th century and was affected by the global evolution of agricultural science. A high yield, good end-use traits, and resistance to abiotic and biotic stresses have always been targets for wheat breeders. The initial approach in durum wheat breeding focused on the exploitation of local genetic resources. Later, the Green Revolution resulted in the release of short, high-yielding durum wheat cultivars from International research institutions, which are used extensively in crosses in almost all national breeding programs.

2.3. The Early Period, before the Green Revolution

At the dawn of the 20th century, the first systematic breeding efforts were made by public research institutes in the countries across the Mediterranean basin that pioneered plant breeding at that time. Initially, landraces were used as the starting materials in breeding programs (Table 2). Later, systematic breeding schemes were gradually applied using parent cultivars with different useful agronomic characteristics, such as high yield, good quality, and resistance to a range of disease.

Table 2. The main Mediterranean wheat landraces.

Country	Reference
Italy	
Biancuccia, Bivona, Castiglione, Ciciredda, Cotrone, Duro Lucano, Farro Lungo, Gioia, Regina, Ruscia, Sammartinara, Timilia	[17]
Greece	
Roussias, Asprostachys, Tsipoura, Asprostaro, Diminitis, Trimini, Gremmenia, Kondouzi, Mavragani, Zochovis, Deves, Rovaki, Arnaouti, Kokkinostaro, Floritsa, Mavrostaro	[18,19]
Turkey	
Uveyik, Sahman, Bagacak, Sorgul, Havrani, Menceki, Iskenderi, Kocabugday, Cambudayi, Kibris bugdayi, Amik, Akbasak, Karabasak, Karakilcik, Kunduru, Sari Bursa, Sari Basak	[20,21]
Algeria	
Tuzelle, Mahon, Biskri, Bidi 17	[22]
Morocco	
ML 19, ML 21, ML 22, ML 23, ML 24, ML 26, ML 28, ML 48, ML 49, ML 32, ML 33, ML 34, ML 35, ML 36, ML 37, ML 38, ML 39, ML 41, ML 42, ML 43, ML 44, ML 45	[23]
Tunisia	
Hamira, Richi, Mahmoudi, Jenah Khotifa	[24]

In Italy, breeding was initially done through public research institutes, universities, and departments from the Ministry of Agriculture [25]; later, private companies such as the Società Italiana Sementi and Società Produttori Sementi Bologna became involved [26]. The pioneer in the modern durum wheat breeding was the Italian Nazareno Strampelli (1866–1942). Strampelli began his efforts with selections of local landraces from southern Italy, the Italian islands, and the Mediterranean region. In 1923, Strampelli released the cultivar "Senatore Capelli", which he had selected in 1915 from the local North African landrace "Jean Retifah". This cultivar was a landmark for the cultivation of durum wheat in Italy, as it covered 60% of Italy's durum wheat acreage for many decades, but also throughout the Mediterranean, where it has been widely used for crosses [16,27]. Casale, an Italian breeder, crossed cv. "S. Capelli" with Palestinian cv. "Eiti 6" in the 1940s and released the cultivar "Capeiti 8" which replaced "S. Capelli" in durum wheat cultivation in 56% of acreages due to its higher yield, although its seed quality was low. Another important breeder, Forlani, saw the possibility of improving durum wheat using interspecific crosses, particularly to introduce resistance to diseases [16,28].

In Greece, modern wheat breeding was started in 1923 by Ioannis (Juan) Papadakis (1903–1997), who founded the Institute of Plant Breeding, the first Research Center in the Balkan Peninsula. Papadakis introduced the new methods of the era, such as selection and crossbreeding, and conducted experiments in different locations using different controls and replications to evaluate the wheat's genetic material. Papadakis started by inserting selections into local breads and durum wheat landraces. In 1924, he made the first crosses by selecting parents from both local and foreign landraces (Table 3) [29]. In 1929, Papadakis recorded and described in detail the genetic material of the wheat found in Greece, according to Professor Percival's classification [30]. This study included 47 local cultivars or landraces of durum wheat cultivated in Greece [18]. The first durum wheat cultivars were released in 1932 [31].

The local durum wheat landraces were cultivated until 1930 were then replaced gradually by the new improved cultivars. The crosses of the plant breeding program of the Institute of Plant Breeding were produced in 1946, resulting in the cultivars "Methoni" and "Electra", which had better quality and featured earlier maturation by one week compared to "Lemnos". After 30 years, the new improved cultivars covered 60% of the durum wheat cultivated area in Greece, with the improved cv. "Lemnos" occupying 42% of the area [32].

Table 3. The main Mediterranean cultivars released before the Green Revolution.

Country	Name	Year of Release	Pedigree, Cross Name, Origin	Agronomic Characteristics	Reference
Italy	Senatore Capelli	1923	Selection from landrace "Jean Retifah"	Waxy, tall, rather late ripening, good quality	[16,17,27]
	Capeiti 8 (=Patrizio 6)	1955	S. Capelli × Eiti 6	Better yield and poorer quality than S. Capelli	[16,17,28,33]
	Sincape 9		Sinai × S. Capelli	Agronomical characteristics superior to S. Capelli	[16,28,33]
	Appulo	1964	(S. Cappelli × Grifoni 235) × Capeiti 8	Optimal grain quality and yield	[16,28,33]
	ISA-1	1971	Patrizio × Sassari 0130	Very early	[28]
Greece	Lemnos	1932	Selection from Landrace Akbasak	High yield, good quality for bread	[31]
	Methoni	1954	Lemnos × 7-B-1231	Better quality and earlier maturing than Lemnos	[32]
	Electra	1957	S. Capelli × [Lemnos × (Florence × Arditto) × Sinai2]	Better quality and earlier maturing than Lemnos	[32]
Spain	Andalucia 344	Before 1940	Selection from Manchón de Alcalá la Real		[34,35]
	Hibrido D	Before 1940	S. Capelli/Colorado dc Jerez		[34]
	Ledesma	Before 1940	S. Capelli/Rubio dc Belalázar		[34]
	Bidi 17	1950	Selection from Bidi or Blé Gounod, from Algeria	Tall, large grains and a weak yellow color index.	[34]
France	Bidi 17	1950	Selection from Algerian population "Oued Zenatti"	Tall, large grains and a weak yellow color index.	[26,36]
	Montferrier, Mandon			Better yield, very sensitive to leaf rust	[28,36]
	Agathe			Better yield, moderately sensitive to leaf rust	[36]
Turkey	Kunduru 1149	1967	Selection from Landrace	Tall, drought tolerant, good winter hardiness	[20,21,37]
	Berkmen 469	1970	Selection from Landrace	Tall, drought tolerant, good winter hardiness	[21,37]
Morocco	Oued.Zenati	1949	Selection from landraces	Tall	[23,38]
	Zeramek	1949	Selection from landraces		[23,38]
	Kyperounda	1956	Selection from landraces	Tall	[23,38]
Tunisia	Chili 931	1953	France		[24]
	Kyperounda	1954	Cyprus		[24]
	INRAT 69	1970	Mahamoudi981/Kyperounda		[24,39]

At this initial stage of breeding, the classification of *Triticum* species was also the subject of extensive study by Spanish researchers. In the early 19th century, Spanish botanists Clemente and Lagasca compiled the "Ceres hispanica", the first herbarium for *Triticum* species. Intensive work on the collection and conservation of durum wheat types in the Iberian Peninsula began in 1930 by Arana and was extended from 1950 to 1956 by Tellez, Prieto, and Garcia-Pozuelo. The first attempt to improve durum wheat in Spain was performed at the Agricultural Research Center of Jerez de la Frontera by Juan Bautista Camacho [40]. Based on the work that had been done in the previous years, selections were made in local durum wheat populations, and crossing programs were carried out, which resulted in the release of important cultivars, such as "Ledesma", "Andalucia 344", and "Hibrido D", which, in 1963, covered 12%, 10%, and 6% of the durum wheat cropping area in Spain, respectively [34].

The characteristic resistance of local landraces to three types of rust has been studied at the Instituto Nacional de Investigaciones Agrarias since 1954 [40].

Concerning France, until the 1960s, production and research programs have mainly been conducted in Algeria, Morocco, and Tunisia. In 1930, breeding work was conducted in Algeria by Ducellier, who identified 29 local cultivars or populations of durum wheat. Wheat research was performed and coordinated by the Institut National de la Recherche Agronomique (INRA) in Montpellier until Algeria's independence in 1962, and then from The Algerian Center for Agronomic, Scientific, and Economic Research (CARASE), or INRA, in Algeria. In the early 1970s, the production of durum wheat increased drastically due to the release of the cultivars "Bidi 17", "Oued Zenati 368", "Hedba 3", and "Mohamed Ben Bachir", which resulted from selections among local landraces [22]. Accordingly, before the 1960s, breeding efforts in Morocco were focused on collecting local durum wheat landraces, from which, after the selection programs, the cultivars "Oued Zenati", "Zeramek", and "Kyperounda" were released (from 1949 to 1956). The latter cultivars were more productive than the local landraces, but their quality was inferior [23]. A similar approach was recorded in Tunisia, where local durum wheat landraces were mainly cultivated in the country until the 1940s. These were followed by cultivars produced from selections within the aforementioned landraces [24].

The Ministry of Agriculture started a breeding program in Libya in the 1950s. Durum wheat landraces, such as "Jlail-Dib" and "Hmira", were used to develop improved varieties. During the period of 1962–1965, nine durum wheat cultivars were collected and characterized [41]. Simultaneously, breeding work for durum wheat took place in Egypt, another important North African country. In 1914, Egyptian breeders released two cultivars with resistance to high temperatures, "Dakar 49" and "Dakar 52", which were selected from local durum wheat landraces. In the 1920s, these breeders released the cultivars "Hindi D", "Hindi 62", and "Tosson" using genetic material imported from India. Since 1942, this breeding program has focused on creating cultivars with resistance to the three aforementioned types of rust [42].

In Turkey, initial wheat research began in 1925 with the establishment of the first "Seed Improvement Station" in Eskisehir. The early durum wheat cultivars that were released resulted from selections within the local durum wheat populations and included "Kunduru 414/44", "Sari Bursa 7113", "Kunduru 1149", "Berkmen 496", "Akbasak 073/44", and "Karakilcik 1133" [20,43]. Finally, in Israel, the onset of genetic improvement in durum wheat was based on selections within local landraces. Local wheat landraces, mainly durum wheat that was previously cultivated in Palestine, were described as early as the 1920s. In 1948, Kostrinsky compiled a descriptive list of cultivars and landraces [44].

2.4. Breeding Work in the Mediterranean during and after Green Revolution

A significant increase in yield was achieved in many national breeding programs through the second half of the 20th century [45]. The key to the dramatic increase in wheat yields during the 'Green Revolution' in the 1960s was the introduction of semi-dwarf genes into bread wheat, which resulted in the replacement of tall cultivars with semi-dwarf cultivars characterized by their responsiveness to inputs (e.g., fertilizers) and resistance to lodging. The primary donor of these genes was the semi-dwarf cultivar "Daruma" from Japan. Initially, the cultivar "Norin 10" was developed as a result of diallelic crosses between "Daruma" and some American cultivars. Norman Borlaug, at the Centro Internacional de Mejoramiento de Maiz y Trigo (CIMMYT), used the cross "Norin 10" x "Brevor 14" for the development of initial genetic material that was first shipped to Latin America and Southeast Asia, where it was rapidly adopted for cultivation with spectacular increases in yield [46]. The cultivar "Norin 10" has also been widely used in breeding programs around the world for the introduction of semi-dwarf genes in durum wheat, which has been recorded since 1956 [47]. This was an important milestone that affected durum wheat breeding efforts worldwide. The CIMMYT durum wheat germplasm continues to play an essential role in increasing the production and genetic gain of national agricultural research systems for developing countries. Indeed, 77% of the cultivars released in developing countries in 1991–1997 originated from crosses between CIMMYT and indigenous genotypes. Moreover, 19% of these included at least one parent from the CIMMYT collection [48].

Another critical event in the evolution of plant breeding was the foundation of the International Center for Agricultural Research in the Dry Areas (ICARDA) in 1977 in Aleppo, Syria. Initially, this center was intended to control the CIMMYT's segregated wheat genetic material and pure lines. Since 1981, the ICARDA Improvement Program has worked on crosses, and, in 2003, this program became independent [49,50]. The first successful cultivar of durum wheat resulting from ICARDA's breeding program was "Waha" [syn. Cham1, Plc/Ruff//Gta/Rtte], which was enlisted in Algeria's National Variety Catalogue in 1984. Another successful cultivar was "Om Rabi 1" [syn. Cham 5, Jori/Haurani], which was released in Morocco in 1989. As of 2018, 130 cultivars that originated from the genetic material of ICARDA's breeding program have been released in 22 countries (Table 4) [50].

A comparison between old and new durum wheat cultivars bred in different periods is useful to understand which of the yield components and other associated traits contribute to the increased productivity of new cultivars. In ICARDA, the impact of 35 years (1977–2011) of public durum wheat breeding resulted in a 0.7 genetic gain per year, mostly based on earlier flowering and an increase in the spike density [50]. In Canada and Italy, durum wheat yield improvement is based on an increased number of grains [51,52]. Similarly, new high yield CIMMYT cultivars increased their numbers of grains per m^2 as a result of the increased number of spikes per m^2 and the grains per spike [48,53]. Royo et al. [54] studied the genetic changes in durum wheat yield components and their associated traits by comparing 24 old (<1945), intermediate (1950–1985), and modern (1988–2000) Italian and Spanish cultivars. As a result, it was found that the number of grains per m^2 increased by 39% and 55% in Italian and Spanish cultivars, respectively. This increase involves a 20% increase of plants per m^2, a 29% increase of spikes per plant, and a 51% increase of grains per spike. The mean rate of increase in the yield per plant was 0.41% per year, 0.11% per year for plant density, 0.55% per year for the number of grains per m^2, 0.48% per year for the harvest index, and 0.81% per year for the plant height [54]. The number of spikelets per spike did not change, so the increased numbers of grains per m^2 was due to the higher grain set in the modern cultivars. Similar results were found in a study on durum wheat cultivars released in Italy during the 20th century, where the genetic gain was mostly based on a higher kernel number per m^2 and spikes per m^2 [52].

Apart from yield, durum wheat breeding programs significantly affect grain quality [52,55]. The evaluation of durum wheat cultivars released during different breeding eras shows that genetic improvement reduces grain protein concentration as a result of improved yields, but without affecting pasta cooking quality [53]. The lower grain protein percentage of modern cultivars is based on the dilution effect caused by their heavier grains or increased amount of carbohydrates [55,56]. However, modern cultivars have increased gluten index, showing an improvement in pasta-making quality [32,52,55,57].

Table 4. The main Mediterranean cultivars released after the Green Revolution.

Country	Name	Year of Release	Pedigree, Cross Name Origin	Agronomic Characteristics	Reference
Italy	Trinakria Creso	1973 1974	(CpB144=Castelfusano) × {[(Yaktana54-Norin10-Brevor) Capelli-63-4] Tehucan}	Good quality Good quality, resistance to *Fusarium graminearum* and brown rust	[16,26] [16,17,33]
	Simeto	1988	Capeiti 8 × Valnova	High yield, low tillering, excellent adaptability	[17,33]
	Svevo	1996	Sel. CIMMYT × Zenit sib	High quality for pasta	[16,26]
Greece	Mexicali-81	1981	Selection from Mexicali 75	High yield	[58]
	Selas	1982	Selection from Stork "S"	Good grain quality	[58]
	Elpida	2010	Sifnos × Mexicali-81	High quality and yield	[58,59]
	Thraki	2014	Simeto × Mexicali-81	High quality and yield	[58,60]

Table 4. *Cont.*

Country	Name	Year of Release	Pedigree, Cross Name Origin	Agronomic Characteristics	Reference
Spain	Claudio	1999	(Sel. Cimmyt × Durango) × (IS193B×Grazia)		[61]
	Mexa	1980	GERARDO-VZ-469/3/JORI(SIB)// ND-61-130/LEEDS		[35,61]
	Vitron	1983	TURCHIA-77/3/JORI-69(SIB)/ (SIB)ANHINGA//(SIB)FLAMINGO		[35,61]
	Cocorit	1978	CIMMYT		[35]
France	Durtal	1972	*Triticum durum* × *T. aestivum*	High yield, short, good earliness, poor quality	[26,28,36]
Turkey	Dicle 74	1975	RAE/4×TC//STW63/3/AA"S" =Cocorit	Spring wheat	[20,37]
	Gediz 75	1976	LD357E/TC2//JO"S"	Spring wheat	[20,37]
	Cakmak 79	1979	UVY162/61.130	Winter wheat, good quality	[20,21,37]
	Kiziltan91	1991	UVY162/61.130	Winter wheat, good quality	[20,21,37]
	Altintac 95	1995	KND//68111/WARD	Irrigated winter wheat	[21,37]
	Selcuklu 97	1997	073/44×2/OVI/3/DF-72//61-130/ UVEYYK 162	Irrigated winter wheat	[21,37]
	Yilmaz98	1998		Irrigated winter wheat	[20,21,37]
	Ankara98	1998		Irrigated winter wheat	[20,21,37]
Algeria	Waha	1984	syn. Cham1, Plc/Ruff//Gta/Rtte	ICARDA genetic material	[50]
Morocco	Cocorit			Yield potential, wide adaptation, low quality	[62]
	Jori			Yield potential, wide adaptation, grain quality	[62]
	Haj-Mouline			Yield potential, wide adaptation, low quality	[62]
	Oum Rabia	1988	INRA 1718, Sel in " Cyprus 3"	High yield potential, better grain quality	[23,38,62]
	Karim	1985	Bittern 'S' or sel in « JO'S'.AA":S'//FG'S' »	High yield potential, better grain quality	[23,38,62]
Tunisia	Karim	1982	(Jori"S"/Anhinga"S"//Flamingo"S") CIMMYT		[24,39]
	Khiar	1992	Chen"S"/Altar 84, CIMMYT		[24]
	Om Rabia	1987	Jori C69/Haurani, ICARDA		[24]
	Nasr	1990	GoVZ512/Cit//Ruff/Fg/3/Pin/Gre//Trob), ICARDA		[24]
	Maali	2003	CMH80A.1016/4/TTURA/CMH74A370/ CMH77.774/3/ YAV79/5/Rassak/6/DACK"S"/YEL3"S"// Khiar, INRAT		[24]
	Salim	2010	ALTAR 84/FD8419-126-1-2/Razzak/3/Krf/ Baladia Hamra, INRAT		[24]

In tandem with the evolution of agricultural science occurring worldwide during the same time, in 1956 in Italy, Gian Tommaso Scarascia Mugnozza and Francesco D'Amato embarked on a pioneering durum wheat breeding program that included both fundamental genetic studies and applied mutation breeding and led to 22 registered varieties, six of which resulted from a direct selection of induced mutants [16,63]. An important cultivar that emerged from this program in 1974 was "Creso". Considerable work has also been done at Italian universities, such as at Palermo University (cv. "Trinakria" in 1973 by G.P. Ballatore), at Sassari and Naples (cvs. "Maristella", "Ichnusa" by R. Barbieri), and at Bari University [cvs. "Messapia", "Salentino", "Salizia" by G.T. Scarascia-Mugnozza, A. Blanco and coworkers). J. Vallega and G. Zitelli, at the Experimental Institute for Cereal Research in Rome, used N. Borlaug's selected genetic material in their crosses in their attempt to produce high-yielding cultivars resistant to lodging and diseases; this material had semi-dwarf genes of "Norin 10" [16]. In 1988, Calcagno released the successful cultivar "Simeto", which was bred at the Experimental Station for wheat at Caltagirone in Sicily. This cultivar was high yielding, low tillering, and had excellent adaptability across different environments. Thus, it has been widely cultivated in all the countries of the Mediterranean basin up to present with a large acreage [26] and is often used as a parent in crosses. In the private sector, seed companies have released many notable cultivars, including "Duilio" (Società Italiana Sementi, S.I.S.) and "Svevo" (Società Produttori Sementi), which are cultivated in several important durum wheat productive countries offering high-quality pasta products.

In Greece in the late 1960s, E. Skorda induced artificial mutations with gamma rays and thermal neutrons to increase genetic variability [64]. In the early 1980s, the breeder's team from the Cereal Institute in Thessaloniki released cultivars that were bred from intra-cultivar selections of genetic material from the Mediterranean basin, including "Mexicali 81", "Kallithea", "Athos", and "Selas". Furthermore, new cultivars selected from CIMMYT segregating germplasm were made available to the farmers. Among the aforementioned released cultivars, "Selas" had an excellent grain quality and was used until 2015 by the pasta industry in Greece. The newest durum wheat cultivars released by the breeders from the Cereal Institute originated either from crossbreeding among different cultivars or from selections among CIMMYT segregating material included in the Greek National and the Common Catalogue of Cultivated Plants [58,60].

At the beginning of the 1960s in France, Pierre Grignac began the first durum wheat breeding program, in which the first crossings among Mediterranean landraces were developed at INRA in Montpellier. In the middle of the same decade, cultivars with good qualities, such as a good yellow-amber color, were imported from North Dakota for cultivation in northern France, and constituted a second genetic pool for the program [26]. Afterwards, Grignac used interspecific crosses with bread wheat to select new lines with improved characteristics [28,36].

In Algeria, the Field Crops Development Institute (IDGC) was founded in 1974. This institute was responsible for durum wheat breeding and has collaborated actively with CIMMYT and ICARDA since 1980. As a result, more than 60 modern durum wheat cultivars have been released, and the peak yields have been obtained with the cultivars "Hoggar" and "Sahel" [22]. In Tunisia, modern cultivars developed from CIMMYT and ICARDA genotypes prevailed during the 1970s and were replaced in the 1980s by more recent cultivars developed at INRAT [24].

In 1970, a result of the collaboration with CIMMYT and, later, with ICARDA was the introduction of foreign durum wheat germplasm into Morocco in an attempt to develop new high-yielding, early-maturing cultivars. The outcome of this program was the release of the cultivars "Marzak", "Karim", "Sebou", and others after 1984 [23]. However, this new germplasm was inferior in quality.

Finally, in Turkey, the agreement between the Turkish government and the Rockefeller Foundation in 1967 resulted in the release of semi-dwarf, high-yielding, and disease-resistant cultivars [20,65].

3. Application of Classical Methodologies of Breeding

3.1. Classical Approaches, New Perspectives, and Tools for Wheat Breeding

The main targets of a durum wheat breeding program established in the Mediterranean basin, where its cultivation is very well adapted, may focus on: (1) grain yield improvement; (2) yield stability and a better understanding of genotype × environment interactions (G × E) and adaptation mechanisms; (3) responsiveness to inputs and the use efficiency of recourses in different cultivation systems; (4) resistance to biotic stresses (pests and diseases), and tolerance to abiotic stresses (drought, salinity, etc.); or (5) improving grain quality.

The above parameters are taken into account by durum wheat breeders in the design of a breeding program for cultivar development. They use common classical breeding methods, such as pedigree, bulk, single-seed, backcross method, pure line selection, and recurrent selection, to develop cultivars with the desired characters mentioned above. These methods can be applied with some alterations to increase efficiency and reduce the duration of the breeding program. As an example, backcross (BC) is a very common method mainly used in durum wheat for the transfer of traits controlled by only one or a few genes, such as resistance to diseases or quality parameters, from one donor parent to an elite line (recurrent parent) [66,67]. Molecular markers can significantly shorten the time needed, compared to the conventional backcross method, to identify the desired plants that have the target gene/genes and apply negative selection for the donor genome to ensure the maximum recovery of a recurrent-parent genome [68]. There are successful results in durum wheat breeding programs regarding the increased disease resistance or quality characteristics. Marker-assisted backcrossing (MABC) through simple sequence repeat (SSR) markers was shown to improve grain protein content in a wheat cultivar [69]; this method can assist the simultaneous selection of multiple stripe rust resistance genes and help avoid escapees during the selection process. In a previous study, the successful targeting of the gene transfer and reconstitution of the genome were completed in a period of four crop cycles, proving the practical application of MABC in developing high grain protein lines in the background of any popular cultivar [69]. In the single-seed descent method (SSD), only one seed from each F_2 selected plant is kept and bulked with all the others to produce an F_3 population. The same procedure is repeated until the F_5–F_6 generation. Two or more generations grow per year in the greenhouse, in winter nurseries, or in a growth chamber. The selection of lines takes place in the F_6 lines, which have increased homozygosity, retaining a large part of the extensive genetic variability from the F_2 generation. The single-seed descent method is considered a tool to exploit durum wheat genetic resources [70]. In the bulk breeding method, some plants are selected from the F_2 population, and their seeds are bulked to form the F_3 population. The same procedure is repeated until the F_5 generation, where the evaluation of lines begins until the F_{10}–F_{11} generation. This is an easy method to apply, thus saving breeders time and effort. Also, this method can increase the frequency of desirable genotypes in a population, but is not very effective for traits with low heritability. However, a modification of this method could be useful in wheat breeding when applied under salinity stress conditions [71]. In this case, using molecular markers through a bulk segregant analysis (BSA) will increase the effectiveness and shorten the needed time for all these processes [72].

Finally, the pedigree breeding method (and its modifications) is the most common method used in breeding programs for the release of durum wheat cultivars. Generally, pedigree breeding includes phenotypic selection in the early generations (until $F_{3\text{-}4}$), and the normal yield test begins in the generations with increased homozygosity (~F_5). The selection for yield during early generations was not very effective when the evaluation was done in normal plant density fields [73–76]; however, it was found to be effective when the evaluation was applied in low plant density fields and the experimental unit involved a single plant, as in the honeycomb methodology [77–80]. Finally, the selected lines from the experiments with replications will be evaluated in large plots (drill strips) over ~2 years and ~5–10 locations to determine yield, stability across locations, maturity, plant height, semolina, pasta and important quality characteristics for cooking [81]. Following a classical approach, a period of 9–12 years is needed from the beginning to the end of a breeding program, but this time could be reduced significantly by using the Marker Assisted Selection (MAS) procedure [82].

It has been recorded that the effectiveness of breeding on durum wheat's yield potential has been remarkable in Spain, Morocco, Turkey, and Italy [20,35,62,83]. The improvement of grain yield (GY) during the past decade has been attributed mostly to the increased number of grains m^{-2} and to the increased number of spikes m^{-2} [83]. Thus, further improvements in these characteristics might improve yield [83]. Important increases in yield were also achieved by increasing the harvest index, which has almost doubled since the beginning of the 20th century [83,84]. Moreover, a further increase in yield was achieved through an increase in biomass and a subsequent increase in yield, which explains the stability of the harvest index over the last three decades [84]. Similarly, an increase in biomass could result in an increase in yield in the future. Other characteristics that contributed to the increased yield in the 20th century include a reduction in the heading date and physiological maturity and an increase in the grain filling period [83]. Similar changes in these characteristics in the future could result in an increase in yield, based on the results of the reduction of the effects of drought and heat stress. Moreover, an increased yield could be achieved through the release of cultivars with higher water use efficiency [85].

Apart from the selection for increased yield potential, further improvements could be achieved through selection for increased tolerance to abiotic and biotic stresses. Moreover, the importance of traits that allow a plant to escape terminal drought and avoid critical stages of seed development (anthesis and seed filling), such as early vigor and an early heading date, has been well recognized [86,87]. All the above classical breeding methodologies have succeeded in making considerable progress in the yield and quality of durum wheat in the Mediterranean basin [24,57,88]. The main problem for the classical methods is that they are time-consuming, and phenotyping procedures are costly. In an effort to aid classical breeding methods, molecular genetics and associated technologies have been developed, and they offer important tools for plant breeders.

The parental selection of wheat lines can be based on phenotypic characterization and biochemical and DNA markers, which can estimate genetic variability even among phenotypically similar genotypes, as identified in several studies undertaken in Mediterranean countries [89–93]. By employing molecular markers in parental selection, the genetic diversity of wild and cultivated wheat can be exploited [94–98]. As an example, SSR markers were proven to be effective in the selection of genetically diverse genotypes with phenotypical similarities [90]. A combination of molecular markers and pedigree data could help in the exploitation of genetic diversity [91,99] and the selection of progenies, significantly increasing the efficiency and precision of plant breeding programs. Molecular markers supply various advantages over morphological markers in the linkage mapping of important agronomic traits. They are also unlimited in number, highly polymorphic, and can be used at any developmental stage without any environmental interference. Molecular markers can increase the precision and speed of selection in a durum wheat program though: (a) selection in the early stages or a simultaneous selection of multiple traits or traits that are difficult or expensive to evaluate; (b) the targeted introgression of useful genes in wide crosses; and (c) accelerated backcrossing. MAS or molecular breeding offers an opportunity to accelerate classical breeding approaches. MAS requires the establishment of a correlation between a desired trait, such as disease resistance, and molecular marker(s); this can be obtained by phenotyping a genetic mapping population followed by a quantitative trait locus (QTL) analysis [81]. For this purpose, several markers that are known to be associated with QTL/genes for some major economic traits are being deployed for MAS in wheat breeding programs. Several examples of the successful use of MAS are now available for wheat, and more examples will become available in the future [100–102]. Furthermore, molecular allele mining can help in broadening the reduced genetic diversity of cultivated wheat through the identification of allelic variation and the isolation of new rare alleles capable of improving tolerance to abiotic and biotic stresses [103,104]. According to Sehgal et al. [104], by using new technologies, unexploited genetic variation can further improve the drought and heat stress tolerance of the elite wheat pool and enrich it with novel drought and heat tolerance genes. This will contribute to achieve adaptability of the released cultivars to high temperature and drought that is for the most important emerging problems emerging in the Mediterranean due to climate change.

3.2. Participatory Plant Breeding

Today, it is recognized that agricultural production requires the adoption of environmentally friendly solutions, the preservation of crop biodiversity, and the release of varieties suitable for low input environments to set new goals for wheat breeding that align with the real needs of farmers and the market that are imprinted in the Mediterranean. Employing a participatory plant breeding (PPB) approach may have many benefits, including increased and more stable productivity, faster release and adoption of wheat varieties, better understanding of farmers' various criteria, enhanced biodiversity, the conservation of crop diversity on farms [105,106], increased cost-effectiveness, the ability to facilitate the learning of farmers, and the empowerment of farmers [107].

Participatory plant breeding (PPB) methods incorporate the involvement of end users in the breeding process [108] and the decentralization of selection sites into farmers' fields [109]. This has been proposed as an alternative to formal plant breeding and is more likely to produce varieties acceptable to farmers in marginal environments [110]. Social studies concerning the related historical and cultural traditions can assess the needs of both farmers and the market for local products [111–113]. Usually, end users value different traits than plant breeders [114]. Plant breeders contribute their expertise in creating genetic variation, in population management, and in designing screening methods that can separate genetic from environmental effects [110]. Participation provides flexibility in the selection program. The objectives could be reoriented to ensure relevant end products in case some changes are necessary during the breeding procedure. Moreover, there is a mistrust of modern varieties among farmers, bakers, and consumers [81,115], and participatory plant breeding could rebuild client trust with improved varieties [116]. In many breeding programs where there are G x E interactions during evaluation, the lines selected under PPB have been found to perform better for farmer priority traits than those selected via formal plant breeding methods [117,118]. PPB programs can also reduce the costs of the breeding process. Cost savings primarily derive from the less frequent testing of advanced lines [119]. Many studies have shown that participatory variety selection (PVS) can improve the adoption of varieties [120–122], and thereby enhance productivity [123].

In Syria, decentralized participatory selection by farmers is significantly more efficient in identifying the highest yielding entries in farmers' fields than any other selection strategy [124]. Farmer-selected populations are not genetically homogenous, which may lead to higher yield stability in varying environments [125]. PPB projects including farmers have resulted in the wider and simpler adoption of new varieties [114,126,127]. Wheat populations after PPB will evolve by adapting gradually and continuously to climate change [128]. A wheat population developed at ICARDA [129] has been evolving for five years at a farm in Tuscany (using evolutionary participatory breeding principals); the name of this population is 'SOLIBAM Tenero Floriddia' [130]. Thanks to the EU Commission Implementing Decision (2014/150/EU), which provides specific derogations for the marketing of wheat populations, this genetically heterogeneous population is now, for the first time, being marketed as a certified seed [130].

3.3. *The Application of Doubled-Haploid Techniques*

The introduction of advanced in-vitro tissue culture techniques, such as androgenesis (anther or microspore culture), chromosome elimination techniques (wide hybridization), and ovule cultures (gynogenesis), in self-pollinating crop species, has helped breeders to accelerate trait fixation in segregating populations of durum wheat in research conducted in Tunisia and ICARDA [131,132]. Of the techniques mentioned above, androgenesis (more precisely, an anther culture) can only be incorporated into breeding programs if they ensure the production of a sufficient number of genetically stable doubled haploid plants from a wide range of genotypes [133]. An anther culture, despite its effectiveness and convenience, has the serious disadvantage of being firmly genotype dependent [134,135]. Furthermore, durum wheat hardly responds at all to this technique (i.e., its embryo production is deficient and most of the plants produced are albinos) [136,137]. For this reason, chromosome elimination techniques are an attractive alternative approach, since they are not genotype dependent [138] and are not influenced by the dominant Kr wheat crossover genes [139]. This technique is mainly used in producing new germplasm, not only in durum but also in bread wheat and triticale. The fourth technique, gynogenesis, is another alternative for producing new germplasm. In gynogenesis, haploid plant development is induced by an unpollinated ovary culture. However, the use of an ovary culture is practiced more rarely in wheat breeding programs [140]. In a recent study in Tunisia, Slama-Ayed et al. [132] compared three doubled haploid techniques and found that gynogenesis is an exciting approach that could be used to produce new durum wheat genotypes as a supplement to maize techniques.

4. Breeding Challenges

Durum wheat breeding is considered to be one of the most cost effective and environmentally safe ways to meet the future challenges that durum wheat productivity will face due to climate change. The durum wheat is cultivated in rain-fed farming systems in Mediterranean basin. This is mainly connected with the high temperatures and drought that are expected to become more severe the next years and affect the cultivation across the Mediterranean [141]. In this context, investment in the productivity of rain-fed areas that cover a significant portion of Mediterranean countries could contribute to food security and rural growth. The UN reports several strategies for agronomic practices based on output and productivity in semi-arid areas, including the use of adaptable varieties, which is considered a very effective practice [142]. Breeding programs must be even more efficient due to the upcoming climate change effects and increased food demands. The identification of genetic resources and the study of genetic variability will provide further information regarding the increased tolerance of durum wheat under abiotic and biotic stresses. This could contribute to the increase and stability of production in future adverse climatic conditions. In this way, genetic studies for the identification of QTLs/genes that control important agronomic traits [82,143–145] and disease resistance [146] could also help. The identification of genomic regions that affect valuable target traits is known as quantitative trait locus (QTL) (or linkage) mapping, and it is a useful tool for the exploitation of loci that are co-segregating with traits of interest in a population [147]. QTL studies have been widely conducted on durum wheat for the genetic dissection of important breeding traits using diverse molecular markers and detailed genetic maps. Maps were used for the identification of QTLs controlling several characteristics, such as grain yield and kernel characters [82,148], grain-milling traits [149], and quality traits like endosperm color [150], grain protein content [151], and other pasta quality traits [152]. In extended experiments, including 249 recombinant inbred durum wheat lines evaluated in 16 environments, it was found that two major QTLs on chromosomes 2BL and 3BS have consistent effects across different environments [82]. Also, a QTL for plant height was identified on chromosomes 1BS, 3AL, and 7AS, and three QTLs for heading date were identified on chromosomes 2AS, 2BL, and 7BS. Moreover, 76 QTLs were identified for yield components along with several morpho-physiological traits (peduncle length, the Normalized Difference Vegetation Index (NDVI), and leaf greenness at the milk-grain stage expressed in Single-photon Avalanche Diode (SPAD) units) [143]. In a study under salinity conditions, four SSR markers were closely linked with grain yield, which could thus be used in the improvement of durum wheat through MAS under abiotic stress [144].

It is also expected that climate change will affect the vulnerability of durum wheat in different diseases [153,154]. Biotic resistance has also been investigated for the identification of QTLs that confer resistance to fungi [155] or pests [156]. Additional QTL studies use even greater genetic diversity, such as multiparental crosses for the identification of yield-related QTLs [157]. Abiotic resistance has also been under investigation for the identification of QTLs. A genome-wide association study of a durum wheat core set using 7652 Single Nucleotide Polymorphism (SNP) markers allowed the identification of major QTLs controlling the adaptation to heat stress [145]. Additionally, 12 loci were found to control the main heat tolerance traits; among these loci, three activated only when heat stress occurred. Moreover, two loci validated in a Kompetitive Allele Specific PCR (KASP) marker, are ready for deployment via MAS and could result in increased productivity in heat-stressed areas and improved resilience to climate change. A haplotype analysis of 208 elite lines confirmed that those with positive allele at all three QTLs had an 8% higher yield in a heat-stressed field environment [145].

Another important parameter is the durum wheat quality in terms of its protein content, endosperm texture, and glutenin content, which cannot be easily measured phenotypically. However, the methods for testing quality are typically costly, time-consuming, and need relatively large amounts of grain, which are available only in the late stages of breeding programs. Thus, markers for wheat quality traits can be very useful to enable the screening of a high number of lines and can be used early in breeding programs [158,159]. The durum wheat breeding programs carried out over the 20th century have focused on an increase of yield in combination with quality characteristics for pasta products [52,160–162] and the achievement of better adaptability under Mediterranean conditions [52]. Little attention has been given to increasing other grain health-promoting components and nutritive constituents of durum wheat, such as dietary fibre (DF), total and soluble arabinoxylan content, and beta-glycan in semolina. There are indications that intense breeding either increases or does not affect these parameters in modern cultivars compared to old cultivars [88], indicating that the breeding process may contribute to a further improvement of durum wheat's nutritive characteristics.

Thus, markers for wheat quality traits can be very useful to enable the screening of a high number of lines and can be used early in breeding programs [158,159]. Six QTLs explained 49%–56% of grain protein variations [163], and seven QTLs explain 62%–91% of the sodium dodecyl sulfate (SDS) volume [160] in durum wheat germplasm. A number of markers targeting different glutenin alleles have been referenced, including markers for *Glu-B1* alleles, based on the sequence variations of Bx type genes [161]. Further, MAS succeeded in increasing pasta-quality-associated properties through the transfer of significant QTLs, such as the *Gli-B1* locus containing *γ-gliadin* 45 and the *Glu-B3* locus containing Low Molecular Weight (LMW)-2 type glutenins [162].

4.1. Seed Storage Proteins and Quality

Seed storage proteins are prolamins that account for 80% of total grain proteins, and their role is crucial in determining the technological properties of durum wheat end products [57]. Prolamins are alcohol-soluble and can be classified according to their electrophoretic mobility in two classes: monomeric gliadins and polymeric glutenins. The former can be further classified as α, β, γ, and ω gliadins or as the high and low molecular weight glutenin subunits (HMW-GS and LMW-GS respectively) [57]. Many reports have discussed the effects of gluten protein composition on durum's end products [164,165]. These effects are either genotypic or environmentally dependent [166,167]. It is well established that certain HMW and LMW glutenin subunits affect the end product quality differently in durum wheat. For example, HMW GS *7+8* alleles are associated with better quality compared to allelic HMW-GS 20 [110]. For LMW-GS, it has also been demonstrated that certain subunits encoded by the loci located on chromosome *1B* (*Glu-*) positively (LMW -2 group of subunits) or negatively (LMW-1 group of subunits) affect pasta-making properties [168].

Wheat gliadin is also characterized by high intervarietal polymorphism, and most individual cultivars show unique electrophoretic patterns [169–171]. In durum wheat, the presence of components *γ-42* and *γ-45* encoded by allelic genes on chromosome *1B* is reported to affect the viscoelastic properties of gluten [172]. Gliadin *γ-45* is associated with a group of LMW–GS subunits termed *LMW-2*, and *γ-42* is associated with LMW-1 glutenin subunits. Gliadin *γ-45* could be used as a genetic marker for high gluten quality, whereas gliadin's *γ-42* component could serve as a genetic marker for poor gluten quality. Also, gliadin alleles were found to be correlated with resistance to cold and stem rust [173]. Finally, in breeding programs, knowledge of the allelic composition at each locus is beneficial in identifying and using the genotypes that carry the most promising qualitative traits.

4.2. Seed Quality Characteristics Connected with Human Health

The durum wheat breeding programs carried out over the 20th century mainly focused on increasing yield in combination with quality characteristics for pasta products [52,174–176] and the achievement of better adaptability to Mediterranean conditions [52]. Little attention was given to other grain health-promoting components. It has been suggested that intensive breeding has led to decreased contents of health-promoting components in modern wheat cultivars [177]. Recently, several researchers have investigated, in detail, the other nutritive constituents of durum wheat, such as dietary fiber (DF), that have many health benefits; it was found that intense breeding has not decreased DF in the modern cultivars compared to the old ones [88]. It was also observed that the total arabinoxylan content in wholemeal or semolina is not differentiated between recent and old genotypes, while modern cultivars have higher proportions of soluble arabinoxylan in wholemeals and of beta-glycan in semolina compared to the old genotypes [88]. These results show that the breeding process could contribute to further improvements in durum wheat's nutritive characteristics.

Apart from its value as a source of nutrients, wheat may cause inflammatory immune reactions and disorders like wheat allergies, celiac disease and non-celiac wheat sensitivity (NCWS), fructose malabsorption, and irritable bowel syndrome (IBS), highlighting the need for less-reactive wheat products that can contribute to quality of life improvements [81,178]. Wheat proteins, including gluten and non-gluten proteins like amylase/trypsin inhibitors (ATI) and others, are characterized as triggering factors. Recent studies have sought to investigate the underlying causes of these immune reactions [178]. According to the types of reactions caused, wheat-related disorders are classified as: (a) allergies, including immunoglobulin E (IgE) and non-IgE mediated allergic reactions; and (b) autoimmune, including celiac disease and herpetiform dermatitis [179]. Several studies have focused on the factors that affect the immunostimulatory capacity of allergic factors present in cereals, since it has been reported that short immunotherapy may represent a valid way to treat the disease [179,180].

Recent comparative studies on the nutritional characteristics of old and modern durum wheat genotypes have found that the breeding process improves durum wheat's gluten quality both in terms of its technological performance in producing high-quality pasta products and its allergenic potential [57]. More specifically, in modern cultivars, a higher gluten index was found to be connected to increased glutenin content. Further, the breeding process contributed to the drastic reduction of a significant allergen in wheat-dependent exercise-induced anaphylaxis (WDEIA), while the old and modern durum cultivars were not different in their α-type and γ-type gliadin content, the former being considered a factor associated with celiac disease toxicity [57]. Despite the existence of allergens in wheat grain, there is genetic variability within wheat's genetic resources, and further research is necessary for the identification and the development of cultivars with lower reactivity and/or higher secondary health-promoting ingredients to meet the different needs of consumers [81,174].

5. Future Prospects

Modern genome-wide association studies (GWASs) offer the advantage of performing association analyses using the association of each marker and the phenotype of interest that has been scored across a large number of unrelated genetic materials. Furthermore, GWASs take advantage of the higher number of gene recombinations used within the panel compared to linkage mapping where meiotic recombinations are limited. As a consequence, the aim of a GWAS is to locate important QTLs for complex characteristics by employing diverse germplasm collections and modern molecular markers. GWASs are complementary to QTL mapping [181]. GWAS studies have been conducted on bread wheat to analyze important characteristics. A genome-wide association study of a durum wheat core set using 7,652 single nucleotide polymorphism (SNP) markers facilitated the identification of a major QTL controlling adaptation to heat stress [145]. Additionally, 12 loci were found to control the main heat tolerance traits; among them, three were activated only when heat stress occurred. Twenty-nine QTLs for three different yield components were identified by a GWAS in a panel of 233 tetraploid wheat accessions, including durum wheat accessions, using SNP markers [182]. GWAS is a valuable tool for breeders since broad genetic resources can be screened for market-trait associations. Germplasm collections that contain a wealth of useful genes for valuable traits such as disease resistance could be used to identify possible sources of resistance.

A GWAS that focuses on drought tolerance and 17 other agronomical traits was conducted for 493 durum wheat accessions; this study identified a putative QTL that controls drought tolerance [183]. Two QTL hotspots related to stress tolerance and yield were identified on chromosomes 2A and 2B using 6211 diversity array technology (DArTseq) SNPs on a panel consisting of 208 durum wheat lines [184]. Other GWAS studies have focused on other traits, such as disease resistance [185–188] and important quality traits. Marcotuli et al. [189] identified 37 marker-trait associations and 19 QTLs, possibly underlining arabinoxylan content in the grains of 104 tetraploid wheat genotypes. Arabinoxylans have been shown to have various health benefits. Furthermore, the co-migration of

QTLs for grain protein content and the candidate genes related to nitrogen metabolism found in a study of a durum wheat germplasm collection show that such approaches can be applied to MAS breeding schemes [190].

The rapid development of next-generation sequencing (NGS) technologies has facilitated the discovery of vast numbers of SNPs across genomes. SNP markers are now popular molecular markers because they are ubiquitous in plant genomes and are very easy and cheap to score. The high-throughput genotyping of wheat varieties is now applied routinely, especially after the construction of specific genotyping arrays and the sequencing of wheat genomes. A recently-developed genotyping array for wheat that includes 90,000 gene-associated SNPs is aiding the fast identification of genetic variation that underlines trait variation in wheat genetic materials [191]. This genotyping array contains mostly bread wheat SNPs, but also includes a large number of durum wheat SNPs. A total of 90 k genotyping wheat arrays have also helped in the construction of a detailed SNP-based genetic map based on 140 RILs developed from a cross between a wild emmer wheat population and a durum wheat cultivar [192]. More genetic maps are available today for durum wheat [193]. The durum wheat genome was only recently sequenced, revealing more valuable information about the crop's genome evolution during domestication and selection [194]. It is expected that this genome will aid in clarifying marker–trait associations and facilitate exploration of the genes underlying important characters. Durum wheat transcriptomes have also become increasingly available, thus aiding MAS breeding [195,196]. Genome sequencing is providing breeders with precise info about the nature of the genome changes in their breeding lines. Furthermore, genetic information obtained by DNA sequencing and extracted with the use of advanced bioinformatics tools will help in the application of new DNA-marker platforms and is expected to help enormously in genomics-assisted breeding for yield and quality. NGS has made possible the development of the first mutant library for wheat, which is now available publicly [197] while plenty of DNA information has been deposited in public databases accessible to scientists working on wheat all over the world.

Genotyping-by-sequencing (GBS) identifies genome polymorphism (SNP) NGS technologies, which facilitate the discovery of genetic variation in natural populations of many plants, including wheat. GBS is a useful tool and has revealed that winter durum wheat lines have significant genetic diversity, which is crucial for breeding [198]. DArTseq and SNP markers based on GBS technology were used to survey the genetic variation and the genomic characterization of 91 durum wheat landraces from Turkey and Syria, revealing extensive mixing of landraces between the two geographical regions [199]. The use of GBS in a large wheat accession collection resulted in the discovery of thousands of new SNP variations for drought and heat stress tolerance [104], which is useful for improving the elite wheat pool and enriching it with novel drought and heat tolerance genes. According to Sehgal et al. [104], this unexploited genetic variation can further improve the drought and heat stress tolerance of the elite wheat pool and enrich it with novel drought and heat tolerance genes. As the ultimate MAS tool, GBS can effectively facilitate breeding.

Furthermore, a very promising modern tool in plant breeding is genomic selection (GS). GS is a strategy used to predict the genetic value of selection candidates based on the estimated genomic breeding value, which is predicted using high-density molecular markers that are dispersed across the genome [200]. GS bases its success on the use of genome-wide markers to ensure that minor to medium effect QTLs cannot be left uncaptured unlike MAS, which focuses on the few markers linked to major genes [201]. GS models have shown high forward prediction accuracies and an enhanced genetic gain for semolina, as well as grain quality characteristics revealing that a combination of MAS and GS can be used effectively to select for quality traits [202].

The development of speed breeding is a very promising technique that could substantially help in this area. Speed breeding entails the use of specifically controlled-environment plant growth conditions and extended photoperiods of 22 h light/2 h dark that accelerates plant development. As a result, plant breeding speeds can reach up to 6–8 generations/year for wheat. Therefore, speed breeding accelerates genetic gain and significantly reduces the length of breeding cycles [203]. Specifically, studies on durum wheat [204] have recently shown that the application of constant light and controlled temperature greenhouse conditions allow the rapid growth of durum wheat seedlings and the quick phenotyping for five important traits. Early selection in the F2 generation of a bi-parental cross has led to the significant improvement of traits like crown rot tolerance, root angle, and root number, thereby proving that a combination of speed breeding with early selection can facilitate the time and efficiency of breeding programs, as recombinant inbred lines can be provided with the desirable alleles [204]. For GWAS, using speed breeding, 393 durum Recombinant Inbred Lines (RILS) and DArT-seq markers have identified a major QTL for the seminal root angle on chromosome 6A [204]. Speed breeding coupled with genomics-based technologies and other advances in phenomics could yield significant progress in the rate of genetic gain in breeding schemes.

Genetically modified wheat has been developed previously [205,206], but today no GMO wheat is cultivated officially in any part of the world. However, new technologies like genome editing and its relevant protocols provide promising tools for the future. Transcription activator-like effector nucleases (TALENs) and clustered regularly interspaced short palindromic repeat (CRISPR)–associated protein 9 (Cas9) systems have been used on bread wheat to modify three homoalleles that code for mildew-resistance locus (MLO) proteins and a TaMLO-A1 allele, respectively, to make the wheat resistant to powdery mildew [207]. Detailed improved protocols for the application of CRISPR/Cas9-mediated mutagenesis are emerging; these protocols would help achieve fast and efficient gene targeting in wheat [208–210]. Furthermore, gene editing could be used in studying gene function. Resequencing of 1526 tetraploid and 1200 hexaploid wheat mutants created a database of 10 million sequenced mutations which, by more than 90%, result in truncations or deleterious amino acid changes [211]. More precise mutations can be introduced in wheat by gene editing and this, coupled with improved transformation technologies now evolving in wheat, would mean that researchers could be further helped in their efforts to introduce novel allelic diversity for breeding durum wheat and better understand basic gene function. However, since genetic modification in Europe is legally tightly regulated and gene editing has recently been ruled out as a form of genetic modification, the production of wheat cultivars based on these techniques has to be carefully considered in the future. Nevertheless, all modern biotechnological approaches (high throughput genome analysis, gene editing, genetic engineering, and proteomics and transcriptomics) are powerful tools to complement the classical methods of breeding. It is now proposed that genome assembly, germplasm characterization, gene function identification, genomic breeding, and gene editing constitute a comprehensive 5G approach in modern breeding that could help develop new varieties with a high yield, good quality, and strong resilience to changing climate conditions [212].

6. Conclusions

Classical breeding approaches will continue to play an important role in durum wheat improvement for the release of cultivars. Advances in DNA sequencing and other technologies, such as bioinformatics, statistics, and other scientific areas, could help breeders increase the efficiency and speed of a breeding program to meet humankind's growing demands for more food that is nutritious and sustainably produced. Ultimately, the use of new molecular biology technologies is essential, but also inexorably coupled with reliable and extensive testing under real field conditions.

Author Contributions: All authors have contributed in writing this review paper. All authors have read and agreed to the published version of the manuscript

References

1. Agriculture and Agri-Food Canada. Canada: Outlook for Principal Field Crops. 19 July 2019. Available online: http://www.agr.gc.ca/eng/industry-markets-and-trade/canadian-agri-food-sector-intelligence/crops/reports-and-statistics-data-for-canadian-principal-field-crops/?id=1378743094676 (accessed on 26 September 2019).
2. EUROSTAT. Available online: https://ec.europa.eu/eurostat/data/database (accessed on 27 September 2019).
3. Agriculture and Agri-Food Canada. Area, Yield, and Production of Canadian Principal Field Crops Report. Available online: https://aimis-simia.agr.gc.ca/rp/index-eng.cfm?action=pR&r=243&lang=EN (accessed on 26 September 2019).
4. United States Department of Agriculture. Wheat Data. Available online: http://www.ers.usda.gov/data-products/wheat-data.aspx (accessed on 26 September 2019).
5. Tidiane Sall, A.; Chiari, T.; Legesse, W.; Seid-Ahmed, K.; Ortiz, R.; van Ginkel, M.; Bassi, F.M. Durum wheat (*Triticum durum* Desf.): Origin, cultivation and potential expansion in Sub-Saharan Africa. *Agronomy* **2019**, *9*, 263. [CrossRef]
6. Tedone, L.; Alhajj Ali, S.; De Mastro, G. Optimization of nitrogen in durum wheat in the Mediterranean climate: The agronomical aspect and greenhouse gas (GHG) emissions. In *Nitrogen in Agriculture—Updates*; Amanullah, K., Fahad, S., Eds.; InTech: London, UK, 2018; Volume 8, pp. 131–162.
7. Shewry, P.R. Wheat. *J. Exp. Bot.* **2009**, *60*, 1537–1553. [CrossRef]
8. Peng, J.H.; Sun, D.; Nevo, E. Domestication evolution, genetics and genomics in wheat. *Mol. Breed.* **2011**, *28*, 281–301. [CrossRef]
9. Ozkan, H.; Brandolini, A.; Schafer-Pregl, R.; Salamini, F. AFLP analysis of a collection of tetraploid wheats indicates the origin of emmer and hard wheat domestication in Southeast Turkey. *Mol. Biol. Evol.* **2002**, *19*, 1797–1801. [CrossRef] [PubMed]
10. Martínez-Moreno, F.; Solís, I.; Noguero, D.; Blanco, A.; Özberk, İ.; Nsarellah, N.; Elias, E.; Mylonas, I.; Soriano, J.M. Durum wheat in the Mediterranean Rim: Historical evolution and genetic resources. *Genet. Resour. Crop. Evol.* **2020**. [CrossRef]
11. Nalam, V.J.; Vales, M.I.; Watson, C.J.W.; Kianian, S.F.; Riera-Lizarazu, O. Map-Based analysis of genes affecting the brittle rachis character in tetraploid wheat (*Triticum turgidum* L.). *Theor. Appl. Genet.* **2006**, *112*, 373–381. [CrossRef]
12. Ben-Abu, Y.; Tzfadia, O.; Maoz, Y.; Kachanovski, D.E.; Melamed-Bessudo, C.; Feldman, M.; Levy, A.A. Durum wheat evolution: A genomic analysis. In *Série A Mediterranean Seminars, No 110, Options Méditerranéennes, Proceedings of the International Symposium on Genetics and Breeding of Durum Wheat, Rome, Italy, 27–30 May 2013*; Porceddu, E., Damania, A.B., Qualset, C.O., Eds.; CIHEAM: Bari, Italy, 2014; pp. 31–45.
13. Christidis, B. *Winter Cereals*, 2nd ed.; Thessaloniki, Greece, 1963; pp. 43–70. (In Greek)
14. Bell, G.D.H. The history of wheat cultivation. In *Wheat Breeding: Its Scientific Basis*; Lupton, F.G.H., Ed.; Springer: Dordrecht, The Netherlands, 1987; pp. 31–49.
15. Valamoti, S.M.; Gkatzogia, E.; Hristova, I.; Marinova, E. Iron age cultural interactions, plant subsistence and land use in Southeastern Europe inferred from archaeobotanical evidence of Greece and Bulgaria. In *Archaeology Across Frontiers and Borderlands: Fragmentation and Connectivity in the North Aegean and the Central Balkans from the Bronze Age to the Iron Age*; Gimatzidis, S., Pieniążek, M., Mangaloğlu-Votruba, S., Eds.; Austrian Academy of Sciences: Vienna, Austria, 2018; pp. 269–290.
16. Scarascia Mugnozza, G.T. The contribution of Italian wheat geneticists: From Nazareno Strampelli to Francesco D'Amato. In *The Wake of the Double Helix: From the Green Revolution to the Gene Revolution, Proceedings of the International Congress, Bologna, Italy, 27–31 May 2003*; Tuberosa, R., Phillips, R.L., Gale, M., Eds.; Avenue Media: Bologna, Italy, 2005; pp. 53–75.
17. Sciacca, F.; Cambrea, M.; Licciardello, S.; Pesce, A.; Sciacca, F.; Cambrea, M.; Licciardello, S.; Pesce, A.; Romano, R.; Spina, A.; et al. Evolution of durum wheat from Sicilian landraces to improved varieties. In *Série A Mediterranean Seminars, No 110, Options Méditerranéennes, Proceedings of the International Symposium on Genetics and Breeding of Durum Wheat, Rome, Italy, 27–30 May 2013*; Porceddu, E., Damania, A.B., Qualset, C.O., Eds.; CIHEAM: Bari, Italy, 2014; pp. 139–145.

18. Papadakis, I. Greek types of wheat. In *Scientific Bulletin of the "Special Station of Plant Breeding in Thessaloniki"*; General Directorate for Settlements of Makedonia: Thessaloniki, Greece, 1929; pp. 6–29. (In Greek and In French)
19. Melas, T.B. *The Finest Varieties of Greek Wheat*; Papaspyrou Printing House: Athens, Greece, 1922; pp. 33–46. (In Greek)
20. Ozberk, I.; Ozberk, F.; Atli, A.; Cetin, L.; Aydemir, T.; Keklikci, Z.; Onal, M.A.; Braun, H.J. Durum wheat in Turkey: Yesterday, today and tomorrow. In *Durum Wheat Breeding: Current Approaches and Future Strategies*; Royo, C., Nachit, M.M., Di Fonzo, N., Araus, J.L., Pfeiffer, W.H., Slafer, G.A., Eds.; The Haworth Press Inc.: Philadelphia, PA, USA, 2005; pp. 981–1010.
21. Palamarchuk, A. Selection strategies for traits relevant for winter and facultative durum wheat. In *Durum Wheat Breeding: Current Approaches and Future Strategies*; Royo, C., Nachit, M.M., Di Fonzo, N., Araus, J.L., Pfeiffer, W.H., Slafer, G.A., Eds.; The Haworth Press Inc.: Philadelphia, PA, USA, 2005; pp. 599–644.
22. Abdelkader, B. The history of wheat breeding in Algeria. In *Série A Mediterranean Seminars, No 110, Options Méditerranéennes, Proceedings of the International Symposium on Genetics and Breeding of Durum Wheat, Rome, Italy, 27–30 May 2013*; Porceddu, E., Damania, A.B., Qualset, C.O., Eds.; CIHEAM: Bari, Italy, 2014; pp. 363–370.
23. Taghouti, M.; Nsarellah, N.; Rhrib, K.; Benbrahim, N.; Amallah, L.; Rochdi, A. Evolution from durum wheat landraces to recent improved varieties in Morocco in terms of productivity increase to the detriment of grain quality. *Rev. Mar. Sci. Agron. Vét.* **2017**, *5*, 351–358.
24. Boukida, F.; Vittadinib, E.; Prandib, B.; Mattarozzid, M.; Marchinib, M.; Sforzab, S.; Sayare, R.; WeonSeof, Y.; Yacoubia, I.; Mejri, M. Insights into a century of breeding of durum wheat in Tunisia: The properties of flours and starches isolated from landraces, old and modern genotypes. *Food Sci. Technol.* **2018**, *97*, 743–751. [CrossRef]
25. Kadkol, G.P.; Sissons, M. Durum wheat: Overview. In *Encyclopedia of Food Grains*, 2nd ed.; Wrigley, C., Corke, H., Seetharaman, K., Faubion, J., Eds.; Academic Press: Oxford, UK, 2016; pp. 117–124.
26. Clarke, J.M.; De Ambrogio, E.; Hare, R.A.; Roumet, P. Genetics and breeding of durum wheat. In *Durum Wheat Chemistry and Technology*, 2nd ed.; Sissons, M., Abecassis, J., Marchylo, B., Carcea, M., Eds.; AACC International: St. Paul, MN, USA, 2012; pp. 15–36.
27. Porceddu, E.; Blanco, A. Evolution of durum wheat breeding in Italy. In *Série A Mediterranean Seminars, No 110, Options Méditerranéennes, Proceedings of the International Symposium on Genetics and Breeding of Durum Wheat, Rome, Italy, 27–30May 2013*; Porceddu, E., Damania, A.B., Qualset, C.O., Eds.; CIHEAM: Bari, Italy, 2014; pp. 157–173.
28. Bagnara, D.; Scarascia Mugnozza, G.T. Outlook in breeding for yield in durum wheat. In *Genetics and Breeding of Durum Wheat, Proceedings of a Symposium on Genetics and Breeding of Durum Wheat, Bari, Italy, 14–18 May 1973*; Scarascia Mugnozza, G.T., Ed.; Eucarpia: Wageningen, The Netherlands, 1975; pp. 249–273.
29. Papadakis, I. *The Plant: Breeding Institute 1923–1933*; Bulletin No. 15; Kastrinaki & Georganta: Thessaloniki, Greece, 1933; pp. 1–36. (In Greek)
30. Percival, J. *The Wheat Plant*; A Monograph; Duckworth: London, UK, 1921; pp. 1–628, A Monograph.
31. Papadakis, I. *List of Seeds: What Seeds Should We Plant*; Central Committee for the Protection of Domestic Seed Production, Plant Breeding Institute: Thessaloniki, Greece, 1934; pp. 1–10. (In Greek)
32. Kokolios, B. Durum wheat in Greece. In *Proceedings of Convegno Internationale del Grano Duro*; Camera di Commercio Industria de Agricoltura: Foggia, Italy, 1962.
33. Di Fonzo, N.; Blanco, A.; Ravaglia, S.; Troccoli, A.; DeAmbrogio, E. Durum wheat improvement in Italy. In *Durum Wheat Breeding: Current Approaches and Future Strategies*; Royo, C., Nachit, M.N., Di Fonzo, N., Araus, J.L., Pfeiffer, W.H., Slafer, G.A., Eds.; The Haworth Press Inc.: Philadelphia, PA, USA, 2005; pp. 825–881.
34. Royo, C.; Briceño-Félix, G.A. Wheat breeding in Spain. In *The World Wheat Book: A History of Wheat Breeding*; Angus, W., Bonjean, A., Van Ginkel, M., Eds.; Lavoisier Publishing Inc.: Paris, France, 2011; Volume 2, pp. 121–154.
35. Royo, C. Durum wheat improvement in Spain. In *Durum Wheat Breeding: Current Approaches and Future Strategies*; Royo, C., Nachit, M.N., Di Fonzo, N., Araus, J.L., Pfeiffer, W.H., Slafer, G.A., Eds.; The Haworth Press Inc.: Philadelphia, PA, USA, 2005; pp. 883–906.

36. Grignac, P. Action of leaf rust (Puccinia recondita f. sp. Tritici) on durum wheat yield and breeding for rust resistance. In *Genetics and Breeding of Durum Wheat, Proceedings of a Symposium on Genetics and Breeding of Durum Wheat, Bari, Italy, 14–18 May 1973*; Scarascia Mugnozza, G.T., Ed.; Eucarpia: Wageningen, The Netherlands, 1975; pp. 489–496.
37. Braun, H.J.; Zencirci, N.; Altay, F.; Atli, A.; Avci, M.; Eser, V.; Kambertay, M.; Payne, T.S. Turkish wheat pool. In *The World Wheat Book: A History of Wheat Breeding*; Bonjean, A.P., Angus, W.J., Eds.; Lavoisier: Paris, France, 2001; pp. 851–879.
38. Taghouti, M.; Nsarellah, N.; Gaboun, F.; Rochdi, A. Multi-Environment assessment of the impact of genetic improvement on agronomic performance and on grain quality traits in Moroccan durum wheat varieties of 1949 to 2017. *Glob. J. Plant Breed. Genet.* **2017**, *4*, 394–404.
39. Byerlee, D.; Moya, P. *Impacts of International Wheat Breeding Research in the Developing World, 1966–1990*; The International Maize and Wheat Improvement Center (CIMMYT): Mexico City, Mexico, 1993; pp. 69–70.
40. Salazar, J.; Branas, M.; Ziteli, G.; Vallega, J. Old Iberian durum wheats as an important source of resistance to stem and leaf rusts. In *Genetics and Breeding of Durum Wheat, Proceedings of a Symposium on Genetics and Breeding of Durum Wheat, Bari, Italy, 14–18May 1973*; Scarascia Mugnozza, G.T., Ed.; Eucarpia: Wageningen, The Netherlands, 1975; pp. 497–508.
41. Al-Idrissi, M.; Sbeita, A.; Jebriel, S.; Zintani, A.; Shreidi, A.; Ghawawi, H.; Tazi, M. *Libya: Country Report to the FAO International Technical Conference on Plant Genetic Resources*; The Food and Agriculture Organization of the United Nations(FAO): Leipzig, Germany, 1996.
42. Ghanem, E.H. Wheat improvement in Egypt with emphasis on heat tolerance. In *Wheat in Heat-Stressed Environments: Irrigated, Dry Areas and Rice-Wheat Farming Systems*; Saunders, D.A., Hettel, G.P., Eds.; The International Maize and Wheat Improvement Center(CIMMYT): Mexico City, Mexico, 1994.
43. Zencirci, N.; Kinaci, E.; Atli, A.; Kalayci, M.; Avci, M. Wheat research in Turkey. In *Wheat: Prospects for Global Improvement. Developments in Plant Breeding*; Braun, H.J., Altay, F., Kronstad, W.E., Beniwal, S.P.S., McNab, A., Eds.; Springer: Dordrech, The Netherlands, 1997; Volume 6, pp. 11–16.
44. Poiarkova, H.; Blum, A. Land-Races of wheat from the northern Negev in Israel. *Euphytica* **1983**, *32*, 257–271. [CrossRef]
45. Isidro, J.; Álvaro, F.; Royo, C.; Villegas, D.; Miralles, D.J.; García del Moral, L.F. Changes in Duration of Developmental Phases of Durum Wheat Caused by Breeding in Spain and Italy during the 20th Century and Its Impact on Yield. *Ann. Bot.* **2011**, *107*, 1355–1366. [CrossRef] [PubMed]
46. Hedden, P. The genes of the Green Revolution. *Trends Genet.* **2003**, *19*, 5–9. [CrossRef]
47. Lebsock, K.L. Transfer of Norin 10 genes for dwarfness to durum wheat. *Crop Sci.* **1963**, *3*, 450–451. [CrossRef]
48. Pfeiffer, W.H.; Sayre, K.D.; Reynolds, M.P. Enhancing genetic grain yield potential and yield stability in durum wheat. In *Durum Wheat Improvement in the Mediterranean Region: New Challenges*; Royo, C., Nachit, M., Di Fonzo, N., Araus, J.L., Eds.; International Centre for Advanced Mediterranean Agronomic Studies(CIHEAM): Zaragoza, Spain, 2000; pp. 83–93.
49. The International Center for Agriculture Research in the Dry Areas(ICARDA). Available online: https://www.icarda.org/ (accessed on 30 August 2019).
50. Bassi, F.M.; Nachit, M.M. Genetic gain for yield and allelic diversity over 35 years of durum wheat breeding at ICARDA. *Crop Breed. Genet. Genom.* **2019**, *1*, 1–19.
51. McCaig, T.N.; Clarke, J.M. Breeding durum wheat in western Canada: Historical trends in yield and related variables. *Can. J. Plant Sci.* **1995**, *75*, 55–60. [CrossRef]
52. De Vita, P.; Matteu, L.; Mastrangelo, A.M.; Di Fonzo, N.; Cattivelli, L. Effects of breeding activity on durum wheat traits breed in Italy during the 20th century. *Ital. J. Agron.* **2007**, 451–462. [CrossRef]
53. Waddington, S.R.; Osmanzai, M.; Yoshida, M.; Ransom, J.K. The Yield of durum wheats released in Mexico between 1960 and 1984. *J. Agric. Sci.* **1987**, *108*, 469–477. [CrossRef]
54. Royo, C.; Álvaro, F.; Martos, V.; Ramdani, A.; Isidro, J.; Villegas, D.; García del Moral, L.F. Genetic changes in durum wheat yield components and associated traits in Italian and Spanish varieties during the 20th century. *Euphytica* **2007**, *155*, 259–270. [CrossRef]
55. Motzo, R.; Fois, S.; Giunta, F. Relationship between grain yield and quality of durum wheats from different eras of breeding. *Euphytica* **2004**, *140*, 147–154. [CrossRef]

56. Guarda, G.; Padovan, S.; Delogu, G. Grain yield, nitrogen-use efficiency and baking quality of old and modern Italian bread-wheat cultivars grown at different nitrogen levels. *Eur. J. Agron.* **2004**, *21*, 181–192. [CrossRef]
57. De Santis, M.A.; Giuliani, M.M.; Giuzio, L.; De Vita, P.; Lovegrove, A.; Shewry, P.R.; Flagella, Z. Differences in gluten protein composition between old and modern durum wheat genotypes in relation to 20th century breeding in Italy. *Eur. J. Agron.* **2017**, *87*, 19–29. [CrossRef] [PubMed]
58. Institute of Plant Breeding and Genetic Resources. Available online: http://http://www.ipgrb.gr/index.php/antikeimena/sitiron (accessed on 27 October 2019).
59. Kotzamanidis, S. Three new cultivars of durum wheat with very high yield potential, quality and excellent adaptation to Greek conditions. *ETHIAGE* **2011**, *43*, 14–15. (In Greek)
60. Korpetis, E.; Bladenopoulos, K.; Kotzamanidis, S. Two new Greek cultivars of Cereal Institute: THRAKI (durum wheat) and ALKMINI (triticale). In Proceedings of the 15th National Conference of Hellenic Scientific Society for Genetics and Plant Breeding, Larisa, Greece, 15–17 October 2014; p. 157. (In Greek)
61. Chairi, F.; Vergara-Diaz, O.; Vatter, T.; Aparicio, N.; Nieto-Taladriz, M.T.; Kefauver, S.C.; Bort, J.; Serret, M.D.; Araus, J.L. Post-Green revolution genetic advance in durum wheat: The case of Spain. *Field Crop. Res.* **2018**, *228*, 158–169. [CrossRef]
62. Nsarellah, N.; Amri, A. Durum wheat genetic improvement in Morocco. In *Durum Wheat Breeding: Current Approaches and Future Strategies*; Royo, C., Nachit, M.N., Di Fonzo, N., Araus, J.L., Pfeiffer, W.H., Slafer, G.A., Eds.; The Haworth Press Inc.: Philadelphia, PA, USA, 2005; pp. 963–980.
63. Scarascia-Mugnozza, G.T.; D'Amato, F.; Avanzi, S.; Bagnara, D.; Belli, M.L.; Bozzini, A.; Brunori, A.; Cervigni, T.; Devreux, M.; Donini, B.; et al. Mutation breeding for durum wheat (*Triticum turgidum* ssp. *durum* Desf.) improvement in Italy. *Mut. Breed. Rev.* **1993**, *10*, 1–28.
64. Skorda, E. The Mutability of Some Greek Varieties of the Genus Triticum. Ph.D. Thesis, Agricultural University of Athens, Athens, Greece, 1971.
65. Klatt, A.R.; Dincer, N.; Yakar, K. Problems associated with breeding spring and winter durums in Turkey. In *Genetics and Breeding of Durum Wheat, Proceedings of a Symposium on Genetics and Breeding of Durum Wheat, Bari, Italy, 14–18May 1973*; Scarascia Mugnozza, G.T., Ed.; Eucarpia: Wageningen, The Netherlands, 1975; pp. 327–337.
66. Yaniv, E.; Raats, D.; Ronin, Y.; Korol, A.B.; Grama, A.; Bariana, H.; Dubcovsky, J.; Schulman, A.H. Evaluation of marker-assisted selection for the stripe rust resistance gene Yr15, introgressed from wild emmer wheat. *Mol. Breed.* **2015**, 35. [CrossRef]
67. Phan, H.T.T.; Rybak, K.; Bertazzoni, S.; Furuki, E.; Dinglasan, E.; Hickey, L.T.; Oliver, R.P.; Tan, K.C. Novel sources of resistance to septoria nodorum blotch in the vavilov wheat collection identified by genome-wide association studies. *Theor. Appl. Genet.* **2018**, *131*, 1223–1238. [CrossRef]
68. Vishwakarma, M.K.; Mishra, V.K.; Gupta, P.K.; Yadav, P.S.; Kumar, H.; Joshi, A.K. Introgression of the high grain protein gene Gpc-B1 in an elite wheat variety of indo-gangetic plains through marker assisted backcross breeding. *Curr. Plant Biol.* **2014**, *1*, 60–67. [CrossRef]
69. Yildirim, A.; Sonmezoglu, O.A.; Sayaslan, A.; Koyuncu, M.; Gulec, T.E.; Kanderim, N. Marker-Assisted breeding of a durum wheat cultivar for γ-gliadin and LMW-glutenin proteins affecting pasta quality. *Turk. J. Agric. For.* **2013**, *37*, 527–533. [CrossRef]
70. Pignone, D.; De Paola, D.; Rapanà, N.; Janni, M. Single seed descent: A tool to exploit durum wheat (*Triticum durum* Desf.) genetic resources. *Genet. Resour. Crop Evol.* **2015**, *62*, 1029–1035. [CrossRef]
71. Farag, H.I.A. Efficiency of three methods of selection in wheat breeding under saline stress conditions. *Egypt. J. Plant Breed.* **2013**, *17*, 85–95. [CrossRef]
72. Aoun, M.; Kolmer, J.A.; Rouse, M.N.; Elias, E.M.; Breiland, M.; Bulbula, W.D.; Chao, S.; Acevedo, M. Mapping of novel leaf rust and stem rust resistance genes in the Portuguese durum wheat landrace. *G3 Genes Genomes Genet.* **2019**, *9*, 2535–2547. [CrossRef] [PubMed]
73. Atkins, R.E.; Murphy, H.C. Evaluation of yield potentialities of oat crosses from bulk hybrid tests. *Agron. J.* **1949**, 41. [CrossRef]
74. Fowler, W.L.; Heyne, E.G. Evaluation of bulk hybrid tests for predicting performance of pure line selections in hard red winter wheat. *Agron. J.* **1955**, *47*, 430–434. [CrossRef]
75. Lupton, F.G.H. Studies in the breeding of self-pollinating cereals. *Euphytica* **1961**, *10*, 209–224.

76. Schut, J.W.; Dourleijn, C.J.; Bos, I. Cross and Line Prediction in Barley Using F_4Small-Plot Yield Trials. Ph.D. Thesis, Wageningen Agricultural University (WAU), Wageningen, The Netherlands, 1998; pp. 21–43.
77. Mitchell, J.W.; Baker, R.J.; Knott, D.R. Evaluation of honeycomb selection for single plant yield in durum wheat. *Crop Sci.* **1982**, *22*, 840–843. [CrossRef]
78. Kotzamanidis, S.T.; Lithourgidis, A.S.; Mavromatis, A.G.; Chasioti, D.I.; Roupakias, D.G. Prediction criteria of promising F3 populations in durum wheat: A comparative study. *Field Crop. Res.* **2008**, *107*, 257–264. [CrossRef]
79. Lungu, D.M.; Kaltsikes, P.J.; Larter, E.N. Honeycomb selection for yield in early generations of spring wheat. *Euphytica* **1987**, *36*, 831–839. [CrossRef]
80. Mylonas, I. Experimentation of the Maximization of Yield Potential in Barley (*Hordeum vulgare* L.) Cultivars. Ph.D. Thesis, School of Agriculture, Aristotle University of Thessaloniki, Thessaloniki, Greece, 2012.
81. Kissing Kucek, L.; Veenstra, L.D.; Amnuaycheewa, P.; Sorrells, M.E. A Grounded Guide to Gluten: How Modern Genotypes and Processing Impact Wheat Sensitivity. *Compr. Rev. Food Sci. Food Saf.* **2015**, *14*, 285–302. [CrossRef]
82. Maccaferri, M.; Sanguineti, M.C.; Corneti, S. Quantitative trait loci for grain yield and adaptation of durum wheat (*Triticum durum* Desf.) across a wide range of water availability. *Genetics* **2008**, *178*, 489–511. [CrossRef]
83. De Vita, P.; Nicosia, O.L.D.; Nigro, F.; Platani, C.; Riefolo, C.; Di Fonzo, N.; Cattivelli, L. Breeding progress in morpho-physiological, agronomical and qualitative traits of durum wheat cultivars released in Italy during the 20th century. *Eur. J. Agron.* **2007**, *26*, 39–53. [CrossRef]
84. Garcia del Moral, L.F.; Royo, C.; Slafer, G.A. Genetic improvement effects on durum wheat yield physiology. In *Durum Wheat Breeding: Current Approaches and Future Strategies*; Royo, C., Nachit, M.N., Di Fonzo, N., Araus, J.L., Pfeiffer, W.H., Slafer, G.A., Eds.; The Haworth Press Inc.: Philadelphia, PA, USA, 2005; pp. 379–396.
85. Araus, J.L.; Slafer, G.A.; Reynolds, M.P.; Royo, C. Plant breeding and drought in C3 cereals: What to breed for? *Ann. Bot.* **2002**, *89*, 925–940. [CrossRef] [PubMed]
86. Richards, R.A. Selectable traits to increase crop photosynthesis and yield of grain crops. *J. Exp. Bot.* **2000**, *51*, 447–458. [CrossRef] [PubMed]
87. Spielmeyer, W.; Hyles, J.; Joaquim, P.; Azanza, F.; Bonnett, D.; Ellis, M.E.; Moore, C.; Richards, R.A. A QTL on chromosome 6A in bread wheat (*Triticum aestivum*) is associated with longer coleoptiles, greater seedling vigour and final plant height. *Theor. Appl. Genet.* **2007**, *115*, 59–66. [CrossRef] [PubMed]
88. De Santis, M.A.; Kozik, O.; Passmore, D.; Flagella, Z.; Shwewry, P.R. Comparison of the dietary fibre composition of old and modern durum wheat (*Triticum turgidum* spp. *durum) genotypes. Food Chem.* **2018**, *244*, 304–310. [CrossRef]
89. Achtar, S.; Moualla, M.Y.; Kalhout, A.; Röder, M.S.; MirAli, N. Assessment of genetic diversity among syrian durum (*Triticum* ssp. *durum*) and bread wheat (*Triticum aestivum* L.) using SSR markers. *Russ. J. Genet.* **2010**, *46*, 1320–1326. [CrossRef]
90. Casassola, A.; Brammer, S.P.; Chaves, M.S.; Wiethölter, P.; Caierão, E. Parental selection of wheat lines based on phenotypic characterization and genetic diversity. *Crop Breed. Appl. Biotechnol.* **2013**, *13*, 49–58. [CrossRef]
91. Marzario, S.; Logozzo, G.; David, J.L.; Zeuli, P.S.; Gioia, T. Molecular genotyping (SSR) and agronomic phenotyping for utilization of durum wheat (*Triticum Durum* Desf.) ex situ collection from Southern Italy: A combined approach including pedigreed varieties. *Genes* **2018**, *9*, 465. [CrossRef]
92. Soriano, J.M.; Villegas, D.; Sorrells, M.E.; Royo, C. Durum wheat landraces from east and west regions of the mediterranean basin are genetically distinct for yield components and phenology. *Front. Plant Sci.* **2018**, *9*, 80. [CrossRef]
93. Abu-Zaitoun, S.Y.; Chandrasekhar, K.; Assili, S.; Shtaya, M.J.; Jamous, R.M.; Mallah, O.B.; Nashef, K.; Sela, H.; Distelfeld, A.; Alhajaj, N.; et al. Unlocking the genetic diversity within a Middle-East panel of durum wheat landraces for adaptation to semi-arid climate. *Agronomy* **2018**, *8*, 233. [CrossRef]
94. Dograr, N.; Akin-Yalin, S.; Akkaya, M. Discriminating durum wheat cultivars using highly polymorphic simple sequence repeat DNA markers. *Plant Breed.* **2000**, *119*, 360–362. [CrossRef]
95. Eujayl, I.; Sorrells, M.; Baum, M.; Wolters, P.; Powell, W. Isolation of EST-derived microsatellite markers for genotyping the A and B genomes of wheat. *Theor. Appl. Genet.* **2002**, *104*, 399–407. [CrossRef] [PubMed]
96. Incirli, A.; Akkaya, M.S. Assessment of genetic relationships in durum wheat cultivars using AFLP markers. *Genet. Resour. Crop Evol.* **2001**, *48*, 233–238. [CrossRef]

97. Pujar, S.; Tamhankar, S.; Rao, V.; Gupta, V.; Naik, S.; Ranjekar, P. Arbitrarily primed-PCR based diversity assessment reflects hierarchical groupings of Indian tetraploid wheat genotypes. *Theor. Appl. Genet.* **1999**, *99*, 868–876. [CrossRef]
98. Soleimani, V.; Baum, B.; Johnson, D. AFLP and pedigree-based genetic diversity estimates in modern cultivars of durum wheat (*Triticum turgidum* L. subsp. *durum* (Desf.) Husn.). *Theor. Appl. Genet.* **2002**, *104*, 350–357. [CrossRef] [PubMed]
99. Babay, E.; Mnasri, S.R.; Mzid, R.; Naceur, M.B.; Hanana, M. Quality selection and genetic diversity of tunisian durum wheat varieties using SSR markers. *Biosci. J.* **2019**, 35. [CrossRef]
100. Bonnett, D.G.; Rebetzke, G.J.; Spielmeyer, W. Strategies for efficient implementation of molecular markers in wheat breeding. *Mol. Breed.* **2005**, *15*, 75–85. [CrossRef]
101. Kuchel, H.; Fox, R.; Hollamby, G.; Reinheimer, J.L.; Jefferies, S.P. The challenges of integrating new technologies into a wheat breeding programme. In Proceedings of the 11th International Wheat Genet Symposium, Brisbane, Australia, 24–29 August 2008; Appels, R., Eastwood, R., Lagudah, E., Langridge, P., Mackay, M., McIntyre, L., Sharp, P., Eds.; Sydney University Press: Sydney, Australia, 2008; pp. 1–5.
102. Jia, Z.; Liu, Y.; Gruber, B.D.; Neumann, K.; Kilian, B.; Graner, A.; von Wirén, N. Genetic dissection of root system architectural traits in spring barley. *Front. Plant Sci.* **2019**, *10*, 400. [CrossRef]
103. Bhullar, N.K.; Zhang, Z.; Wicker, T.; Keller, B. Wheat gene bank accessions as a source of new alleles of the powdery mildew resistance gene Pm3: A large scale allele mining project. *BMC Plant Biol.* **2010**, *10*, 88. [CrossRef]
104. Sehgal, D.; Vikram, P.; Sansaloni, C.P.; Ortiz, C.; Pierre, C.S.; Payne, T.; Ellis, M.; Amri, A.; Petroli, C.D.; Wenzl, P.; et al. Exploring and mobilizing the gene bank biodiversity for wheat improvement. *PLoS ONE* **2015**, *10*, e0132112. [CrossRef]
105. Cleveland, D.A.; Soleri, D.; Smith, S.E. Do folk crop varieties have a role in sustainable agriculture? Incorporating folk varieties into the development of locally based agriculture may be the best approach. *Bioscience* **1994**, *44*, 740–751. [CrossRef]
106. Love, B.; Spaner, D. Agrobiodiversity: Its value, measurement, and conservation in the context of sustainable agriculture. *J. Sustain. Agric.* **2007**, *31*, 53–82. [CrossRef]
107. Sperling, L.; Ashby, J.A.; Smith, M.E.; Weltzien, E.R.; McGuire, S. A framework for analyzing participatory plant breeding approaches and results. *Euphytica* **2001**, *136*, 21–35.
108. Witcombe, J.R.; Joshi, K.D.; Gyawali, S.; Musa, A.M.; Johansen, C.; Virk, D.S.; Sthapit, B.R. Participatory plant breeding is better described as highly client-oriented plant breeding. I. Four indicators of client-orientation in plant breeding. *Exp. Agric.* **2005**, *41*, 299–319. [CrossRef]
109. Ceccarelli, S. Efficiency of plant breeding. *Crop Sci.* **2015**, *55*, 87–97. [CrossRef]
110. Atlin, G.N.; Cooper, M.; Bjornstad, A. A comparison of formal and participatory breeding approaches using selection theory. *Euphytica* **2001**, *122*, 463–475. [CrossRef]
111. Pretty, J.; Vodouhê, S.D. Using rapid or participatory rural appraisal. In *Improving Agricultural Extension*; A Reference Manual; The Food and Agriculture Organization of the United Nations (FAO): Rome, Italy, 1997.
112. Soleri, D.; Cleveland, D. Breeding for quantitative variables, Part 1: Farmers' and scientists' knowledge and practice in variety choice and plant selection. In *Plant Breeding and Farmer Participation*; Ceccarelli, S., Guimarães, E.P., Weltzien, E., Eds.; The Food and Agriculture Organization of the United Nations (FAO): Rome, Italy, 2009; pp. 324–366.
113. Organic Seed Alliance (OSA). *Participatory Plant Breeding Toolkit*; Organic Seed Alliance (OSA): Port Townsend, WA, USA, 2012.
114. Ashby, J.A. The impact of participatory plant breeding. In *Plant Breeding and Farmer Participation*; Ceccarelli, S., Guimarães, E.P., Weltzien, E., Eds.; The Food and Agriculture Organization of the United Nations (FAO): Rome, Italy, 2009; pp. 649–671.
115. Davis, W. *Wheat Belly: Lose the Wheat, Lose the Weight, and Find: Your Path Back to Health*; Emmaus, P.A., Ed.; Rodale Press: Emmaus, PA, USA, 2011.
116. Kissing Kucek, L.; Darby, H.; Mallory, E.; Dawson, J.; Davis, M.; Dyck, E.; Lazor, J.; O'Donnell, S.; Mudge, S.; Kimball, M.; et al. Participatory breeding of wheat for organic production. In Proceedings of the Organic Agriculture Research Symposium, LaCrosse, WI, USA, 25–26 February 2015; pp. 25–26.

117. Goldringer, I. Co-Constructiuon of a French program for participatory breeding of wheat cultivars adapted to organic cultivation and transformation. In Proceedings of the EGOSGN Conference, Montreal, QC, Canada, 4–6 July 2014.
118. Kirk, A.P.; Fox, S.L.; Entz, M.H. Comparison of organic and conventional selection environments for spring wheat. *Plant Breed.* **2012**, *131*, 687–694. [CrossRef]
119. Mangione, D.; Senni, S.; Puccioni, M.; Grando, S.; Ceccarelli, S. The cost of participatory barley breeding. *Euphytica* **2006**, *150*, 289–306. [CrossRef]
120. Ceccarelli, S. Specific adaptation and breeding for marginal conditions. *Euphytica* **1994**, *77*, 205–219. [CrossRef]
121. Ceccarelli, S. Adaptation to low/high input cultivation. *Euphytica* **1996**, *92*, 203–214. [CrossRef]
122. Ceccarelli, S.; Erskine, W.; Hamblin, J.; Grando, S. Genotype by environment interaction and international breeding programmes. *Exp. Agric.* **1994**, *30*, 177–187. [CrossRef]
123. Ceccarelli, S.; Grando, S.; Baum, M. Participatory plant breeding in water-limited environments. *Exp. Agric.* **2007**, *43*, 411–435. [CrossRef]
124. Ceccarelli, S.; Grando, S.; Bailey, E.; Amri, A.; El-Felah, M.; Nassif, F. Farmer participation in barley breeding in Syria, Morocco and Tunisia. *Euphytica* **2001**, *122*, 521–536. [CrossRef]
125. Entz, M.H.; Kirk, A.P.; Carkner, M.; Vaisman, I.; Fox, S.L. Evaluation of lines from a farmer participatory organic wheat breeding program. *Crop Sci.* **2018**, *58*, 2433. [CrossRef]
126. Ortiz-Perez, R.; Rios-Labrada, H.; Miranda-Lorigados, S.; Ponce-Brito, M.; Quintero-Fernandez, E.; Chaveco-Perez, O. Avances del mejoramiento genetico participativo del frijol en Cuba. *Agron. Mesoam.* **2006**, *17*, 337–346. [CrossRef]
127. Mustafa, Y.; Grando, S.; Ceccarelli, S. *Assessing the Benefits and Costs of Participatory and Conventional Barley Breeding Programs in Syria*; The International Development Research: Ottawa, ON, Canada, 2006.
128. Ceccarelli, S. Drought. In *Plant Genetic Resources and Climate Change*; Jackson, M., Ford-Lloyd, B.V., Parry, M.L., Eds.; CAB International: Boston, MA, USA, 2014; pp. 221–235.
129. Ceccarelli, S. Landraces: Importance and use in breeding and environmentally friendly agronomic systems. In *Agrobiodiversity Conservation: Securing the Diversity of Crop Wild Relatives and Landraces*; Maxted, N., Dulloo, M.E., Ford-Lloyd, B.V., Frese, L., Iriondo, J., Pinheiro de Carvalho, M.A.A., Eds.; CABI Publishing: New York, NY, USA, 2012; pp. 103–117.
130. Petitti, M.; Bocci, R.; Bussi, B.; Ceccarelli, S.; Spillane, C.; McKeown, P. Future proofing decentralised evolutionary wheat populations' seed systems in Italy using a climate analogues approach: The example of Tuscany. In Proceedings of the EUCARPIA Symposium on Breeding for Diversification, Witzenhausen, Germany, 19–21 February 2018.
131. Tasesse, W.; Sanchez-Garcia, M.; Tawkaz, S.; Baum, M. Doubled haploid production in wheat. In *Advances in Breeding Techinques for Cereal Crops*; Ordon, F., Friedt, W., Eds.; Burleigh Dodds Science Publishing: Cambridge, UK, 2019; pp. 93–116.
132. Slama-Ayed, O.; Bouhaouel, I.; Ayed, S.; Jacques De Buyser, J.; Picard, E.; Hajer Slim Amara, H.S. Efficiency of three haplomethods in durum wheat (*Triticum turgidum* subsp. *durum* Desf.): Isolated microspore culture, gynogenesis and wheat × maize crosses. *Czech J. Genet. Plant Breed.* **2019**, *55*. [CrossRef]
133. Snape, J.; Simpson, E. The genetical expectations of doubled haploid lines derived from different filial generations. *Theor. Appl. Genet.* **1981**, *60*, 123–128. [CrossRef]
134. Almouslem, A.B.; Jauhar, P.P.; Peterson, T.S.; Bommineni, V.R.; Rao, M.B. Haploid durum wheat production via hybridization with maize. *Crop Sci.* **1998**, *38*, 1080–1087. [CrossRef]
135. Niroula, R.K.; Bimb, H.P. Overview of wheat x maize system of crosses for dihaploid induction in wheat. *World Appl. Sci. J.* **2009**, *7*, 1037–1045.
136. Foroughi-Wehr, B.; Zeller, F.J. In-Vitro microspore reaction of different German wheat cultivars. *Theor. Appl. Genet.* **1990**, *79*, 77–80. [CrossRef]
137. Loschenberger, F.; Hansel, H.; Heberle-Bors, E. Plant formation from pollen of tetraploid wheat species and Triticum tauschii. *Cereal Res. Commun.* **1993**, *21*, 133–139.
138. Hussain, M.; Niaz, M.; Iqbal, M.; Iftikhar, T.; Ahmad, J. Emasculation techniques and detached tiller culture in wheat x maize crosses. *J. Agric. Res.* **2012**, *50*, 1–19.
139. Laurie, D.A. Factors affecting fertilization frequency in crosses of *Triticum aestivum* cv. "Highbury" x *Zea mays* cv. "Seneca 60". *PlantBreed.* **1989**, *103*, 133–140. [CrossRef]

140. Sibi, M.L.; Kobaissi, A.; Shekafandeh, A. Green haploid plants from unpollinated ovary culture in tetraploid wheat (*Triticum durum* Defs.). *Euphytica* **2001**, *122*, 351–359. [CrossRef]
141. Gibbon, D. Rainfed farming systems in the Mediterranean region. *Plant Soil* **1981**, *58*, 59–80. [CrossRef]
142. United Nations (UN). *Global Drylands: A UN System-Wide Response*; United Nations (UN): New York, NY, USA, 2011.
143. Graziani, M.; Maccaferri, M.; Royo, C.; Salvatorelli, F.; Tuberosa, R. QTL dissection of yield components and morpho-physiological traits in a durum wheat elite population tested in contrasting thermo-pluviometric conditions. *Crop Pasture Sci.* **2014**, *65*, 80–95. [CrossRef]
144. Dura, S.; Duwayri, M.; Nachit, M.; Sheyaboudi, F.A. Detection of molecular markers associated with yield and yield components in durum wheat (*Triticum turgidum* L. Var. *durum*Desf.) under drought conditions. *Crop Pasture Sci.* **2013**, *64*, 957–964. [CrossRef]
145. Hassouni, K.E.; Belkadi, B.; Sall, A.T.; Filali-Maltouf, A.; Al-Abdallat, A.; Nachit, M.M.; Bassi, F.M. Loci controlling adaptation to heat stress occurring at the reproductive stage in durum wheat. *Agronomy* **2019**, *9*, 414. [CrossRef]
146. Kthiri, D.; Loladze, A.; MacLachlan, P.R.; N'Diaye, A.; Walkowiak, S.; Nilsen, K.; Dreisigacker, S.; Ammar, K.; Pozniak, C.J. Characterization and mapping of leaf rust resistance in four durum wheat cultivars. *PLoS ONE* **2018**, *13*, e0197317. [CrossRef]
147. Alqudah, A.M.; Sallam, A.; Baenziger, P.S.; Börner, A. GWAS: Fast-Forwarding gene identification and characterization in temperate Cereals: Lessons from Barley—A review. *J. Adv. Res.* **2020**, *22*, 119–135. [CrossRef]
148. Patil, R.M.; Tamhankar, S.A.; Oak, M.D.; Raut, A.L.; Honrao, B.K.; Rao, V.S.; Misra, S.C. Mapping of QTL for agronomic traits and kernel characters in durum wheat (*Triticum durum* Desf.). *Euphytica* **2013**, *190*, 117–129. [CrossRef]
149. Elouafi, I.; Nachit, M.M. A genetic linkage map of the durum×triticum dicoccoides backcross population based on SSRs and AFLP markers, and QTL analysis for milling traits. *Theor. Appl. Genet.* **2004**, *108*, 401–413. [CrossRef] [PubMed]
150. Pozniak, C.J.; Knox, R.E.; Clarke, F.R.; Clarke, J.M. Identification of QTL and association of a phytoene synthase gene with endosperm colour in durum wheat. *Theor. Appl. Genet.* **2007**, *114*, 525–537. [CrossRef] [PubMed]
151. Peleg, Z.; Cakmak, I.; Ozturk, L.; Yazici, A.; Jun, Y.; Budak, H.; Saranga, Y. Quantitative trait loci conferring grain mineral nutrient concentrations in durum wheat×wild emmer wheat RIL population. *Theor. Appl. Genet.* **2009**, *119*, 353–369. [CrossRef] [PubMed]
152. Zhang, W.; Chao, S.; Manthey, F.; Chicaiza, O.; Brevis, J.C.; Echenique, V.; Dubcovsky, J. QTL analysis of pasta quality using a composite microsatellite and SNP map of durum wheat. *Theor. Appl. Genet.* **2008**, *117*, 1361–1377. [CrossRef] [PubMed]
153. Royo, C.; Soriano, J.M.; Alvaro, F. Wheat: A crop in the bottom of the Mediterranean diet pyramid. *Mediterr. Identities Environ. Soc. Cult.* **2017**, 381–399. [CrossRef]
154. Nachit, M.M.; Elouafi, I. Durum wheat adaptation in the Mediterranean dryland: Breeding, stress physiology, and molecular markers. In *Challenges and Strategies of Dryland Agriculture*; Rao, S.C., Ryan, J., Eds.; CSSA Special Publication: Madison, WI, USA, 2004; pp. 203–218.
155. Prat, N.; Guilbert, C.; Prah, U.; Wachter, E.; Steiner, B.; Langin, T.; Buerstmayr, H. QTL mapping of Fusarium head blight resistance in three related durum wheat populations. *Theor. Appl. Genet.* **2017**, *130*, 13–27. [CrossRef]
156. Varella, A.C.; Zhang, H.; Weaver, D.K.; Cook, J.P.; Hofland, M.L.; Lamb, P.; Talbert, L.E. A novel QTL in durum wheat for resistance to the wheat stem sawfly associated with early expression of stem solidness. *G3 Genes Genomes Genet.* **2019**, *9*, 1999–2006. [CrossRef]
157. Milner, S.G.; Maccaferri, M.; Huang, B.E.; Mantovani, P.; Massi, A.; Frascaroli, E.; Salvi, S. A multiparental cross population for mapping QTL for agronomic traits in durum wheat (*Triticum turgidum* ssp. *durum*). *Plant Biotechnol. J.* **2016**, *14*, 735–748. [CrossRef]
158. Peña, R.J. Wheat for bread and other foods. In *Bread Wheat Improvement and Production*; Curtis, B.C., Rajaram, S., Gómez Macpherson, H., Eds.; The Food and Agriculture Organization of the United Nations (FAO): Rome, Italy, 2002; pp. 483–542.
159. Gale, K.R. Diagnostic DNA markers for quality traits in wheat. *J. Cereal Sci.* **2005**, *41*, 181–192. [CrossRef]

160. Blanco, A.; Bellomo, M.P.; Lotti, C.; Maniglio, T.; Pasqualone, A.; Simeone, R.; Troccoli, A.; Fonzo, N.D. Genetic mapping of sedimentation volume across environments using recombinant inbred lines of durum wheat. *Plant Breed.* **1998**, *117*, 413–417. [CrossRef]
161. Ma, W.; Zhang, W.; Gale, K.R. Multiplex-PCR typing of high molecular weight glutenin alleles in wheat. *Euphytica* **2003**, *134*, 51–60. [CrossRef]
162. Yıldırım, A.; Sönmezoğlu, Ö.A.; Sayaslan, A.; Kandemir, N.; Gökmen, S. Molecular breeding of durum wheat cultivars for pasta quality. *Qual. Assur. Saf. Crop.* **2019**, *11*, 15–21. [CrossRef]
163. Blanco, A.; de Giovanni, C.; Laddomada, B.; Sciancalepore, A.; Simeone, R.; Devos, K.M.; Gale, M.D. Quantitative trait loci influencing grain protein content in tetraploid wheats. *Plant Breed.* **1996**, *115*, 310–316. [CrossRef]
164. Edwards, N.M.; Gianibelli, M.C.; McCaig, T.N.; Clarke, J.M.; Ames, N.P.; Larrique, O.R.; Dexter, G.E. Relationship between gluten strength, polymeric protein quantity and composition for diverse durum wheat genotypes g. *Cereal Sci.* **2007**, *45*, 140–149. [CrossRef]
165. Sissons, M. Role of durum wheat composition on the quality of pasta and bread. *Food* **2008**, *2*, 75–90.
166. Giuliani, M.M.; De Santis, M.A.; Pompa, M.; Giuzio, I.; Flagella, Z. Analysis of gluten proteins composition during grain filling in two durum wheat cultivars submitted to two water regoms. *Ital. J. Agron.* **2014**, *9*, 558.
167. Giuliani, M.M.; Palermo, C.; De Santis, M.A.; Mentana, A.; Pompa, M.; Giuzio, I.; Masci, S.; Centonze, D.; Flagella, Z. Differential expression of durum wheat glutenproteome under water stress during grain filling. *J. Agric. Food Chem.* **2015**, *63*, 6501–6512. [CrossRef]
168. D'Ovidio, R.; Masci, S. The wheat low-molecular-weight glutenin subunits. *J. Cereal Sci.* **2004**, *39*, 321–339. [CrossRef]
169. Sosinov, A.A.; Poperelya, F.A. Prolamin polymorphism and plant breeding. *Bull. Agric. Sci.* **1979**, *10*, 21–34.
170. Wrigley, C.W.; Autran, J.C.; Bushuk, W. Identification of cereal varieties by gel electrophoresis of grain proteins. *Adv. Cereal Sci. Technol.* **1982**, *5*, 211–259.
171. Melnikova, N.V.; Kudryavtseva, A.V.; Kudryatsev, A.M. Catalogue of alleles of gliadin-coding loci in durum wheat (*Triticum durum* Desg.). *Biochimie* **2011**, *94*, 551–557. [CrossRef] [PubMed]
172. Damideaux, R.; Autran, J.C.; Grignak, P.; Feillet, P. Mise en evidence de relations applicables en selection 921 entre l' electrophoregramme des gliadines et les properties viscoelastiques du gluten de *Triticum durum* 922 Desf. *C. R. Acad. Sci.* **1978**, *287*, 701–704.
173. Poperelya, F.A.; Babayants, I.G. Gliadin component block Gld-183 as a marker of gene determining stem rust resistance. *Dokl. Vaskhnil* **1978**, *6*, 6–7.
174. Raciti, C.N.; Doust, M.A.; Lombardo, G.M.; Boggini, G.; Pecetti, L. Characterization of durum wheat mediterranean germplasm for high and low molecular weight glutenin subunits in relation with quality. *Eur. J. Agron.* **2003**, *19*, 373–382. [CrossRef]
175. Rossini, F.; Provenzano, M.E.; Sestili, F.; Ruggeri, R. Synergistic effect of sulfur and nitrogen in the organic and mineral fertilization of durum wheat: grain yield and quality traits in the Mediterranean environment. *Agronomy* **2018**, *8*, 189. [CrossRef]
176. Li, L.; Niu, Y.; Ruan, Y.; DePauw, R.M.; Singh, A.K.; Gan, Y. Agronomic advancement in tillage, crop rotation, soil health, and genetic gain in durum wheat cultivation: A 17-year Canadian story. *Agronomy* **2018**, *8*, 193. [CrossRef]
177. Morris, C.E.; Sands, D.C. The breeder's dilemma-yield or nutrition? *Nat. Biotechnol.* **2006**, *24*, 1078–1080. [CrossRef]
178. Scherf, K.A. Immunoreactive cereal proteins in wheat allergy, non-celiac gluten/wheat sensitivity (NCGS) and celiac disease. *Curr. Opin. Food Sci.* **2019**, *25*, 35–41. [CrossRef]
179. Ziegler, K.; Neumann, J.; Liu, F.; Fröhlich-Nowoisky, J.; Cremer, C.; Saloga, J.; Reinmuth-Selzle, K.; Pöschl, U.; Schuppan, D.; Bellinghausen, I.; et al. Nitration of wheat amylase trypsin inhibitors increases their innate and adaptive immunostimulatory potential In Vitro. *Front. Immunol.* **2019**, *9*, 3174–3184. [CrossRef]
180. Cianferoni, A. Wheat allergy: Diagnosis and management. *J. Asthma Allergy* **2016**, *9*, 13–25. [CrossRef]
181. Korte, A.; Farlow, A. the advantages and limitations of trait analysis with gwas: A review. *Plant Methods* **2013**, *9*, 29. [CrossRef] [PubMed]

182. Mangini, G.; Gadaleta, A.; Colasuonno, P.; Marcotuli, I.; Signorile, A.M.; Simeone, R.; De Vita, P.; Mastrangelo, A.M.; Laidò, G.; Pecchioni, N.; et al. Genetic dissection of the relationships between grain yield components by genome-wide association mapping in a collection of tetraploid wheats. *PLoS ONE* **2018**, *13*, e0190162. [CrossRef] [PubMed]
183. Wang, S.; Xu, S.; Chao, S.; Sun, Q.; Liu, S.; Xia, G. A genome-wide association study of highly heritable agronomic traits in durum wheat. *Front. Plant Sci.* **2019**, *10*, 919. [CrossRef] [PubMed]
184. Sivakumar, S.; Matthew, P.R.; Sansaloni, C. Genome-Wide association analyses identify QTL hotspots for yield and component traits in durum wheat grown under yield potential, drought, and heat stress environments. *Front. Plant Sci.* **2018**, 81. [CrossRef]
185. Aoun, M.; Breiland, M.; Turner, K.M.; Loladze, A.; Chao, S.; Xu, S.; Acevedo, M. Genome-Wide association mapping of leaf rust response in a durum wheat worldwide germplasm collection. *Plant Genome* **2016**, *9*. [CrossRef]
186. Liu, W.; Maccaferri, M.; Bulli, P.; Rynearson, S.; Tuberosa, R.; Chen, X.; Pumphrey, M. Genome-Wide association mapping for seedling and field resistance to *Puccinia striiformis* f. sp. *tritici in elite durum wheat. Theor. Appl. Genet.* **2017**, *130*, 649–667. [CrossRef]
187. Liu, W.; Maccaferri, M.; Rynearson, S.; Letta, T.; Zegeye, H.; Tuberosa, R.; Pumphrey, M. Novel sources of stripe rust resistance identified by genome-wide association mapping in Ethiopian durum wheat *(Triticum turgidum* ssp. *durum*). *Front. Plant Sci.* **2017**, *8*, 774. [CrossRef]
188. Kidane, Y.G.; Hailemariam, B.N.; Mengistu, D.K.; Fadda, C.; Pè, M.E.; Dell'Acqua, M. Genome-Wide association study of septoria tritici blotch resistance in ethiopian durum wheat landraces. *Front. Plant Sci.* **2017**, *8*, 1586. [CrossRef]
189. Marcotuli, I.; Houston, K.; Waugh, R.; Fincher, G.B.; Burton, R.A.; Blanco, A.; Gadaleta, A. Genome wide association mapping for arabinoxylan content in a collection of tetraploid wheats. *PLoS ONE* **2015**, *10*, e0132787. [CrossRef]
190. Nigro, D.; Gadaleta, A.; Mangini, G.; Colasuonno, P.; Marcotuli, I.; Giancaspro, A.; Blanco, A. Candidate genes and genome-wide association study of grain protein content and protein deviation in durum wheat. *Planta* **2019**, *249*, 1157–1175. [CrossRef]
191. Wang, S.; Wong, D.; Forrest, K.; Allen, A.; Chao, S.; Huang, B.E.; Mastrangelo, A.M. Characterization of polyploid wheat genomic diversity using a high-density 90,000 single nucleotide polymorphism array. *Plant Biotechnol. J.* **2014**, *12*, 787–796. [CrossRef] [PubMed]
192. Avni, R.; Nave, M.; Eilam, T.; Sela, H.; Alekperov, C.; Peleg, Z.; Distelfeld, A. Ultra-Dense genetic map of durum wheat× wild emmer wheat developed using the 90K iSelect SNP genotyping assay. *Mol. Breed.* **2014**, *34*, 1549–1562. [CrossRef]
193. Maccaferri, M.; Ricci, A.; Salvi, S.; Milner, S.G.; Noli, E.; Martelli, P.L.; Ammar, K. A high-density, SNP-based consensus map of tetraploid wheat as a bridge to integrate durum and bread wheat genomics and breeding. *Plant Biotechnol. J.* **2015**, *13*, 648–663. [CrossRef] [PubMed]
194. Maccaferri, M.; Harris, N.S.; Twardziok, S.O.; Pasam, R.K.; Gundlach, H.; Spannagl, M.; Himmelbach, A. Durum wheat genome highlights past domestication signatures and future improvement targets. *Nature Genet.* **2019**, *51*, 885. [CrossRef]
195. Krasileva, K.V.; Buffalo, V.; Bailey, P.; Pearce, S.; Ayling, S.; Tabbita, F.; Dubcovsky, J. Separating homeologs by phasing in the tetraploid wheat transcriptome. *Genome Biol.* **2013**, *14*, R66. [CrossRef] [PubMed]
196. Vendramin, V.; Ormanbekova, D.; Scalabrin, S.; Scaglione, D.; Maccaferri, M.; Martelli, P.; Tuberosa, R. Genomic tools for durum wheat breeding: De novo assembly of Svevo transcriptome and SNP discovery in elite germplasm. *BMC Genom.* **2019**, *20*, 278. [CrossRef] [PubMed]
197. Krasileva, K.V.; Vasquez-Gross, H.A.; Howell, T.; Bailey, P.; Paraiso, F.; Clissold, L.; Fosker, C. Uncovering hidden variation in polyploid wheat. *Proc. Natl. Acad. Sci. USA* **2017**, *114*, 913–921. [CrossRef]
198. Sieber, A.N.; Longin, C.F.H.; Würschum, T. Molecular characterization of winter durum wheat (*Triticumdurum*) based on a genotyping-by-sequencing approach. *Plant Genet. Res.* **2017**, *15*, 36–44. [CrossRef]
199. Baloch, F.S.; Alsaleh, A.; Shahid, M.Q.; Çiftçi, V.; de Miera, L.E.S.; Aasim, M.; Hatipoğlu, R. A whole genome DArTseq and SNP analysis for genetic diversity assessment in durum wheat from central fertile crescent. *PLoS ONE* **2017**, *12*, e0167821. [CrossRef]
200. Newell, M.A.; Jannink, J.L. Genomic selection in plant breeding. In *Crop Breed*; Humana Press: New York, NY, USA, 2014; pp. 117–130.

201. Hayes, B.J.; Goddard, M.E. Prediction of total genetic value using genome-wide dense marker maps. *Genetics* **2001**, *157*, 1819–1829.
202. Fiedler, J.D.; Salsman, E.; Liu, Y.; Michalak de Jiménez, M.; Hegstad, J.B.; Chen, B.; Li, X. Genome-Wide association and prediction of grain and semolina quality traits in durum wheat breeding populations. *Plant Genome* **2017**, *10*. [CrossRef] [PubMed]
203. Watson, A.; Ghosh, S.; Williams, M.J.; Cuddy, W.S.; Simmonds, J.; Rey, M.D.; Adamski, N.M. Speed breeding is a powerful tool to accelerate crop research and breeding. *Nat. Plants* **2018**, *4*, 23. [CrossRef] [PubMed]
204. Alahmad, S.; Dinglasan, E.; Leung, K.M.; Riaz, A.; Derbal, N.; Voss-Fels, K.P.; Hickey, L.T. Speed breeding for multiple quantitative traits in durum wheat. *Plant Methods* **2018**, *14*, 36. [CrossRef] [PubMed]
205. Altpeter, F.; Vasil, V.; Srivastava, V.; Vasil, I.K. Integration and expression of the high-molecular-weight glutenin subunit 1Ax1 into wheat. *Nat. Biotechnol.* **1996**, *14*, 1155–1159. [CrossRef] [PubMed]
206. He, G.Y.; Rooke, L.; Steele, S.; Bekes, F.; Gras, P.; Tatham, A.S.; Fido, R.; Barcelo, P.; Shewry, P.R.; Lazzeri, P.A. Transformation of pasta wheat (*Triticum durum* L. var. *durum*) with high-molecular-weight glutenin subunit genes and modification of dough functionality. *Mol. Breed.* **1999**, *5*, 377–396. [CrossRef]
207. Wang, H.; Yang, H.; Shivalila, C.S.; Dawlaty, M.M.; Cheng, A.W.; Zhang, F.; Jaenisch, R. One-Step generation of mice carrying mutations in multiple genes by CRISPR/Cas-mediated genome engineering. *Cell* **2013**, *153*, 910–918. [CrossRef]
208. Shan, Q.; Wang, Y.; Li, J.; Gao, C. Genome editing in rice and wheat using the CRISPR/Cas system. *Nat. Protoc.* **2014**, *9*, 2395–2410. [CrossRef]
209. Gil-Humanes, J.; Wang, Y.; Liang, Z.; Shan, Q.; Ozuna, C.V.; Sánchez-León, S.; Baltes, N.J.; Starker, C.; Barro, F.; Gao, C.; et al. High-Efficiency gene targeting in hexaploid wheat using DNA replicons and CRISPR/Cas9. *Plant J.* **2017**, *89*, 1251–1262. [CrossRef]
210. Zhang, Y.; Liang, Z.; Zong, Y.; Wang, Y.; Liu, J.; Chen, K.; Qiu, J.-L.; Gao, C. Efficient and transgene-free genome editing in wheat through transient expression of CRISPR/Cas9 DNA or RNA. *Nat. Commun.* **2016**, *7*, 1–8. [CrossRef]
211. Uauy, C.; Paraiso, F.; Colasuonno, P.; Tran, R.K.; Tsai, H.; Berardi, S.; Comai, L.; Dubcovsky, J. A modified TILLING approach to detect induced mutations in tetraploid and hexaploid wheat. *BMC Plant Biol.* **2009**, *9*, 115. [CrossRef]
212. Varshney, R.K.; Sinha, P.; Singh, V.K.; Kumar, A.; Zhang, Q.; Bennetzen, J.L. 5Gs for crop genetic improvement. *Curr. Opin. Plant Biol.* **2020**. [CrossRef] [PubMed]

13

Molecular and Morphological Divergence of Australian Wild Rice

Dinh Thi Lam [1,2], Katsuyuki Ichitani [3], Robert J. Henry [4] and Ryuji Ishikawa [5,*]

1 United Graduate School of Agricultural Sciences, Iwate University, Morioka, Iwate 020-8550, Japan; lamiasvn@gmail.com
2 Institute of Agricultural Science for Southern Vietnam, District 1, Ho Chi Minh City 121, Vietnam
3 Faculty of Agriculture, Kagoshima University, 1-21-24 Korimoto, Kagoshima, Kagoshima 890-0065, Japan; ichitani@agri.kagoshima-u.ac.jp
4 Queensland Alliance for Agriculture and Food Innovation, University of Queensland, Brisbane QLD 4072, Australia; robert.henry@uq.edu.au
5 Faculty of Agriculture and Life Science, Hirosaki University, 3 Bunkyo-cho, Hirosaki, Aomori 036-8561, Japan
* Correspondence: ishikawa@hirosaki-u.ac.jp.

Abstract: Two types of perennial wild rice, Australian *Oryza rufipogon* and a new taxon Jpn2 have been observed in Australia in addition to the annual species *Oryza meridionalis*. Jpn2 is distinct owing to its larger spikelet size but shares *O. meridionalis*-like morphological features including a high density of bristle cells on the awn surface. All the morphological traits resemble *O. meridionalis* except for the larger spikelet size. Because Jpn2 has distinct cytoplasmic genomes, including the chloroplast (cp), cp insertion/deletion/simple sequence repeats were designed to establish marker systems to distinguish wild rice in Australia in different natural populations. It was shown that the new taxon is distinct from Asian *O. rufipogon* but instead resembles *O. meridionalis*. In addition, higher diversity was detected in north-eastern Australia. Reproductive barriers among species and Jpn2 tested by cross-hybridization suggested a unique biological relationship of Jpn2 with other species. Insertions of retrotransposable elements in the Jpn2 genome were extracted from raw reads generated using next-generation sequencing. Jpn2 tended to share insertions with other *O. meridionalis* accessions and with Australian *O. rufipogon* accessions in particular cases, but not Asian *O. rufipogon* except for two insertions. One insertion was restricted to Jpn2 in Australia and shared with some *O. rufipogon* in Thailand.

Keywords: *Oryza*; speciation; divergence; life history; phylogenetic relation; Australian continent

---

1. Introduction

The *Oryza* genus is comprised of 23 species with varying genome compositions and ploidy levels [1]. The two cultivated species, *Oryza sativa* and *Oryza glaberrima*, belong to AA genome species, and their progenitors were wild *Oryza rufipogon* and *Oryza barthii*, respectively. The AA genome species are dispersed across the major continents and were once classified as a single species, *Oryza perennis*, comprising Asian, American, African, and Oceanian forms [2]. The Asian species, *O. rufipogon* represents different life histories and varies from annual to perennial. Their life history is a continuum with annual, intermediate, and perennial forms [3,4]. The American species, *Oryza glumaepatula* also varies from annual to perennial. African species, *O. barthii* and *Oryza longistaminata*, however, are exclusively annual and perennial types, respectively.

Oceanian species had been known as *O. perennis* (later changed to the current species nomenclature) including annual and perennial types as a continuum within a single species [3]. After rearrangement of the species classification, an annual type was defined as an Oceanian endemic species, *Oryza meridionalis* and the perennial form as *O. rufipogon* [4]. Their distributions in Australia are well studied [5,6]. Speciation of these species has been confirmed using retrotransposon insertions [7,8] and crossing ability [9–11].

In general, annual and perennial species have different adaptive strategies to allocate their energy resources [3,12]. Annual species tend to have higher seed productivity than perennial species. *O. meridionalis*, the Australian annual species, produces plenty of seeds and disperses these seeds. *O. meridionalis* inhabits ponds or the periphery of ponds, ditches, or lakes during the rainy season. Water levels in wild rice habitats recede and water in the peripheral areas of annual species disappears during the dry season [13]. Annual species produce large amounts of seed for the next generation. In contrast, the life history of Australian perennial species is similar to Asian perennial species except for a unique taxon known as Jpn2 or taxon B [6,14]. In addition, Jpn2 type wild rice exhibits different morphological and genetic characteristics [14]. Including the new wild rice type, Australian perennial and annual rice chloroplast (cp) genomes have been completely sequenced in order to understand the uniqueness in evolutionary relationships among other wild rice [15,16]. This showed that the cp genome of Australian *O. rufipogon*, Jpn1 (taxon A) has a closer relationship to *O. meridionalis* than to Asian *O. rufipogon*, although its nuclear type tended to show higher similarity to Asian *O. rufipogon*. Another perennial species, Jpn2 (taxon B), also shared similarity not only with the cp genome to *O. meridionalis* but also the nuclear type [14,17,18]. This analysis showed that all Australian wild rice shared some cp genetic similarity with *O. meridionalis*. Nuclear genomes in Australia showed huge variation never seen in Asian wild rice. These findings with ecological observations confirmed that there were two types of perennial rice. Their distribution in northern Queensland and their unique morphological traits were also reported [6,14].

In this paper, we further characterized these two taxa at morphological and reproductive levels, which enabled us to determine how they have diverged at the species level. Cytoplasmic markers to distinguish them were developed and variation among natural populations was evaluated. These findings will help to distinguish these taxa in field research for further analysis and also give clues to their evolutionary origins. Retro-transposable elements were also used to screen the species examined in this study. Some of these provide clear evidence of phylogenetic relationships because of the unique mechanism of transposition insertion.

2. Results

2.1. Morphological Features

Two types of Australian perennial wild rice were collected (Table 1). Based on our previous report [14], identifying two types of perennials: Jpn1 (taxon A) and Jpn2 (taxon B), morphological traits were able to be discriminated between the Australian perennials. Bristle cells have a thorn-like architecture along the awns (Figure 1). SEM enabled us to compare the density of these cells. They varied from 2.33 to 5.33 per 200 μm square among Asian wild rice (Table 2). In *O. meridionalis*, W1299 and W1300 had 12.67 and 14.67 per 200 μm^2, respectively. The density in Jpn1 was similar to that in Asian wild rice. That in Jpn2 was similar to *O. meridionalis*. There were significant differences between the two groups, W1299/W1300/Jpn2 and W106/W0120/W0137/Jpn1. Other traits, such as anther length, suggested that Jpn2 shared short anthers with other annual accessions such as W0106 in *O. rufipogon*, and W1299 and W1300 in *O. meridionalis*.

Table 1. Samples collected in Australia and control core collections developed in NBRP.

Sites	Populations	No. of Plants	GPS Data S	GPS Data E	Year
P26	P26a	8	S16.5332	E145.2138	2011
	P26b	8	S16.5529	E145.2136	2011
	P26c	8	S16.5541	E145.2130	2011
	P26d	8	S16.5539	E145.2128	2011
	P26e	8	S16.5536	E145.2125	2011
	P26f	8	S16.5536	E145.2124	2011
	P26g	8	S16.5535	E145.2123	2011
	P26h	8	S16.5534	E145.2122	2011
	P26i	8	S16.5533	E145.2122	2011
Jpn1	Jpn1	8	S16.3809	E145.1936	2011
Jpn2	Jpn2	8	S15.2622	E144.1239	2011
Jpn3	Jpn3	8	S15.0431	E143.4321	2011
P6	P6	8	S15 41519	E145 02473	2011
P7	P7E	8	S15 42003	E145 04219	2011
	P7L	8	S15 42003	E145 04219	2011
P8	P8W	4	S15 41302	E145 07237	2011
	P8R	4	S15 41302	E145 07237	2011
P10	P10H	8	S15.41416	E145.09040	2011
	P10L	8	S15.41416	E145.09040	2011
P12	P12	8	S15.31486	E144.22564	2011
P17	P17	8	S15.09256	E143.49481	2011
P21	P21	8	S15.23047	E144.07228	2011
P22	P22	8	S15.78099	E144.14333	2011
P23	P23	8	S15.31198	E144.22079	2011
P26j	P26j	8	S16.55118	E145.2136	2011
P26PL	P26PL	8	S16.15579	E145.2060	2011
P27	P27				
P5	P5O	8	S15 45329	E144 59419	2010
	P5N	8	S15 45329	E144 59419	2010
	P5W	8	S15 45329	E144 59419	2010
Core collection of NBRP					
Oryza meridinoalis		19*			
Oryza rufipogon		32			

*W1299 noted as no rank in Oryza database, was added to the core collection in this study.

Table 2. Comparison of morphological traits, anther length, pancle length, paniclewidth, and density of bristle cells.

				Anther Length (mm)				Spikelet Length (mm)				Spikelet Width (mm)				Density of Bristle Cells (per 200 μm²)			
Species	**Accession**	**Life History**	**n=**	**mean**	**±**	**SD**		**mean**	**±**	**SD**		**mean**	**±**	**SD**		**mean**	**±**	**SD**	
O.rufipogon																			
	W0106	Annual	5	1.50	±	0.07	**	7.01	±	0.63	**	2.42	±	0.20		4.67	±	1.92	**
	W0120	Perennial	5	1.88	±	0.10	**	7.06	±	0.16	**	2.34	±	0.11		2.67	±	1.92	**
	W0137	Perennial	5	2.31	±	0.05	**	6.95	±	0.25	**	2.60	±	0.11		3.67	±	1.92	**
O.meridionalis																			
	W1299	Annual	5	1.42	±	0.05	**	6.47	±	0.27	**	2.32	±	0.12		12.67	±	3.42	
	W1300	Annual	5	ND				6.91	±	0.19	**	2.13	±	0.05	**	14.67	±	1.60	
	Australian wild rice																		
Greenhouse																			
	Jpn1	Perennial	5	3.48	±	0.20		7.13	±	0.15	**	2.02	±	0.16	**	2.67	±	3.42	**
	Jpn2	Perennial	5	1.64	±	0.07	**	8.28	±	0.14	*	2.62	±	0.11		12.00	±	2.34	
Field																			
	Jpn1	Perennial	5	3.79	±	0.20		7.62	±	0.17	**	2.00	±	0.07	**	ND			
	Jpn2	Perennial	5	1.74	±	0.19	**	8.60	±	0.23		2.40	±	0.23		ND			

*,**: Significant differences compared with the longest anther of Jpn1, the largest panicle of Jpn2, the widest width of W0137, and density of bristle cells of W1300 at 5 and 1% levels, respectively.

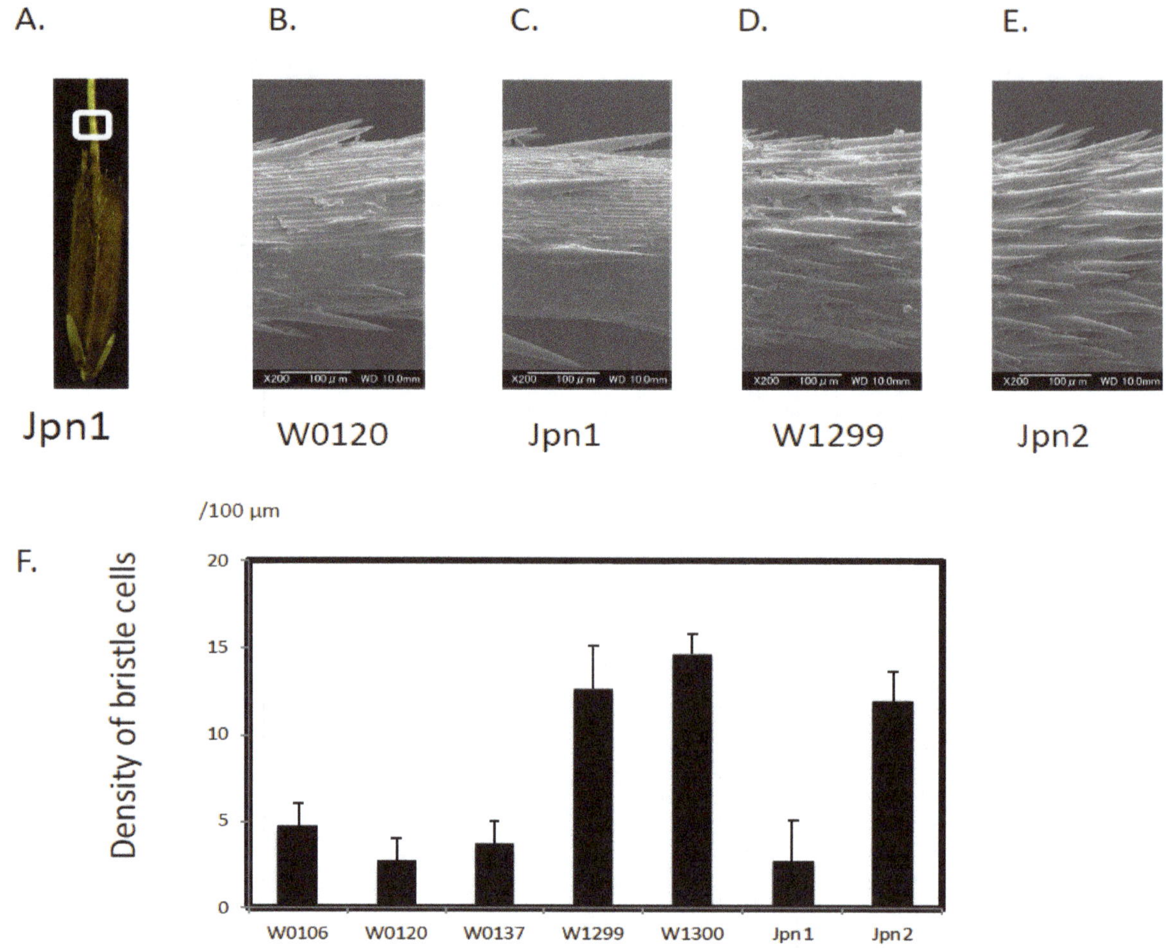

Figure 1. Variation in the density of bristle cells in awns. Panel **A**: spikelet of Jpn1, Panels B to E: enlarged SEM photos of W0120 (Panel **B**), Jpn1 (Panel **C**), W1299 (Panel **D**), Jpn2 (Panel **E**). Panel **F**: density of bristle cells per 200 μm^2. Bars indicating standard error (n = 3).

2.2. Maternal Lineages

In order to trace maternal lineages, next-generation sequencing data obtained from Jpn1 and Jpn2 were used for re-sequencing and comparison with the Nipponbare complete cp genome sequence. More than 53 million reads were obtained from the two accessions. Two genome sequences of *O. meridionalis*, and *O. rufipogon* were added for comparison. In all cases, 100% coverage was achieved with 733 to 2002 mean depth. When the nuclear genome was used as a reference genome, 66%–88% coverage with 7.6 to 11.4 mean depth was obtained.

Simple sequence repeats were found at 20 loci in the cp genomes. Simple insertions or deletions (INDELs) were also found at 21 loci (Table 3). Two loci were not amplified, and six loci were not confirmed because of difficulty of primer design for these fragments. One region ranging from nucleotide 17,336 to 17,392 of the Nipponbare cp genome was amplified as a single amplicon because of its short size. In total, 29 insertions/deletions (INDELs)/simple sequence repeats (SSRs) in the cp genome were polymorphic. Australian rice accessions including *O. meridionalis*, Jpn1, and Jpn2 shared the same genotype at 26 out of the 29 loci developed by plastid INDELs and SSRs.

Table 3. INDEL and SSR markers in chloroplast genomes and developed markers.

Marker Type	Genotype (Based on Relative Migration Distance)								INDEL	INDEL Start Position in Nipponbare cp Genome	Forward	Reverse
	Nipponbare	W0106	W0120	W0137	W1299	W1300	Jpn1	Jpn2				
INDEL1	2	2	2	2	2	2	1	2	-C	1605	CTATTCCGAAGAGGAAGTCTAC	TCTCCGTATCAATGATCTGGTG
SSR1	1	1	1	1	2	2	3	2	+AA	3535	CTTTTGACTTTGGGATACAGTC	GATTAGTGCCTGATGTAGGG
INDEL2	2	2	2	2	1	1	1	1	-CAATC	5852	GGAATTTCCATCCTCAACAGA	GTTTTGTTACGGAAAAATGGTATG
SSR2	1	1	1	1	2	2	2	2	+A	6098	TTCTCGTATTTCTTCGACTCG	GATAAGAACTGCTCGTTAGATAG
INDEL3	1	1	1	1	2	2	2	2	+AGAAA	8192	GCCGCTTTAGTCCACTCAGCCATC	TCAATGCCTTTTTTCAATGGTCTC
SSR3	1	1	1	1	2	2	2	2	+A	11441	CTGGCTCGGTTATTCTATC	GAAAACCGGTATAGTTCTAGG
INDEL4	2	1	2	1	1	1	1	1	-AGGG	12669	GCAACAGGGTTCCCTAAACCG	GCCAAATTGAGCAGGTTGCG
INDEL5	2	2	2	2	1	1	1	1	-T	13566	GCTTCGCGACTCTGTACTCA	TACTTAAGGCGTCCTTAAGG
INDEL6	2	1	2	1	1	1	1	1	-AC	14011	GAAATCTGGGCCATAGAGAA	CTAAGCAGAGACATTCAGAATC
INDEL7	NA								-TATTTCTAAGA	14527	-	-
SSR4	2	2	2	2	1	1	1	1	-A	17099	GAAAAAATCCATGGAGGGAGAG	CCCAACATATCGCACATTTTCC
SSR5	2	1	1	1	3	3	3	3	-TTTCTA	17336	GGTCGCTTCTAGTAGCGATTATG	TGCCGAACTTTATTCTTTCTCTC
SSR6	including SSR5 to INDEL10								-TTTCTA	17358		
INDEL8									-ATAGAA	17379		
INDEL9									+AGAATTAT	17385		
INDEL10									+GAATTATATAGAAC	17392		
INDEL11	1	1	1	1	2	2	1	2	+TGG	19001	GAATATCATAAACTGTAAGTGGCAG	CACATGAAATTCTCGGGAACTCC
SSR7	1	1	1	1	2	2	2	2	+T	41464	GAGGCAAGTGTTCGGATCTATTATG	CTATATTATGCTCAAGGAAAGTAGA
SSR8	1	1	2	1	1	1	1	1	-TATAT	46086	CTCTAATTCGCAAATCTATTTTTC	CAAGAAATTCGCATGTTCTC
SSR9	2	1	2	1	1	1	1	1	-T	46174	GAGAACATGCGAATTTCTTG	CATACTATAACGCTTGATATTC
INDEL12	1	2	1	2	2	2	2	2	+T	47211	GTCGTGAGGGTTCAAGT	CGAGTTAATAATCGACATTCCTTGCC
INDEL13	1	1	1	1	2	2	1	2	+AGGAC	50351	GCCTGTCCAGTCTATAAACAAG	GGGTCTTTGAAACAGTTCG
SSR10	1	1	1	1	2	2	2	2	+T	53999	CATAGAATGTACACAGGGTGTACCC	CTCACAACGACAGGGTCTAC
INDEL14	INDEL14 and SSR11								-CTTTTTTTTTAGAATA	57017	GGATAGAAAGGCCGCGAG including	GACTATTGTATTTTTGAGTTTGC
SSR11									-A	57061		
INDEL15	2	2	2	2	1	1	1	1	-CTTTTCAAT	64815	CCAGATGCTTTGTCATTCCC	TCATGACTCTAAGGTCCAACC
INDEL16	1	1	1	1	2	2	2	2	+TTCCTATTTAATA	65452	GTCGTTATTGTCGTAAGCATACGA	GATGAATACCCTCGATACATATG
SSR12	NA								+T	65615	-	-
INDEL17	1	1	1	1	2	2	2	2	+t	66896	CCAATGGCTTTTGCTACTATAACC	GAAAGAAAGGGCTCCGGTG
SSR13	1	1	1	1	2	2	2	2	+A	71377	GCACCTGTTATCTCTATCAAG	GTCTGGTTGCGAGGTCTGAATAG
SSR14	2	2	2	2	1	1	1	1	-TTTCTA	75980	GATATCCGTTTCAGGGTAAA	CTGATTCGTAGGCGTGGAC
SSR15	2	1	2	1	1	1	1	1	-A	76232	CAAATTTTACGAACAGAAGCTC	CCGAAGACTCGAAGGATACC

Table 3. *Cont.*

Marker Type	Genotype (Based on Relative Migration Distance)									INDEL Start Position in Nipponbare cp Genome		
	Nipponbare	W0106	W0120	W0137	W1299	W1300	Jpn1	Jpn2	INDEL		Forward	Reverse
SSR16				NA					-T	76574	CATAACTAAACCCTCGAAAGTAA	CCCGCCTATAGCGGTAATC
INDEL18	2	3	2	3	1	1	1	1	-T	77728	GCTACATTTAAAAGGGTCTGAGG	CTGCCAGCAAAATGCCC
SSR17				NA					-T	78423	-	-
INDEL19	1	1	1	1	2	2	2	2	+T	80090	GGGTTGTACCAAGTCTGAA	GCTCGAGGACGTAGTTCTCCCATAA
SSR18				NA					-C	93004	-	-
SSR19	1	1	1	1	2	2	2	2	+C	93534	GTTCGTCCTCAATGGGAAAATG	GGGAAGTCCTATTGATTGCTG
INDEL20	1	2	2	2	2	2	2	2	+AACA	104530	GATCATTTTCTGGCGTCAGCG	GAATATTGTACCGAGGAATTCG
INDEL21	1	1	1	1	2	2	2	2	+G	121618	AAGGCTCGAATGGTACGATC	CTTCTCGAGAATCCATACATCCC
SSR20				NA					-G	122142	-	-

NA: not amplified.

Five chloroplast markers, INDEL1, INDEL11, INDEL13, INDEL18, and INDEL19, represented polymorphisms among natural populations (Table S2 (Supplementary Materials)). Plastid types were defined as distinct combinations of each genotype. In total, nine plastid types (Type 1 to 9, r1, and r2) with r1 and r2 types in the control *O. rufipogon*, were recognized. Asian *O. rufipogon* and *O. sativa* accessions, were obviously different from the Australian wild rices.

Three accessions in PNG *O. rufipogon*, W1235, W1238, and W1239, and W2109 in Australian *O. rufipogon* shared the Type 5 plastid type with *O. meridionalis*. W1230 in Papua New Guinea *O. rufipogon* shared the r2 plastid type with the Asian type. W1236 carried a unique plastid type. Jpn2 shared Type 1 with *O. meridionalis*. Other *O. meridionalis* in the core collection divided into three types, Types 1, 5, and 8. Only two types, Types 1 and 2, were detected in the Northern Territory and in Western Australia. Newly collected accessions from Queensland carried seven types. Five of them were newly detected.

2.3. Reproductive Isolation

Biological species can be detected by the pollen fertility of hybrids. Jpn1 and Jpn2 were crossed with Asian wild rice and *O. meridionalis*. Each F_1 plant was grown in a greenhouse, and leaf samples were used to check whether they were hybrids originating from the cross. Anthers were taken to check pollen fertility by staining with I_2–KI. Well-stained pollen grains were counted.

Seed fertility was also assessed but this may not reflect reproductive ability of the respective plants (Table 4). W0106, W0120, and W1299 showed more than 95% pollen fertility. However, except for W0120, they showed lower seed fertility of 19.5% in W0106 and 22.3% in W1299. The panicles were bagged to prevent out-crossing and this might explain the low seed fertility. In combinations with Jpn1 and Asian *O. rufipogon*, F_1 plants with W0106 and W0120 had more than 90% pollen fertility. However, seed fertility was relatively low, similar to self-pollination of W0106 and W1299. We relied on data from pollen fertility rather than seed fertility and concluded that by this criterion, Jpn1 is related to Asian *O. rufipogon*, and that Jpn2 is not close to either *O. rufipogon* or *O. meridionalis*.

Table 4. Pollen and seed fertility of self pollinated plants and F_1 plants among Asian *O. rufipogon*, *O. meridionalis* and alternative perennials in Australia.

		Pollen and Seed Fertilitty (%)							
Female	**Male**	**Self Pollinated**		**Crossed with Jpn1**		**Crossed with Jpn2**		**Crossed with W1297**	
		Pollen	**Seed**	**Pollen**	**Seed**	**Pollen**	**Seed**	**Pollen**	**Seed**
W0106		97.2	30.1	87.2	15.7	10.2	0.0	0.1	0.0
		98.8	8.9	95.7	18.3	0.8	0.0	0.1	0.0
	Mean*	98.0	19.5	91.4	17.0	5.5	0.0	0.1	0.0
W0120		98.1	96.8	84.0	24.5	10.2	0.0	21.3	0.0
		98.7	74.5	78.3	21.6	12.7	0.0	20.2	0.0
	Mean	98.4	85.6	81.2	23.1	11.5	0.0	20.8	0.0
W1299		95.7	20.0	0.0	0.0	29.4	0.0	99.3	42.9
		96.1	24.5	4.5	0.0	39.8	0.3	-	-
	Mean	95.9	22.3	2.2	0.0	34.6	0.1	-	-
W1297		95.7	54.4	5.3	0.0	53.2	0.0	-	-
		96.9	75.9	6.4	0.0	33.8	0.0	-	-
	Mean	96.3	65.2	2.2	0.0	34.6	0.1	-	-

*Mean: data obtained from multiple plants were averaged and noted the mean.

2.4. Unique Insertion of Retrotransposable Element in Jpn2

In total, six presumed insertions were confirmed only in the Jpn2 genome but not in Nipponbare (Table 5). Two *pSINE1* insertions were shared among Jpn2 and 19 *O. meridionalis* accessions. Another insertion amplified with Chr3-10559212-r (w/L) and pSINE1-L showed an insertion shared among Jpn1, Jpn2, and 19 *O. meridionalis* accessions (Figure 2). Chr1-4067055-f (w/L) and pSINE1-L amplified the same amplicons not only from Jpn2 and 19 *O. meridionalis* accessions but also with W0106, which originated in India, suggesting that some parts of the Jpn2 genome share the insertion with wild rice from India. No *O. rufipogon* accessions in the core collection except for W2266 and W2267 were tested because of lack of DNA, and 19 *O. meridionalis* showed these insertions. Results suggested that the insertion was probably shared among *O. meridionalis* and W0106. Chr3-10203820-f (w/L) can amplify with pSINE1-L only in Jpn2 and no other *O. meridionalis* showed any amplicons. In screening for the insertion among 30 *O. rufipogon* accessions in the core collection, W0180 and W1921, both of which originated from Thailand, showed amplicons. The insertion sequence in Jpn2 was screened from the raw reads and 53 bp were recovered. When aligned with *pSINE1*, 92.4% high similarity was retained. When *pSINE3* insertions were examined, three of the presumed insertions were amplified only among Jpn2 and 19 *O. meridionalis* accessions.

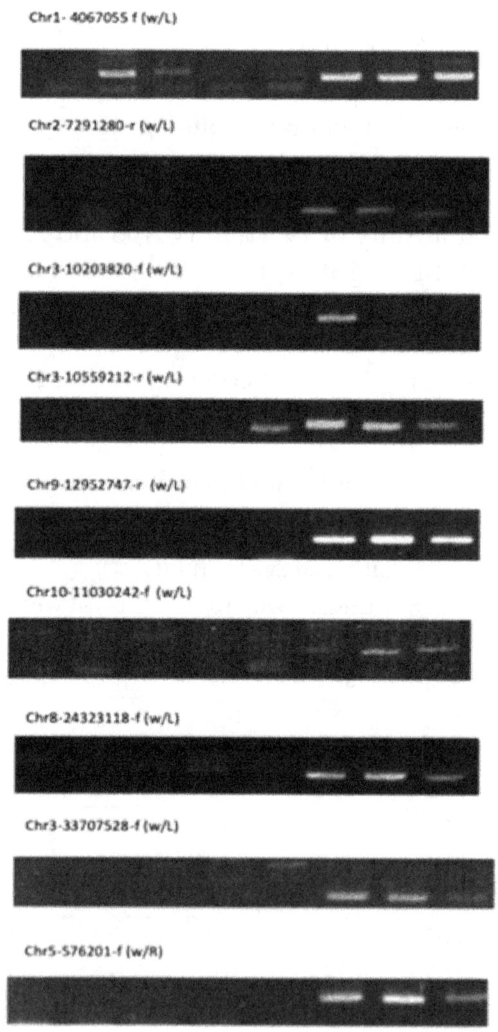

Figure 2. *pSINE1* and *pSINE3* insertions amplified with flanking primers and outward primers from *pSINE* consensus sequences. From lane 1 to 8, Nipponbare, W0106, W0120, W0137, Jpn1, Jpn2, W1299, and W1300 were used as each DNA template.

Table 5. Screening retrotransposable element insertions.

Flanking Primers to Confirm *pSINE1/3* Insertions		Genome	Insertion (+) / no-Insertion (-)								
with Inner Primers toward Outside	**Sequence**	**Position**	**Nipponbare**	**W0106**	**W0120**	**W0137**	**Jpn1**	**Jpn2**	**W1299**	**W1300**	**Remarks**
pSINE1											
Jpn2-meridionalis class											
Chr2-7291280-r (w/L)	TCTCTCTACAGATAATGCTC	7291535	-	-	-	-	-	+	+	+	Other *O. meridionalis*
Chr9-12952747-r (w/L)	CACACCCATCTACATCGATG	12953007	-	-	-	-	-	+	+	+	Other *O. meridionalis*
Chr10-11030242-f (w/L)	GATTGCCGGCTTCTTTACTAG	11029860									
Australia class											
Chr3-10559212-r (w/L)	ACCTATAACAACTGAGAGAC	10559538	-	-	-	-	+	+	+	+	Other *O. meridionalis*
Jpn2-meridionalis-W0106 class											
Chr1-4067055-f (w/ L)	GAAAGAGATCACAGGTAAAC	4066713	-	+	-	-	-	+	+	+	Other *O. meridionalis*
Jpn2-W0180,W1921 class											
Chr3-10203820-f (w/L)	TCCACCGACTTATAAATCAC	10203450	-	-	-	-	-	+	-	-	No other *O. meridionalis*, but two *O. rufipogon*,W0180, W1921
Inner primer toward outside											
pSINE1-L	GAAGACCCCTGGGCGTTTCT										
(paired primer to amplify *pSINE1* insertion)											
pSINE3											
Chr3-33707528-f (w/L)	GTGTAAATATGTATTGTACC	33702014	-	-	-	-	-	+	+	+	Other *O. meridionalis*
Chr8-24323118-f (w/L)	GCCTATTACTATCAATCACC	24322813	-	-	-	-	-	+	+	+	Other *O. meridionalis*
Chr5-576201-f (w/R)	GATAACTAGGGTAAATGAC	575868		-	-	-	-	+	+	+	Other *O. meridionalis*
Inner primer toward outside											
pSINE3-R	TCCTTCCTAGATTGGTCCC										
pSINE3-L	TGCTAGCCGGGAAGACC										
(paired primer to amplify *pSINE3* insertion)											

3. Discussion

3.1. Unique Morphological Traits in Australian Wild Rice

O. rufipogon is composed of a continuum of annual and perennial strains in Asia. They represent different life history traits related to the r-K strategy to maximize fitness [4,12]. Perennial and annual types are regarded as K- and r-strategists, respectively. Intermediates represented the r–K continuum. K selection works for individuals to increase their life span and r selection works to produce more offspring. Thus, perennials spend more energy on vegetative organs before the flowering stage. Annuals spend energy to produce more panicles and seeds. Because anther size is related to preference for outcrossing, perennials tend to carry longer anthers than annuals and produce more pollen to maximize the chance of outcrossing [3,4,12]. Such resource allocation was also confirmed in three Asian *O. rufipogon*, the Australian perennial Jpn1, and the Oceanian annual *O. meridionalis*. In order to adapt to the dry season, *O. meridionalis* plants produce many seeds and die after scattering their seeds. *O. meridionalis* has short anthers and slender panicles. The appearance of Jpn1 is similar to Asian *O. rufipogon*, with similar long anthers and open panicles. Our measurements also suggested a trend. In our previous paper, we reported that Jpn2 represents a perennial life history [14]. It generated shoots and roots from its stems to follow water in peripheral areas, growing to the inner side because of the shrinking water mass during the dry season. The morphological appearance of Jpn2 was quite different to Jpn1. Anthers length is a unique characteristic in the morphology of this type, being shorter in *O. meridionalis* [14]. *O. rufipogon* W0106, an annual type, also shared this short anther characteristic. In this study, we also demonstrated another morphological trait characteristic of Jpn2. Jpn2 has a high density of bristle cells along the awn. Genome sequencing also suggested that Jpn2 shared higher similarity to *O. meridionalis* than *O. rufipogon* [16]. These characteristics of Jpn2 infer that this species/taxon has diverged from *O. meridionalis*.

3.2. Maternal Variation

Cytoplasmic marker systems can be developed for the mitochondrial genome as suggested in this report. Other markers were also developed based on whole cp genome sequences. Whole cp genome sequences have been determined for several Australian accessions [15–19]. The maternal genome data clearly showed Jpn1 and Jpn2 shared high similarity to *O. meridionalis* with some variation. INDELs and SSRs were designed based on the cp genome sequences. Core collections and natural populations were examined to determine the distribution of cp variation. Higher variation was found among accessions in Queensland compared with others accessions collected from the Northern Territory and Western Australia. Variations in cp genomes were distinguished at high resolution using single nucleotide polymorphisms [17,18]. This study showed that easily scored INDELs and SSRs also detected higher diversity in the northern Queensland accessions. These marker systems with whole cp genome screening will provide more clues about the maternal relationships among these related species/taxa and how they diverged.

3.3. Reproductive Barriers Among Australian Wild Rice

Reproductive barriers among *Oryza species* including the Australian species have been confirmed and numerical characteristics supported speciation reproductive barriers [2]. *O. meridionalis* already has high genetic reproductive barriers and sterility detected in F_1 lines of crosses with Asian wild rice. Even among *O. meridionalis*, some sterile lines were reported [9]. In our study, F_1 between Jpn2 and *O. meridionalis* displayed reproductive sterility. Jpn2 in particular, developed a reproductive barrier with both Asian wild rice and *O. meridionalis*. Because the extent of the reproductive barrier corresponded to that of two different organisms, it is concluded that Jpn2 does not belong to *O. rufipogon* or *O. meridionalis*. We have not yet determined when they diverged from each other. Clade analysis of cp genomes suggested that a clade including *O. meridionalis* diverged at a date estimated as 0.86–11.99 million years ago [18]. Similar estimation has also been reported based on sequences among *Oryza* genus [19–22]. Such a long time since divergence has allowed the accumulation of quite diverse genomes in the north-eastern part of Australia and created Jpn2 and various wild rice found at the P5 site.

F_1 hybrids between Jpn2 and *O. meridionalis* showed relatively high pollen fertility with *O. meridionalis*, although the F_1 showed complete seed sterility. It was suggested that the divergence between these two plants is a relatively recent event compared with the divergence from other species.

3.4. Retrotransposable Elements

Retrotransposable elements are well known markers for examining evolutionary pathways. This is mainly due to the unique mechanisms of transpositions. Two retrotransposable elements, *pSINE1* and *pSINE3*, were recognized in species divergence among AA genome and between Asian wild rice and *O. meridionalis* [3,6,23]. These have offered researchers a powerful tool for phylogenetic analysis. On the other hand, when there are no genome sequences, new markers to distinguish particular genomes are not available. In fact, there was no genomic data on insertions in a novel taxon such as Jpn2. Thus, we established a screening methodology to extract retrotransposable elements from raw reads. Recent developments in sequencing technology offer huge numbers of reads to increase target sites. Even with our limited volume of data, we succeeded in picking up insertions in the Jpn2 genome. The uniqueness of Jpn2 was also found with an insertion of the *pSINE1* retrotransposon, which was detected in Jpn2 only and in none of the other accessions of *O. meridionalis*. Two accessions of *O. rufipogon* in Thailand may provide key information on how Australian wild rice originated. This tool will open the way to draw a more precise evolutionary pathway and to understand valuable genetic resources among wild rice.

4. Materials and Methods

4.1. Plant Materials

Wild rice was collected in Australia with permission from the Queensland government, EcoAccess. We developed these collections as de novo resources, which can be accessed repeatedly from the same site with accurate GPS data allowing us to reconfirm their life cycles. Successive observations were made from 2009 until 2011, and the life history traits at the collection sites were reconfirmed year by year. This field research was supported by overseas scientific research funds (JSPS) and collaborative research with the Queensland Herbarium and Queensland Alliance for Agriculture and Food Innovation (QAAFI), University of Queensland. Thirty populations were collected from their natural habitat. Observation of the ecological habitats and life cycle of each population in April 2008, August 2009, and September 2009 were used to determine their life history such as annual or perennial behavior especially for Jpn1 and Jpn2 populations. Jpn1 and Jpn2 were typical perennial sites and individuals survived as living plants in a swamp (Jpn1) or a pond (Jpn2). In order to compare these accessions with cultivated rice, *Oryza sativa*, and wild species, *O. rufipogon* (W0106, W0120, and W0137) and *O. meridionalis* (W1297, W1299 and W1300) were compared. All plant materials were grown in greenhouse conditions at Kagoshima University. Samples collected from nature were compared with Jpn1 and Jpn2 grown from seeds collected to compare environmental effects on anther length and lemma size. Jpn1 and Jpn2 were crossed with W0106, W0120, W1297, and W1299 to test these relationships. The density of bristle cells on the surface of awns per 100 μm^2 was counted using a scanning electric microscope (JSM-7000F, JEOL co., Japan). A core collection derived from the National Bio-Resource (NBR) Project in Japan was kindly provided, as shown in Table S1 (Supplementary Materials) [24].

4.2. Crossing and Fertility Test

Jpn1 and Jpn2 were crossed with Asian wild rice, W106 and W120, and to Australian *O. meridionalis*, W1299 and W1300. F_1 plants were grown in a greenhouse at Kagoshima University. Anthers were taken and stained with I_2–KI solution and well-stained grains were counted as fertile pollen. The remaining panicles were wrapped in paraffin bags to prevent outcrossing. Fully filled grains were counted to calculate seed fertility.

4.3. Data Mining from Whole Genome Sequences

Whole genome sequences of Jpn1 and Jpn2 were obtained using Illumina GAIIx to develop INDEL markers of the cpDNA and retrotransposon INDEL markers. The total numbers of pair-end reads were 52,087,744 for Jpn1 and 54,749,858 for Jpn2. Total nucleotides sequenced were 3.9 Gb for Jpn1 and 4.1 Gb for Jpn2. Mean depth was 2022 in Jpn1 and 2002.5 in Jpn2. With our draft data, we aligned these raw reads to the cpDNA of Nipponbare (GenBank: GU592207.1) using CLC-work bench genomics version 6.0. Several INDELs were grouped together to screen for using single PCR reactions with the designed markers listed in Table 5.

Retrotransposable elements *pSINE1* and *pSINE3* have been reported to be uniquely found in either species, *O. rufipogon* or *O. meridionalis* [7,8,23]. Consensus sequences of the 5′ and 3′ termini were used to design a consensus probe to screen the draft sequence data. Based on the alignment of *pSINE1* elements, consensus sequences were presumed [8]. Two probe sequences were applied for *pSINE1* to screen the data: CCA.CA.CTTGTGGAGCTAGCCGG, in which the periods indicate degenerate nucleotides, for the 5′ termini, and TAGGT.TTCCCTAATATTCGCG for the 3′ termini. These degenerate probes were applied to screen raw reads of Jpn1 and Jpn2. 5′ termini which were confirmed to carry AAGACCCCTGGGCATTTCTC as the complementary sequence. Then, internal sequences of the 5′ probe were obtained from the read to confirm whether it shares homology, ranging from 74% to 82%. In these cases, we adopted the outside of 5′ termini as flanking sequences of *pSINE1* insertions. 3′ termini carried TAG followed by a poly T stretch. We adopted the downstream sites as flanking sequences of *pSINE1* insertions.

Based on *pSINE3* family elements, *r3004*, *r3005*, *r3012*, and *r3024*, the 5′ terminal consensus probe GCCGGGAAGACCCCGGGCC was used to screen internal sequences. The internal sequences were used to design an internal probe, CTAGCTCAGCTTGTGCTA. In order to examine the flanking sequences of insertions, the consensus probe was applied. After confirming the 5′ end shared the 5′ terminus of *pSINE3*, the outside sequences from TTTCTC were regarded as pSINE3 insertions.

When multiple reads were obtained as single locations, we specified the genome position based on the Nipponbare genome and detected 14 and 16 insertions that did not overlap. Of these, 21 could be aligned to the Nipponbare genome without *pSINE1* insertions at the site. Flanking sequences in the Nipponbare genome were applied to design primers to amplify the presumed insertions of either *pSINE1* or *pSINE3*. Outward primers inside *pSINE1* or *pSINE3* were also designed as shown in Table 5. Preliminary screening was performed with Nipponbare, three *O. rufipogon*, W0106, W0120, and W0137, two *O. meridionalis*, W1299 and W1300, and Jpn1, and Jpn2.

4.4. Data Analysis

Dendrograms were constructed using the neighbor-joining method based on Nei's unbiased genetic distances by Populations1.2.30 beta2 program, which was downloaded from http://bioinformatics.org/~{}tryphon/populations/#ancre_bibliographie. All dendrograms were drawn by the TreeExplorer software used to show and edit population dendrograms as supplied with MEGA [25].

5. Conclusions

These data suggested that Jpn2 (taxon B) may be a distinct new species belonging to the *Oryza* genus and isolated from other species by reproductive barriers.

Author Contributions: Conceptualization, R.I., R.J.H., and K.I.; methodology, R.I.; software, R.I.; validation, D.T.L., R.I., R.J.H., and K.I.; formal analysis, D.T.L., R.J.H., K.I., and R.I.; investigation, D.T.L., R.J.H., K.I., and

R.I.; resources, R.I., R.J.H., K.I., and R.I.; writing—original draft preparation, R.I.; writing—review and editing, D.T.L., R.I., R.J.H., and K.I.; visualization, D.T.L. and R.I.; supervision, R.I.; project administration, R.I.; funding acquisition, R.I. All authors have read and agreed to the published version of the manuscript.

Acknowledgments: We acknowledge Bryan Simon, who supported our mission in Australia.

Abbreviations

Cp	chloroplast
INDEL	insertion/deletion
SSR	simple sequence repeat

References

1. Brar, D.S.; Khush, G.D. Wild relative of rice: A valuable genetic resources for genomics and breeding research. In *The Wild Oryza Genomes*; Mondal, T.K., Henry, R.J., Eds.; Springer International Publishing AG: Basel, Switzerland, 2018; pp. 1–25.
2. Morishima, H. Phenetic similarity and phylogenetic relationships among strains of *Oryza perennis*, estimated by methods of numerical taxonomy. *Evolution* **1969**, *23*, 429–443. [CrossRef]
3. Oka, H.-I.; Morishima, H. Variations in the breeding systems of a wild rice, *Oryza perennis*. *Evolution* **1967**, *21*, 249–258. [CrossRef]
4. Morishima, H.; Oka, H.I.; Chang, W.T. Directions of differentiation in populations of wild rice, *O. perennis* and O. sativa f. spontanea. *Evolution* **1961**, *15*, 326–339. [CrossRef]
5. Ng, N.Q.; Hawkes, J.G.; William, J.T.; Chang, T.T. The recognition of a new species of rice (Oryza) from Australia. *Bot. J. Linn Soc.* **1981**, *82*, 327–330. [CrossRef]
6. Henry, R.J.; Rice, N.; Waters, D.L.E.; Kasem, S.; Ishikawa, R.; Hao, Y.; Dillon, S.; Crayn, C.; Wing, R.; Vaughan, D. Australian Oryza: Utility and Conservation. *Rice* **2010**, *3*, 235–241. [CrossRef]
7. Cheng, C.; Tsuchimoto, S.; Ohtsubo, H.; Ohtsubo, E. Evolutionary relationships among rice species with AA genome based on SINE insertion analysis. *Genes Genet. Syt.* **2002**, *77*, 323–334. [CrossRef]
8. Xu, J.-H.; Osawa, I.; Tsuchimoto, S.; Ohtsubo, H.; Ohtsubo, E. Two new SINE elements, *p-SINE2* and *p-SINE3*, from rice. *Genes Genet. Syst.* **2005**, *80*, 161–171. [CrossRef]
9. Juliano, A.B.; Elizabeth, M.; Naredo, B.; Lu, B.-R.; Jackson, M.T. Genetic differentiation in *Oryza meridionalis* Ng based on Molecular and Crossability Analyses. *Genet. Resour. Crop Evol.* **2005**, *52*, 435–445. [CrossRef]
10. Elizabeth, M.; Naredo, B.; Juliano, A.B.; Lu, B.-R.; Jackson, M.T. Hybridization of AA genome rice species from Asia and Australia, I. Crosses and development of hybrids. *Genet. Resour. Crop Evol.* **1997**, *44*, 17–23.
11. Lu, B.-R.; Elizabeth, M.; Naredo, B.; Juliano, A.B.; Jackson, M.T. Hybridization of AA genome rice species from Asia and Australia II. Meiotic analysis of *Oryza meridionalis* and its hybrids. *Genet. Resour. Crop Evol.* **1997**, *44*, 25–31. [CrossRef]
12. Sano, Y.; Morishima, H. Variation in resources allocation and adaptive strategy of a wild rice, *Oryza perennis* Moench. *Bot. Gaz.* **1982**, *143*, 518–523. [CrossRef]
13. Vaughan, D.A. *The Wild Relatives of Rice-A Genetic Resources Handbook*; IRRI: Manila, Philippines, 1994; p. 64.
14. Sotowa, M.; Ootsuka, K.; Kobayashi, Y.; Hao, Y.; Tanaka, K.; Ichitani, K.; Flowers, J.M.; Purugganan, M.D.; Nakamura, I.; Sato, Y.-I.; et al. Molecular relationships between Australian annual wild rice, Oryza meridionalis, and two related perennial forms. *Rice* **2013**, *6*, 26. [CrossRef]
15. Waters, D.; Nock, C.J.; Ishikawa, R.; Rice, N.; Henry, R.J. Chloroplast genome sequence confirms distinctiness of Australian and Asian wild rice. *Ecol. Evol.* **2012**, *2*, 211–217. [CrossRef]
16. Brozynska, M.; Omar, E.S.; Furtado, A.; Crayn, D.; Simon, B.; Ishikawa, R.; Henry, R.J. Chloroplast Genome of Novel Rice Germplasm Identified in Northern Australia. *Trop. Plant Biol.* **2014**, *7*, 111–120. [CrossRef]
17. Brozynska, M.; Copetti, D.; Furtado, A.; Wing, R.; Crayn, D.; Fox, G.; Ishikawa, R.; Henry, R.J. Sequencing of Australian wild rice genomes reveals ancestral relationships with domesticated rice. *Plant Biot. J.* **2016**, *15*, 765–774.

18. Moner, A.M.; Furtado, A.; Chivers, I.; Fox, G.; Crayn, D.; Henry, R.J. Diversity and evolution of rice progenitors in Australia. *Ecol. Evol.* **2018**, *8*, 4360–4366. [CrossRef]
19. Wambugu, P.W.; Brozynska, M.; Furtado, A.; Waters, D.; Henry, R.J. Relationships of wild and domesticated rices (Oryza AA genome species) based upon whole chloroplast genome sequences. *Sci. Rep.* **2015**, *5*, 13957. [CrossRef]
20. Tang, L.; Zou, X.H.; Achoundong, G.; Potgieter, C.; Second, G.; Zhang, D.Y.; Ge, S. Phylogeny and biogeography of the rice tribe (*Oryzeae*): Evidence from combined analysis of 20 chloroplast fragments. *Mol. Phylogenetics Evol.* **2010**, *54*, 266–277. [CrossRef]
21. Kim, K.; Lee, S.C.; Lee, J.; Yu, Y.; Yang, K.; Choi, B.S.; Koh, H.-J.; Waminal, N.E.; Choi, H.-I.; Kim, N.-H.; et al. Complete chloroplast and ribosomal sequences for 30 accessions elucidate evolution of Oryza AA genome species. *Sci. Rep.* **2015**, *5*, 15655. [CrossRef]
22. Zhang, Q.-J.; Zhua, T.; Xiaa, E.-H.; Shia, C. Rapid diversification of five Oryza AA genomes associated with rice adaptation. *Proc. Natl. Acad. Sci. USA* **2014**, *46*, E4954–E4962. [CrossRef]
23. Xu, J.-H.; Kurta, N.; Akimoto, M.; Ohtsubo, H.; Ohtsubo, E. Identification and characterization of Australian wild strains of *Oryza meridionalis* and *Oryza rufipogon* by SINE insertion polymorphism. *Genes Genet. Syst.* **2005**, *80*, 129–134. [CrossRef]
24. Nonomura, K.I.; Morishima, H.; Miyabayashi, T.; Yamaki, S.; Eiguchi, M.; Kubo, T.; Kurata, N. The wild Oryza collection in National BioResource Project (NBRP) of Japan: History, biodiversity and utility. *Breed. Sci.* **2010**, *60*, 502–508. [CrossRef]
25. Kumar, S.; Dudley, J.; Nei, M.; Tamura, K. MEGA: A biologist-centric software for evolutionary analysis of DNA and protein sequences. *Brief Bioinform.* **2008**, *9*, 299–306. [CrossRef]

14

Rice Novel Semidwarfing Gene *d60* can be as Effective as Green Revolution Gene *sd1*

Motonori Tomita [1,*] and Keiichiro Ishimoto [2]

1 Research Institute of Green Science and Technology, Shizuoka University, 836 Ohya, Suruga-ku, Shizuoka City, Shizuoka 422-8529, Japan
2 Faculty of Agriculture, Tottori University, 4-101 Koyama Minami, Tottori 680-8550, Japan; kei-ishimoto@pref.nagasaki.lg.jp
* Correspondence: tomita.motonori@shizuoka.ac.jp

Abstract: Gene effects on the yield performance were compared among promising semidwarf genes, namely, novel gene *d60*, representative gene *sd1* with different two source IR8 and Jukkoku, and double dwarf combinations of *d60* with each *sd1* allele, in a Koshihikari background. Compared with the culm length of variety Koshihikari (mean, 88.8 cm), that of the semidwarf or double dwarf lines carrying Jukkoku_*sd1*, IR8_*sd1*, *d60*, Jukkoku_*sd1* plus *d60*, or IR8_*sd1* plus *d60* was shortened to 71.8 cm, 68.5 cm, 65.7 cm, 48.6 cm, and 50.3 cm, respectively. Compared with the yield of Koshihikari (mean, 665.3 g/m^2), that of the line carrying Jukkoku_*sd1* allele showed the highest value (772.6 g/m^2, 16.1% higher than Koshihikari), while that of IR8_*sd1*, *d60* and IR8_*sd1* plus *d60*, was slightly decreased by 7.1%, 5.5%, and 9.7% respectively. The line carrying Jukkoku_*sd1* also showed the highest value in number of panicles and florets/panicle, 16.2% and 11.1% higher than in Koshihikari, respectively, and these effects were responsible for the increases in yield. The 1000-grain weight was equivalent among all genetic lines. Except for the semidwarf line carrying Jukkoku_*sd1*, semidwarf line carrying *d60* was equivalent to line carrying IR8_*sd1*in the yield of unpolished rice, and yield components such as panicle length, panicle number, floret number /panicle. Therefore, the semidwarfing gene *d60* is one of the best possible choices in practical breeding.

Keywords: rice; semidawarf gene; *d60*; *sd1*; yield component; phenotyping; growth

---

1. Introduction

Semidwarfing prevents plants from lodging at their full-ripe stage, making them lodging-resistant to wind and rain, enhances their adaptability for heavy manuring and markedly improved the global productivity of rice and wheat between 1960–1990 (up to double yields of rice and quadruple yields of wheat) [1,2]. Semidwarf rice contributes stable production in the monsoonal regions of Asia, where typhoons frequently occur during the yielding season and also brings benefits such as erect leaf angle, reduced photoinhibition, and possibility to plant at higher densities to japonica varieties grown in California and also in South America [3]. However, gene source of semidwarfness is limited. The International Rice Research Institute (IRRI) developed a semidwarf rice variety IR8 in 1966 by using Taiwanese native semidwarf variety Dee-geo-woo-gen (DGWG). IR8 called as Miracle Rice, has been improved with lodging resistance and light-reception attitude, and it brought the Green Revolution in tropical Asia [2]. In Japan, semidwarf cultivars in the Kyushu region were developed in the 1960s using the native semidwarf variety Jukkoku [4]. In the Tohoku region, semidwarf cultivars were developed in the 1970s using the semidwarf mutant Reimei induced by Fujiminori-gamma-ray irradiated [5]. In the United States, Calrose 76 was developed in 1976 by Calrose-gamma-ray irradiated [6,7].

Genetic study has also been devoted on the genes responsible for semidwarfism in rice. First, a recessive semidwarf gene *d47* was identified in DGWG, the parental line of IR8 [8,9]. Next, the semidwarf gene *sd1* in Calrose 76 was shown to be allelic to *d47* [10,11]. Finally, semidwarf genes in Taichung Native 1 descend from DGWG, Shiranui from Jukkoku, and *d49* in the mutant cultivar Reimei were attributed to the same allele by allelism examination [12–14]. Therefore, only a single semidwarf gene, *sd1*, has been commonly used across the world. A little genetic source of current semidwarf rice cultivars have a risk for environmental change. Thus, it is an emerging subject to find a novel semidawrf gene to replace *sd1* and to utilize it to diversify genetic variations of semidwarf rice worldwide.

A novel semidwarf gene, *d60*, which was found in the mutant Hokuriku 100 induced by irradiation of 20 kR of gamma-ray to Koshihikari, is thus of particular importance [15]. While *sd1* is on rice chromosome 1 [16,17], *d60* is located on chromosome 2 (Tomita et al., submitted to Genes). *sd1* is a defective allele encoding GA20-oxidase gene in a late step in the GA biosynthesis pathway [18–20]. Moreover, unlike *sd1*, *d60* complements the gamete lethal gene, *gal*. Therefore, in the cross between Hokuriku 100 (*d60d60GalGal*) and Koshihikari (*D60D60galgal*), male and female gametes, in which *gal* and *d60* coexistent, become lethal and the pollen and seed fertility in the F_1 (genotype *D60d60Galgal*) breakdown to 75%. As a result, the F_2 progeny exhibits a unique genotype ratio of 6 fertile long-culm (4*D60D60*:2*D60d60GalGal*: 2 partially fertile long-culm (*D60d60Galgal* = F_1 type):1 dwarf(*d60d60GalGal*) [15].

Although there are multiple alleles in *sd1* locus of DGWG, Jukkoku, Reimei, and Calrose 76, the differences in their influences on the yield performance have not been reported. Therefore, investigating the differences in phenotypic traits among the different *sd1* allele-carrying plants, *d60*-carrying plant and their double dwarf plants, will be beneficial for practical selection of *d60* and *sd1*alleles. In this study, semidwarf or double dwarf lines, which were integrated with *sd1* of Jukkoku, *sd1* of IR8, *d60*, or both gene combinations in the genetic background of Koshihikari, were used for investigating the influence of these semidwarf genes on phenotypic traits, especially related to yield performance.

2. Results

2.1. *Effects of Semidwarf and Double Dwarf Genes on Growth*

The trends in full-length growth, depicted by growth curves, were comparable among all lines. (Figure 1). The full length in lines carrying one or two semidwarf genes was already shorter than that of Koshirikari lines at the time of transplanting (June 7, 28 days after sowing), and the differences became prominent around 64–70 days after sowing (July 13 and 19) (Figure 1, Table 1). The full length of *d60*-carrying line was longer than that of *sd1*-carrying lines at the time of transplanting. However, the full length of line carrying Jukkoku_*sd1* and that of line carrying IR8_*sd1* exceeded that of line carrying *d60* on June 23 (43 days after sowing) and on July 13 (64 days after sowing), respectively: full length in lines carrying either Jukkoku_*sd1* or IR8_*sd1* was longer than that in line carrying *d60* at the time of final measurement (August 23, 103 days after sowing). Days to heading ranged from 86.5 days of line carrying IR8_*sd1* to 90.5 days of those carrying *d60*. Such a four-day difference was thought to be little. Therefore, the differences appeared in morphological traits, such as culm length and panicle length, were attributed to genetic reason.

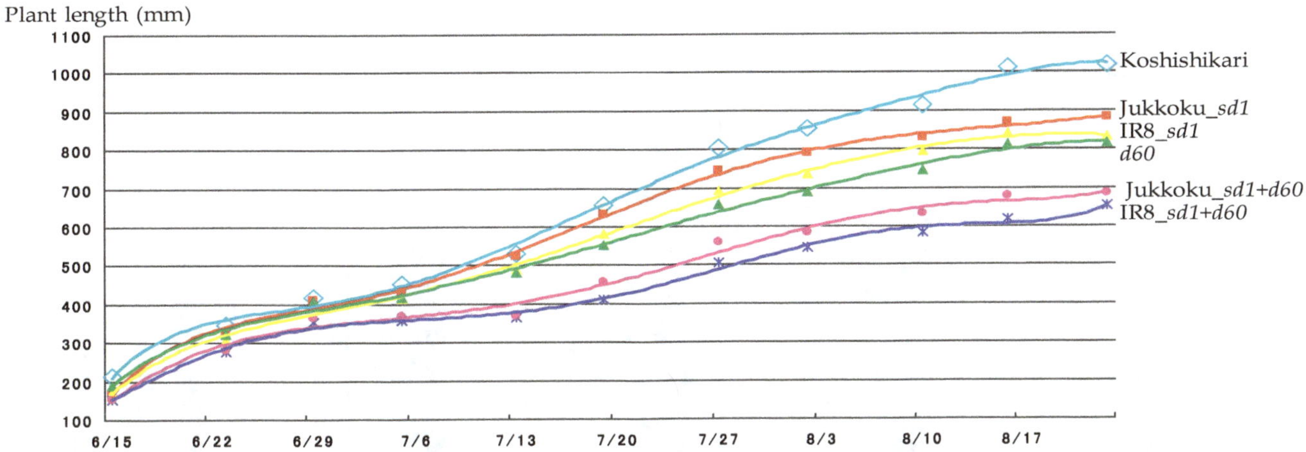

Figure 1. Effect of growth of semidwarf and double dwarf gene lines. Ten plants were randomly selected, and the distance between the ground and the highest standing point (i.e., the full length) was measured every week for approximately three months until the panicle emerged. The full length of *d60*-carrying line was longer than that of *sd1*-carrying lines at the time of transplanting. However, the full length of line carrying Jukkoku_*sd1* and that of line carrying IR8_*sd1* exceeded that of line carrying *d60* on June 23 (43 days after sowing) and on July 13 (64 days after sowing), respectively: full length in lines carrying either Jukkoku_*sd1* or IR8_*sd1* was longer than that in line carrying *d60* at the time of final measurement (August 23, 103 days after sowing).

Table 1. Plant length of semidwarf and double dwarf gene lines.

Days after Sowing	35	43	49	55	64	70	78	82	90	96	103
Koshihikari	212.6	345.0	415.4	452.4	529.5	657.0	805.6	852.6	913.1	1011.6	1020.3
Jukkoku_*sd1*	166.8*	330.2	408.9	430.7	523.8	631.0*	743.0*	791.7*	828.7*	865.9*	880.8*
IR8_*sd1*	170.3*	308.9*	393.2	414.9*	488.8*	578.2*	693.2*	734.4*	794.6*	837.9*	831.0*
d60	189.1*	321.0*	406.5	415.9*	480.5*	550.4*	656.8*	689.3*	744.1*	810.6*	814.3*
Jukkoku_*sd1*+*d60*	153.1*	275.4*	352.8*	357.8*	364.1*	409.2*	506.4*	543.4*	581.9*	616.6*	651.7*
IR8_*sd1*+*d60*	153.7*	279.2*	364.8*	368.5*	369.1*	454.5*	558.4*	580.9*	630.9*	679.2*	684.2*

The full length in lines carrying one or two semidwarf genes was already shorter than that of Koshirikari lines at the time of transplanting 28 days after sowing, and the differences became prominent around 64–70 days after sowing. Finally, the full length of semidwarf and double dwarf lines were significantly shorter than Koshihaikri. Color in the boxes of genetic lines coincide the color of growth curve in Figure 1. *: statistically significant at the 5% level.

Integration of a semidwarf gene (or genes) resulted in a reduction in culm length: the mean culm length of Koshihikari was 88.8 cm, while that of lines carrying Jukkoku_*sd1*, IR8_sd1, *d60*, Jukkoku_*sd1* plus *d60*, or IR8_*sd1* plus *d60* was 71.8 cm, 68.5 cm, 65.7 cm, 48.6 cm, or 50.2 cm, respectively. Leaf length was shorter in line carrying Kinuhikari_*sd1* (9–16% reduction compared with Koshihikari) or *d60* (9–18% reduction compared with Koshihikari) than in those carrying Jukkoku_*sd1* (1–9% reduction compared with Koshihikari (Figure 2). Furthermore, leaves of the semidwarf and double dwarf lines were slightly shorter and straighter (pointing upwards) than in Koshihikari (Figure 3), indicating improved light-reception attitude by the integration of semidwarf gene (or genes). Panicle length was slightly longer (by 2.5%) in line carrying Jukkoku_*sd1* and slightly shorter in lines carrying Kinuhikari_*sd1* (by 2.4%) or *d60* (by 3.0%), compared with Koshihikari (Table 2). However, the reduction in panicle length was quite less than that in culm length (22.8% decrease in lines carrying Kinuhikari_*sd1* vs a 26.1% decrease in lines carrying *d60*). Therefore, the negative effects of semidwarf genes *sd1* and *d60* on panicle length were negligible.

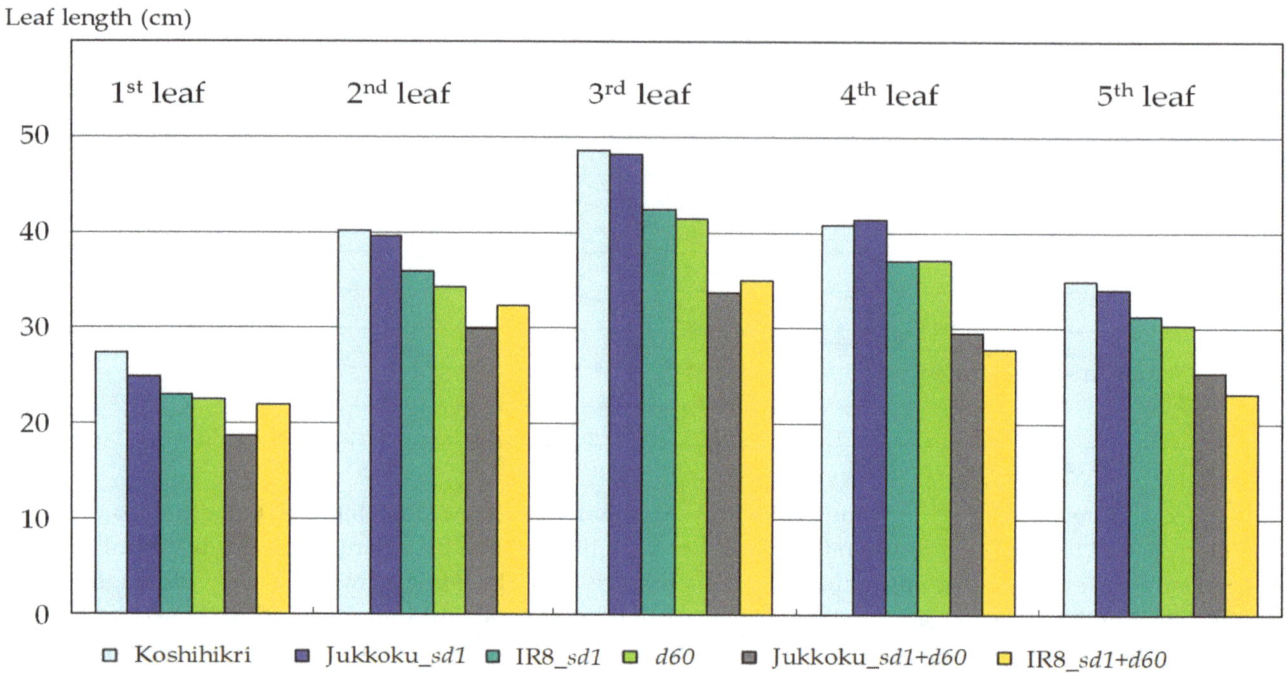

Figure 2. Effect of semidwarf and double dwarf genes to leaf length. Upper five leaves, arising from the main culm, were measured. Except for Jukkoku_*sd1* line. leaves of the semidwarf and double dwarf lines were slightly shorter than that of Koshihikari.

Figure 3. Plant phenotype of semidwarf and double dwarf gene lines. Leaves of the semidwarf and double dwarf lines were straighter (pointing upwards) than in Koshihikari, indicating improved light-reception attitude by the integration of semidwarf gene (or genes).

2.2. *Effects of Semidwarf and Double Dwarf Genes on Yield*

The yield components of each genotype are summarized in Table 2. The weight of unpolished rice/1000 grains and the proportions of fertile florets differed only slightly between lines. The effect of these genes on the proportion of fertile florets and the weight of unpolished rice/1000 grains were negligible. The number of panicles/plants was 17.9 in Koshihikari: 20.8 in line carrying Jukkoku_*sd1* (+16.2% vs Koshihikari), and 15.4 in Jukkoku DW line (−14.0% vs Koshihikari) (Figure 4, Table 3). In addition, the floret number/panicle was 87.3 in line carrying Jukkoku_*sd1* (+11.1% vs Koshihikari) and 72.1 in Jukkoku DW line (−8.3% vs Koshihikari) (Figure 5, Tables 2 and 4). The number of panicles was larger in line carrying Jukkoku_*sd1*, while floret density was larger in all semidwarf varieties than in Koshihikari (Figure 6, Table 5). Thus, an increase in both the number of panicles/plant and the floret number/panicle resulted in an increase in the number of panicles/m^2 and a consequent increase in yield (Figure 7, Table 6).

Table 2. Effect of semidwarf and double dwarf genes to yield components.

	Koshihikari	**Jukkoku_*sd1***	**IR8_*sd1***	***d60***	**Jukkoku _*sd1*+*d60***	**IR8_*sd1*+*d60***
Weight of unpolished rice/1000 grains (g)	22.1	20.2	20.4	20.9	21.3	20.5
Panicles /m^2	397.4	461.8 *	391.8	397.4	341.9 *	399.6
Floret number/panicle	78.6	87.3 *	82.8	79.3	72.1 *	76.2
Seed fertility (%)	96.6	94.9	93.4	95.4	95.6	96.2
Yield of unpolished rice (g/m^2=kg/a)	665.3	772.6 *	617.9 *	628.5	502.1 *	600.5 *

Compared with the yield of Koshihikari (mean, 665.3 g/m^2), that of the line carrying Jukkoku_*sd1* was highest value 772.6 g/m^2 increased by 16.1%, while that of IR8_*sd1*, *d60* and IR8_*sd1* plus *d60*, was slightly decreased by 7.1%, 5.5%, and 9.7%, respectively. The line carrying Jukkoku_*sd1* also showed highest value in number of panicles and florets/panicle, each 16.2% and 11.1% higher than in Koshihikari, which were responsible for the increases in yield. The weight of rice/1000 grains was equivalent among all genetic lines. Except for the semidwarf line carrying Jukkoku_*sd1*, semidwarf line carrying *d60* was equivalent to line carrying IR8_*sd1*in the yield of unpolished rice, and yield components such as panicles/m^2, floret number /panicle. *: statistically significant at the 5% level.

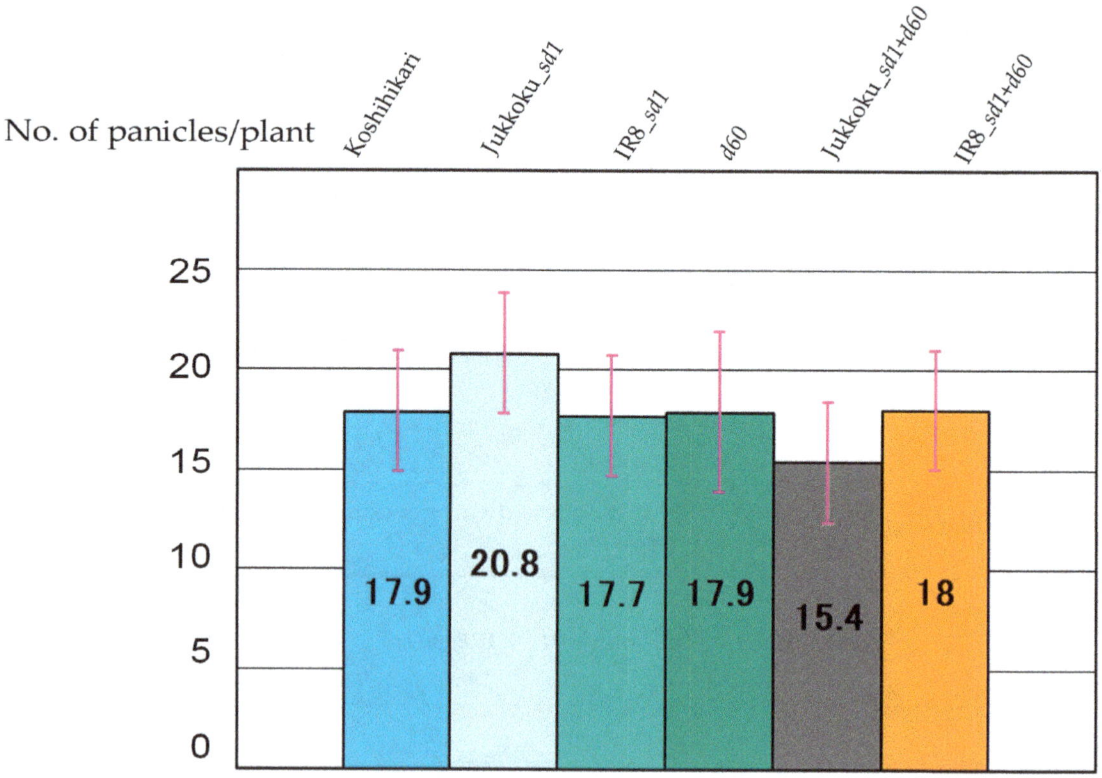

Figure 4. No. of panicles/plant in semidwarf and double dwarf gene lines. The number of panicles/plants was highest at 20.8 in line carrying Jukkoku_*sd1* (+10.2% vs Koshihikari).

Table 3. Effect of semidwarf and double dwarf genes to No. of panicles/plant.

	Koshihikari	Jukkoku_*sd1*	IR8_*sd1*	*d60*	Jukkoku _*sd1+d60*	IR8_*sd1+d60*
No. of panicles/plant	17.9	20.8 *	17.7	17.9	15.4 *	18.0
Percent change (%)	-	+16.2	−1.4	±0	−14.0	+0.6

The number of panicles/plants in line carrying *d60* (17.9) was comparable to that in line carrying IR8_*sd1* (17.7). *: statistically significant at the 5% level.

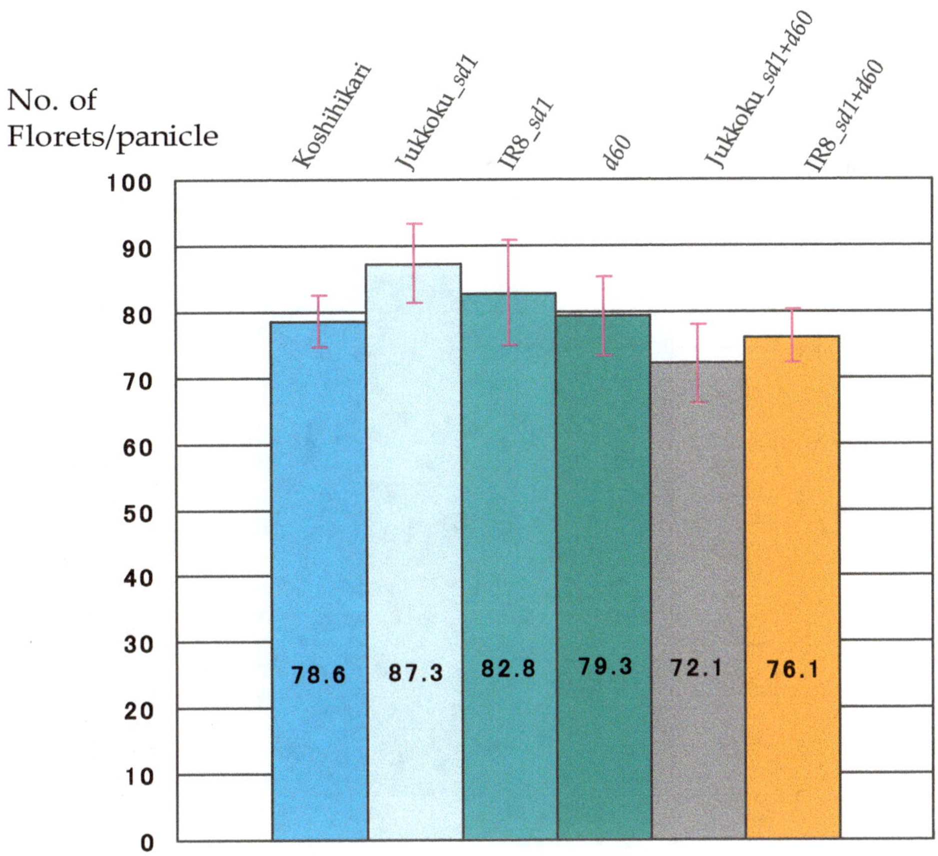

Figure 5. Floret number/panicle of semidwarf and double dwarf gene lines. The number of panicles/plants in line carrying *d60* was comparable to that in line carrying IR8_*sd1*.

Table 4. Effect of semidwarf and double dwarf genes to floret number/panicle.

	Koshihikari	Jukkoku_*sd1*	IR8_*sd1*	*d60*	Jukkoku _*sd1*+*d60*	IR8_*sd1*+*d60*
No. of Florets/panicle	78.6	87.3 *	82.8	79.3	72.1 *	76.1
Percent change (%)	-	+11.1	+5.3	+0.9	−8.3	−3.1

The floret number/panicle was highest at 87.3 in line carrying Jukkoku_*sd1* (+11.1% vs Koshihikari). *: statistically significant at the 5% level.

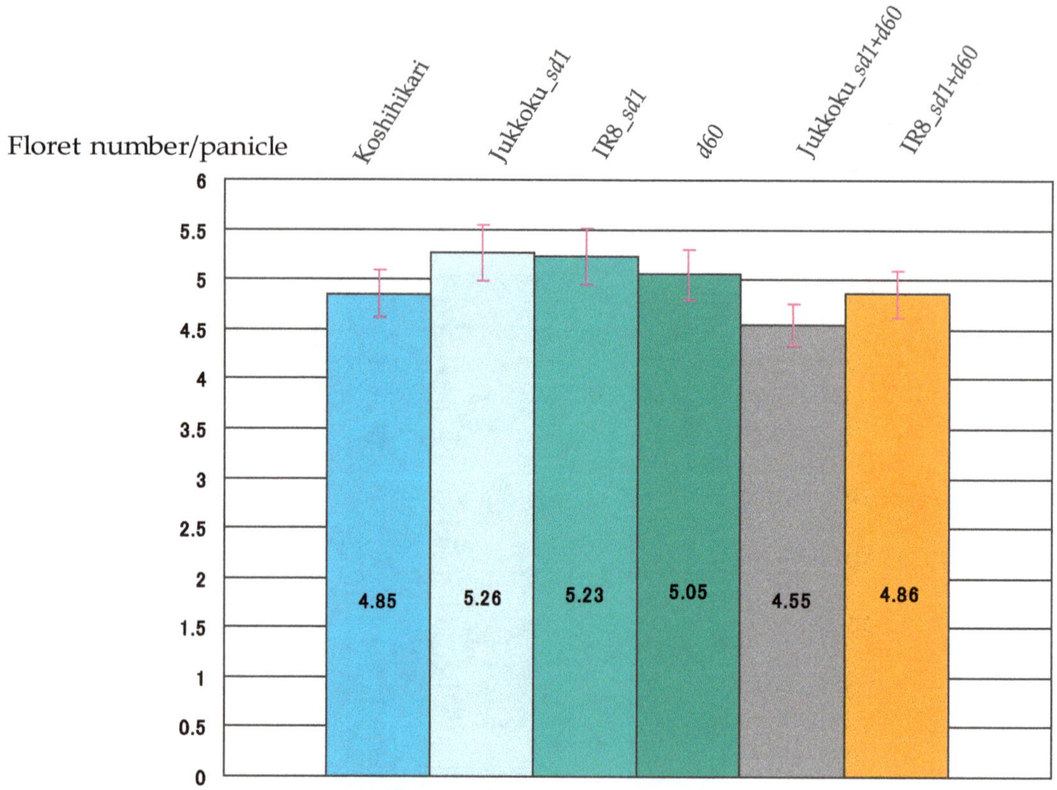

Figure 6. Effect of semidwarf and double dwarf genes to floret density. The floret density was larger in all lines carrying one semidwarf gene than that of Koshihikari.

Table 5. Effect of semidwarf and double dwarf genes to panicle.

	Koshihikari	Jukkoku_*sd1*	IR8_*sd1*	*d60*	Jukkoku _*sd1*+*d60*	IR8_*sd1*+*d60*
Floret number/panicle	78.6	87.3 *	82.8	79.3	72.1 *	76.1
Panicle length (cm)	16.2	16.6	15.8	15.7	15.8	15.7
Floret density (/cm)	4.85	5.26 *	5.23 *	5.05	4.55 *	4.86
Percent change of floret density (%)	-	+8.53	+7.95	+4.11	−6.15	+0.32

Panicle length was slightly longer (by 2.5%) in line carrying Jukkoku_*sd1* and slightly shorter in lines carrying IR8_*sd1* (by 2.4%) or *d60* (by 3.0%), compared with Koshihikari. The reduction in panicle length was quite less than that in culm length (22.8% decrease in line carrying IR8_*sd1* vs a 26.1% decrease in line carrying *d60*).

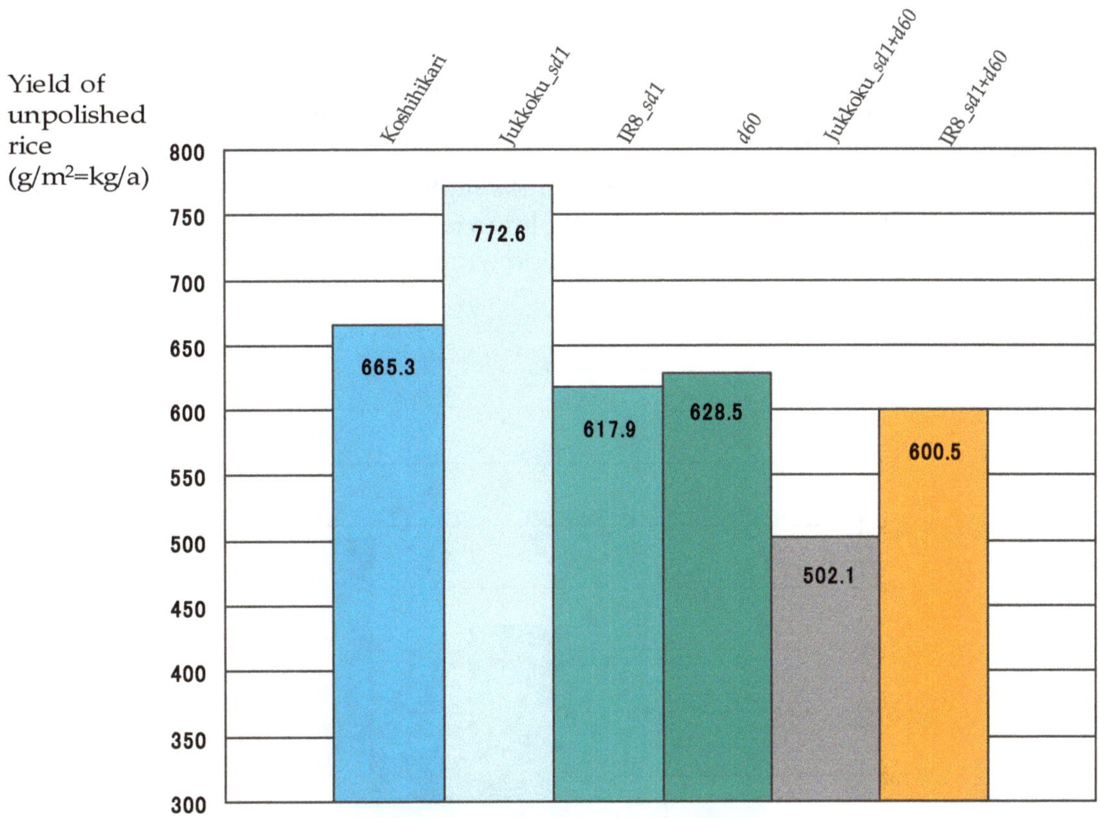

Figure 7. Yield of semidwarf and double dwarf gene lines. The yield of unpolished rice was 665.3 g/m^2 in Koshihikari, 772.6 g/m^2 in line carrying Jukkoku_*sd1* (+15.9% vs Koshihikari), 617.9 g/m^2 in line carrying Kinuhikari_*sd1* (−7.1% vs Koshihikari), and 628.5 g/m^2 in line carrying *d60* (−5.5% vs Koshihikari).

Table 6. Effect of semidwarf and double dwarf genes to yield.

	Koshihikari	Jukkoku_*sd1*	IR8_*sd1*	*d60*	Jukkoku _*sd1*+*d60*	IR8_*sd1*+*d60*
Yield of unpolished rice (g/m^2=kg/a)	665.3	772.6 *	617.9 *	628.5	502.1 *	600.5 *
Percent change (%)	-	+16.1	−7.1	-5.5	−24.5	−9.7

The yield of line carrying *d60* was comparable to that in line carrying IR8_*sd1*. *: statistically significant at the 5% level.

The yield of unpolished rice was 665.3 g/m^2 in Koshihikari, 772.6 g/m^2 in line carrying Jukkoku_*sd1* (+15.9% vs Koshihikari), 617.9 g/m^2 in line carrying Kinuhikari_*sd1* (−7.1% vs Koshihikari), and 628.5 g/m^2 in line carrying *d60* (−5.5% vs Koshihikari) (Figure 7, Table 6). The introduction of Kinuhikari_*sd1* or *d60* into Koshihikari appears to cause a slight reduction in yield. On the other hand, the yield of DW lines was markedly lower than that of Koshihikari: for Jukkoku DW line (−24.5% vs Koshihikari) and 600.5 g/m^2 for IR8 DW line (−9.7% vs Koshihikari) (Figure 7, Table 6). When using the alternative equation, the yield index was higher in all semidwarf-gene-carrying lines than in Koshihikari (Figure 8). The high yield index and lodging resistance of semidwarf varieties suggest that introduction of *sd1* and *d60* into non-dwarf genomes will be beneficial for increasing crop yield. Moreover, only minor differences in the grain appearance were observed among lines including Koshihikari, indicating that the grain quality of semidwarf lines is equivalent to that of Koshihikari. Taken together, semidwarf genes *sd1* and *d60* are useful in the agricultural industry.

The yield index was higher in line carrying IR8_*sd1* than in those carrying Jukkoku_*sd1* (Figure 8, Table 7). Although line carrying Jukkoku_*sd1* gave a higher yield than those carrying IR8_*sd1*, the higher yield index associated with IR8_*sd1* than Jukkoku_*sd1* suggests that the efficiency of the distribution to sink organs (e.g., seeds) is higher. Thus, the yield favors a gain in dry matter, which may be also higher in plants carrying IR8_*sd1* than in those carrying Jukkoku_*sd1*. Furthermore, the yield index for Jukkoku DW and IR8 DW are high—-45.6% and 46.5% higher, respectively than that of Koshihikari (Figure 8, Table 7). In order to increase a markedly low yield in DW lines, the use of conditions that favor a gain in dry matter, such as intensive cultivation with heavy fertilization to increase the number of tillers, may be effective.

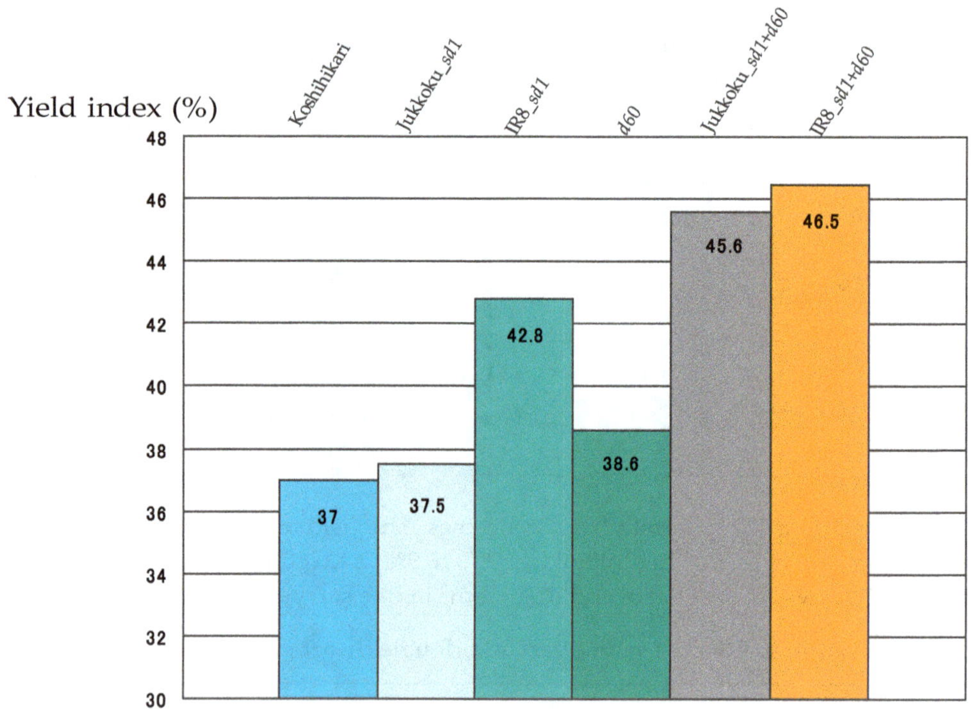

Figure 8. Yield index of semidwarf and double dwarf gene lines. The yield index was higher in line carrying IR8_*sd1* than in those carrying Jukkoku_*sd1*. The yield index for Jukkoku DW and IR8 DW are 45.6% and 46.5%, respectively higher than that of Koshihikari.

Table 7. Effect of semidwarf and double dwarf genes to yield index.

	Koshihikari	Jukkoku_*sd1*	IR8_*sd1*	*d60*	Jukkoku _*sd1*+*d60*	IR8_*sd1*+*d60*
Weight of winnowed paddy (g)	26.9	27.3	26.4	22.7 *	20.3 *	23.9 *
Weight of the plant part above the ground (g)	72.7	72.8	61.5 *	58.9 *	44.6 *	51.4 *
Yield index (%)	37.0	37.5	42.8 *	38.6	45.6 *	46.5 *

The high yield index and lodging resistance of semidwarf varieties suggest that introduction of *sd1* and *d60* into non-dwarf genomes will be beneficial for increasing crop yield. *: statistically significant at the 5% level.

3. Discussion

As exemplified by IR8, which was the variety behind the Green Revolution, many of the rice varieties cultivated worldwide commonly carry the semidwarf gene *sd1*. Another semidwarf gene *d60* is non-allelic to *sd1* and is of particular interest as a different source of semidwarfism to give genetic diversity among the semidwarf varieties. In this study, semidwarf lines, namely Jukkoku (Jukkoku _sd1), *sd1* of Kinuhikari (IR8_*sd1*: Kinuhikari maintains *sd1* of IR8 origin), *d60* or *sd1* plus *d60* into the Koshihikari background, were used to investigate influence of these semidwarf genes on phenotypic traits, in relation to yield.

This study showed that all tested semidwarf lines had shorter culm lengths than Koshihikari, indicating improved lodging resistance. The effect on culm length carrying *d60*(65.7 cm) is slight shorter than in those carrying *sd1* (Jukkoku_*sd1*, 71.8 cm; IR8_*sd1*, 68.5 cm), Among the genetic lines, line carrying Jukkoku_*sd1*showed the highest yield of unpolished rice 772.6 g/m^2, which is 16.1% higher than in Koshikikari. The Jukkoku_*sd1* line also showed highest value in the number of panicles, the number of florets per panicle than in Koshihikari. Therefore, it was highly possible that the increasing yield of Jukkoku_*sd1* line was ascribed to the increasing numbers of panicles and florets. Although the yield of unpolished rice of *d60* line, 628.5 g/m^2 was 5.5% lower than that of Koshihikari (665.3 g/m^2), but this is almost equivalent yield performance of IR8_*sd1* (617.9 g/m^2). Ogi et al. (1993) [21] and Murai et al. (2004) [22] reported characteristics of isogenic line carrying *sd1* derived from DGWG, the source of IR8 *sd1*. These isogenic lines showed almost same number of panicles as that of original varieties, 'Norin 29' and 'Shiokari'. Therefore, it was concluded that Jukkoku_*sd1* especially has potential increasing panicle numbers compared to IR8_*sd1*. Hence, Jukkoku_*sd1* appears to confer a pleiotropic effect of increasing panicle number very well in the Koshihikari genetic background. The difference of such as effect between *sd1* alleles may be ascribed to that IR8_*sd1* suffered 383 bp deficit in the region exon 1-2 of *GA20-ox* [18], whereas, Jukkoku_*sd1* has only a SNP against the wild type *GA20-ox* [18] and the transcripts existed [23].

This study demonstrated that *d60* confers slightly shorter culm length than IR8_*sd1*, but almost equivalent yield performance with IR8_*sd1* together with effects on yield-related phenotypic traits comparable to IR8_*sd1*, which actually contributed to green revolution [2]. Although many dwarf genes are accompanied with a reduction in panicle length, yield of unpolished rice, and grain thresh ability (which is likely attributed to excessive dwarfing effects), *d60* does not exert such negative effects on yield-related phenotypic traits of rice plant. In conclusion, *d60* is applicable to practical breeding and one of choice for expanding genetic diversity of rice varieties.

4. Materials and Methods

4.1. Genetic Lines

The following rice semidwarf or double dwarf lines, Koshihikari, Koshihikari carrying Jukkoku_*sd1* [Koshihikari*6//(Kanto 79/Jukkoku F_4) B_6F_4], Koshihikari carrying IR8_*sd1* [Koshihikari/Kinuhikari F_5)], Koshihikari carrying *d60* [Koshihikari Koshihikari*7//(Koshihikari/Hokuriku 100) B_7F_3], Koshihikari carrying *d60* and Jukkoku_*sd1* [Jukkoku_DW, Koshihikari carrying Jukkoku_*sd1*/Koshihikari carrying_*d60* F_7), and Koshihikari carrying *d60* and IR8_*sd1* (IR8 DW: Koshihikari carrying IR8_*sd1*/ Koshihikari carrying_*d60* F_7) were used in this study. Koshihikari carrying Jukkoku_*sd1* was developed by six times of backcrosses with Koshihiakri as a recurrent parent using the short stemmed *sd1* homozygous fixed F_4 strain in Kanto No. 79 × Jukkoku (*sd1*) as a non-recurrent parent [24]. Koshihikari carrying *d60* was developed by seven times of backcrosses with Koshihikari as the recurrent parent using the short-stemmed F_2 plant as a non-recurrent parent [15]. Koshihikari carrying IR8__*sd1* was the short stemmed *sd1* homozygous fixed F_7 strain derived from Koshihikari × Kinuhikari. Koshihikari carrying *d60* and Jukkoku_*sd1*(Jukkoku DW) was double dwarf *d60sd1* homozygous fixed F_7 strain derived from Koshihikari carrying Jukkoku_*sd1* × Koshihikari carrying_*d60* [15]. Koshihikari carrying *d60* and IR8_*sd1*(IR8 DW) was double dwarf *d60sd1* homozygous fixed F_7 strain derived from Koshihikari carrying IR8_*sd1* × Koshihikari carrying_*d60* [15]. Kinuhikari has *sd1* derived from IR8 [25], which suffer 383 bp deficit in the region exon 1-2 from wild type *GA20-ox* [18]. Koshihikari carrying Jukkoku_*sd1*was plant-variety registered via further 7–8th backcrosses and it was designated as Hikarishinseiki [24], whose *sd1* has only a SNP against wild type *GA20-ox* [18,26]. Genomic *sd1*allele and the RNA transcript in Hikarishinseiki are detectable by diagnosis targeting the SNP [23,27].

4.2. Cultivation

Rice seeds were taken from stocks kept in a refrigerator. Seeds of each line were immersed in enough water just to cover the seeds. Water was exchanged every day for seven days (May 2 to May 8) during seed soaking and stimulation of germination. Seedlings were grown in nursery boxes (30 × 15 × 3 cm) for approximately 20 days. Seedlings were then individually transplanted into a paddy field (120 m^2: 6.0 × 20.0 m) of the University Farm on June 8. Two 4-m^2 plots (2 × 2 m) with transplanting densities 22.2 seedlings/m^2 (one seedling per 30 × 15 cm, 78 seedlings per field) were prepared for each genetic line (two instances). The paddy field was fertilized by 4.0 kg of basal fertilizer containing nitrogen, phosphorus, and potassium (weight ratio, nitrogen:phosphorus:potassium = 2.6:3.2:2.6) at the rate with 4.3 g/m^2 nitrogen, 5.3 g/m^2 phosphorus, and 4.3 g/m^2 potassium evenly across the field. A herbicide (Joystar L Floable, Kumiai Chemical Industry, Tokyo, Japan) was applied on June 20 to kill weeds grown uncontrollably, and the water level was then kept at a high enough level to cover the weeds for a week.

4.3. Growth Analysis

Ten seedlings were randomly selected for each line at the time of transplantation, and the full length was measured individually. After transplantation, ten plants were randomly selected, and the distance between the ground and the highest standing point (i.e., the full length) was measured every week for approximately three months until the panicle emerged. The time when the tip of the panicle first emerged from the flag leaf sheath was recorded as the heading time for all plants.

4.4. Plant Phenotyping

After ripening, ten plants typical of each genotype were sampled twice. Sampled plants were air-dried, and were assessed or measured the following traits. Culm length: the length between the ground surface and the panicle base of the main culm was measured at the time of sampling. Leaf length: the lengths of the upper five leaves, arising from the main culm, were individually measured. Length and weight of panicle: the length between the panicle base and the tip of the panicle, and the weight of the panicle, were measured. Total panicle number: the number of panicles were counted by sampled individuals and panicle numbers per 1 m^2 area (panicles/m^2) were counted twice in each plot of the paddy field. Total floret number: florets were counted to obtain total floret number. Floret number/panicle: the total floret number (including both sterile and fertile florets) was divided by the total panicle number. Proportion of fertile florets: each floret was assessed to determine its fertility. Floret density (floret number/cm): the number of florets per panicle was divided by the length of the panicle. Presence of awns: florets with an awn were counted when counting the florets. Grain threshability: was manually tested during examination of phenotypic traits. Appearance of grains: the size, color, and presence of an awn were observed for assessment of grain quality. Total weight of winnowed paddy: the total weight of winnowed paddy was weighed after grain selection using the salt solution (salt content of 1.06 g/m^3). Weight of sieved unpolished rice/1000 grains: obtained by multiplying the total weight of winnowed paddy by 0.84. Weight of plant parts above the ground: the weight of the plant parts above the ground was measured. Yield index: the winnowed paddy weight was divided by the weight of the plant part above the ground to obtain the yield index. The means of traits were statistically compared using the t-test.

4.5. Yield

Yield of unpolished rice was calculated using the following equation.

- Yield of unpolished rice (g/m^2) = (number of panicles/m^2) × (floret number/panicle) × (proportion of fertile florets) × (weight of unpolished rice/grain)

The following alternative equation (see below) was also used to calculate the comparison of yields:

- Yield = (yield index) × (weight of the plant parts above the ground)

Author Contributions: Conceptualization—M.T.; methodology—M.T.; investigation—K.I., M.T.; resources—M.T.; writing—original draft preparation—M.T.; writing—review and editing—M.T.; project administration, M.T.; funding acquisition—M.T.

References

1. Khush, G.S. Green revolution: Preparing for the 21st century. *Genome* **1999**, *42*, 646–655. [CrossRef] [PubMed]
2. Hergrove, T.; Coffman, W.R. Breeding history. In *Rice that Changed the World: Cerebrating 50 Years of IR8*; Special supplement focusing on IR8; International Rice Research Institute: Los Baños, Philippines, 2016; pp. 6–10. Available online: http://books.irri.org/RT_Supplement-IR8.pdf (accessed on 25 September 2019).
3. Jennings, P. Rice revolutions in Latin America. In *Rice that Changed the World: Cerebrating 50 Years of IR8*; Special supplement focusing on IR8; International Rice Research Institute: Los Baños, Philippines, 2016; p. 19. Available online: http://books.irri.org/RT_Supplement-IR8.pdf (accessed on 25 September 2019).
4. Okada, M.; Yamakawa, Y.; Fujii, K.; Nishiyama, H.; Motomura, H.; Kai, S.; Imai, T. On the new varieties of paddy rice, 'Hoyoku', 'Kokumasari' and 'Shiranui'. *J. Agric. Exp. Stn. Kyushu* **1967**, *2*, 187–224.
5. Futuhara, Y. Breeding of new rice variety Reimei by gamma ray irradiation. *Gamma. Field Symp.* **1968**, *7*, 87–109.
6. Rutger, J.N.; Peterson, M.L.; Hu, C.H.; Lehman, W.F. Induction of usefil short stature and early maturing mutants in two japonica rice cultivars. *Crop Sci.* **1976**, *16*, 631–635. [CrossRef]
7. Rutger, J.N.; Peterson, M.L.; Hu, C.H. Registration of Calrose76 rice. *Crop Sci.* **1977**, *17*, 978. [CrossRef]
8. Aquino, R.C.; Jennings, P.R. Inheritance and significance of dwarfism in an indica rice variety. *Crop Sci.* **1966**, *6*, 551–554. [CrossRef]
9. Heu, M.; Chang, T.T.; Cabanilla, V.L. The inheritance of culm length, panicle length, duration to heading and bacterial leaf blight reaction in rice cross: Sigadis×Taichung (Native)1. *Jpn. J. Breed.* **1968**, *18*, 7–11.
10. Foster, K.W.; Rutger, J.N. Inheritance of semidwarfism in rice, *Oryza sativa* L. *Genetics* **1978**, *88*, 559–574.
11. Mackill, D.J.; Rutger, J.N. The inheritance of induced-mutant semidwarfing genes bin rice. *J. Hered.* **1979**, *70*, 335–341. [CrossRef]
12. Ikehashi, H.; Kikuchi, F. Genetic analysis of semidwarfness and their significance for breeding of high-yielding varieties in rice. *JARQ* **1982**, *15*, 231–235.
13. Kikuchi, F.; Itakura, N.; Ikehashi, H.; Yokoo, M.; Nakane, A.; Maruyama, K. Genetic analysis of semidwarfism in high-yielding rice varieties in Japan. *Bull. Natl. Inst. Agr. Sci. Ser. D* **1985**, *36*, 125–145.
14. Kikuchi, F.; Futsuhara, Y. Inheritance of morphological characters. 2. Inheritance of semidwarf. In *Science of the Rice Plant*, 3rd ed.; Matsuo, T., Shimizu, S., Tsunoda, S., Murata, Y., Kumazawa, K., Futsuhara, Y., Hoshikawa, K., Yamaguchi, H., Kikuchi, F., Eds.; Tokyo Food and Agricultural Policy Research Center: Tokyo, Japan, 1997; Volume 3, pp. 309–317.
15. Tomita, M. Combining two semidwarfing genes *d60* and *sd1* for reduced height in 'Minihikari', a new rice germplasm in the 'Koshihikari' genetic background. *Genet. Res. Camb.* **2012**, *94*, 235–244. [CrossRef] [PubMed]
16. Suh, H.S.; Heu, M.H. The segregation mode of plant height in the cross of rice varieties. VI. Linkage analysis of the semi-dwarfness of the rice variety Tongil. *Korean J. Breed.* **1978**, *10*, 1–6.
17. Maeda, H.; Ishii, T.; Mori, H.; Kuroda, J.; Horimoto, M.; Takamure, I.; Kinoshita, T.; Kamijima, O. High density molecular map of semidwarfing gene, *sd-1*, in rice (*Oryza sativa* L.). *Breed. Sci.* **1997**, *47*, 317–320. [CrossRef]
18. Sasaki, A.; Ashikari, M.; Ueguchi-Tanaka, M.; Itoh, H.; Nishimura, A.; Swapan, D.; Ishiyama, K.; Saito, T.; Kobayashi, M.; Khush, G.S.; et al. Green revolution: A mutant gibberellin-synthesis gene in rice. *Nature* **2002**, *416*, 701–702. [CrossRef] [PubMed]
19. Monna, L.; Kitazawa, N.; Yoshino, R.; Suzuki, J.; Masuda, H.; Maehara, Y.; Tanji, M.; Sato, M.; Nasu, S.; Minobe, Y. Positional cloning of rice semidwarfing gene, *sd-1*: Rice "Green revolution gene" encodes a mutant enzyme involved in gibberellin synthesis. *DNA Res.* **2002**, *9*, 11–17. [CrossRef]
20. Spielmeyer, W.; Ellis, M.H.; Chandler, P.M. Semidwarf (*sd-1*), "green revolution" rice, contains a defective gibberellin 20-oxidase gene. *Proc. Natl. Acad. Sci. USA* **2002**, *99*, 9043–9048. [CrossRef]

21. Ogi, Y.; Kato, H.; Maruyama, K.; Kikuchi, F. The effects of the culm length andother agronomic characters cause by semidwarfing genes at the sd-1 locus inrice. *Jpn. J. Breed.* **1993**, *43*, 267–275. [CrossRef]
22. Murai, M.; Komazaki, T.; Sato, S. Effects of *sd1* and *Ur1* (Undulate rachis-1) on lodging resistance and related traits in rice. *Breed. Sci.* **2004**, *54*, 333–340. [CrossRef]
23. Tomita, M.; Matsumoto, S. Transcription of rice Green Revolution *sd1* gene is clarified by comparative RNA diagnosis using the isogenic background. *Genom. Appl. Biol.* **2011**, *2*, 29–35.
24. Tomita, M. Introgression of Green Revolution *sd1* gene into isogenic genome of rice super cultivar Koshihikari to create novel semidwarf cultivar 'Hikarishinseiki' (Koshihikari-sd1). *Field Crops Res.* **2009**, *114*, 173–181. [CrossRef]
25. Tabuchi, H.; Hashimoto, N.; Takeuchi, A.; Terao, T.; Fukuta, Y. Genetic analysis of semidwarfism of the japonica rice cultivar Kinuhikari. *Breed. Sci.* **2000**, *50*, 1–7. [CrossRef]
26. Tomita, M.; Ishii, K. Genetic performance of the semidwarfing allele *sd1* derived from a Japonica rice cultivar and minimum requirements to detect its single-nucleotide polymorphism by MiSeq whole-genome sequencing. *BioMed Res. Int.* **2018**, *2018*, 4241725. [CrossRef] [PubMed]
27. Naito, Y.; Tomita, M. Identification of an isogenic semidwarf rice cultivar carrying the Green Revolution *sd1* gene using multiplex codominant ASP-PCR and SSR markers. *Biochem. Genet.* **2013**, *51*, 530–542. [CrossRef] [PubMed]

15

Genetic Advance of Durum Wheat under High Yielding Conditions: The Case of Chile

Alejandro del Pozo [1,*], Iván Matus [2], Kurt Ruf [1,2], Dalma Castillo [2], Ana María Méndez-Espinoza [1] and María Dolores Serret [3]

1 Centro de Mejoramiento Genético y Fenómica Vegetal, Facultad de Ciencias Agrarias, Universidad de Talca, 3460000 Talca, Chile
2 CRI-Quilamapu, Instituto de Investigaciones Agropecuarias, 3800062 Chillán, Chile
3 Centre de Recerca en Agrotecnologia (AGROTECNIO), 2519825198 Lleida, Spain
* Correspondence: adelpozo@utalca.cl.

Abstract: In Chile, durum wheat is cultivated in high-yielding Mediterranean environments, therefore breeding programs have selected cultivars with high yield potential in addition to grain quality. The genetic progress in grain yield (GY) between 1964 and 2010 was 72.8 kg ha^{-1} per year. GY showed a positive and significant correlation with days to heading, kernels per unit ground area and thousand kernel weight. The gluten and protein content tended to decrease with the year of cultivar release. The correlation between the $\delta^{13}C$ of kernels and GY was negative and significant (−0.62, $p < 0.05$, for all cultivars; and −0.97, $p < 0.001$, excluding the two oldest cultivars). The yield progress (genetic plus agronomic improvements) of a set of 40–46 advanced lines evaluated between 2006 and 2015 was 569 kg ha^{-1} per year. Unlike other Mediterranean agro-environments, a longer growing cycle together with taller plants seems to be related to the increase in the GY of Chilean durum wheat during recent decades.

Keywords: agronomic traits; carbon isotope; days to heading; grain quality; yield components

---

1. Introduction

Durum wheat (*Triticum turgidum* L. ssp. *durum*) covers ~17 million hectares worldwide, which is less than 10% of the total wheat area. However, its importance for human consumption is very high because it is used for making pasta, couscous, burghul and firik [1]. According to the International Grain Council, the largest producers of wheat in the world are the European Union, Canada, the United States, Turkey and Algeria.

For the production of high-quality durum wheat, dry environments are necessary, with warm days and cold nights during the growing season so that large grains are obtained with yellow color, vitreous kernels (more than 95%), hard texture and high test weight (about 82 kg hL^{-1}), alongside high protein content (greater than 10%) and strong gluten (greater than 30% wet gluten), which gives elasticity to dough for industrial use [2]. In Chile, durum wheat is grown in Mediterranean climate environments from the Valparaíso Region (32 °S) to the Biobio Region (37 °S), but mostly under irrigation conditions or in areas where rainfall is sufficient to satisfy most or all of the crop potential evapotranspiration. The sowing area has increased from 9600 ha in 2001 to 27,000 ha in 2015, and the average yield for 2011–2015 was 6.7 Mg ha^{-1} [3].

Wheat yields in different regions of the world have increased greatly since the 1960s as a result of genetic improvement and better agronomic practices [4]. With the Green Revolution, breeding programs have seen the introduction of semi-dwarfing genes that interfere with the action or production of gibberellin [5], leading to a reduction in plant size and an increase in the partitioning of the above-ground biomass towards spikes and grains [6,7]. In bread wheat, the genetic gain in grain yield (GY) was positively correlated with harvest index and the number of grains per spike and per m^2 [6,8]. In durum wheat, a Spanish study conducted with 12 cultivars from Italy and 12 from Spain released between 1930 and 2000 showed that the changes in grain yield were also associated with increases in the harvest index and number of grains per m^2 [9]. Another study carried out on 14 cultivars released in Italy between 1900 and 2000 indicated that the total aerial biomass had not changed and that the increase in yield was associated with a reduction in plant height and an increase in the harvest index and number of grains per m^2 [10]. These studies in durum wheat have been conducted under rain fed conditions and yields were below 6 Mg ha^{-1}, however, there is no information about the genetic progress in durum wheat in high yield potential environments (>10 Mg ha^{-1}) and how grain quality traits have been affected. Moreover, these works have not focused on studying exclusively the trends in breeding advances of post-Green-Revolution (i.e., semi-dwarf) durum wheat cultivars during the last half-century. This is despite the importance of this issue in the context of climate change and the fact that at least for bread wheat, there are studies reporting a stagnation in yields (or at least a drastic decrease in genetic advance) during the last decade [11].

In bread wheat, grain protein content, sedimentation value and wet gluten have increased in modern cultivars [8,12–14]. In durum wheat, modern dwarf and semi-dwarf cultivars have a higher gluten index compared to landraces or traditional Mediterranean cultivars [15]. Subira et al. [16] also reported significant changes in grain quality traits in a historical series of 24 durum wheat cultivars released in Italy and Spain in different periods of the 20th Century, particularly in gluten strength, sedimentation index and yellow color index. High protein content and 'strong' gluten are necessary to process semolina into a suitable final pasta product.

Physiological changes associated with breeding advances have also been reported for bread and durum wheat. For instance, modern bread wheat presented higher stomatal conductance (on an area basis) and carbon isotope discrimination ($\Delta^{13}C$; or a lower carbon isotope composition, $\delta^{13}C$), and lower oxygen isotope composition ($\delta^{18}O$) than older varieties [8,17–19]. In durum wheat, modern varieties have higher $\Delta^{13}C$ (or lower $\delta^{13}C$) compared to landraces [20–22], although no clear differences were found for $\delta^{18}O$ [21]. However, no information exists for durum wheat growing in a high-yielding Mediterranean environment.

The aim of this work was to analyze a) the changes in agronomic traits, grain quality and isotope composition in a set of ten durum wheat cultivars released in Chile between 1964 and 2010; and b) the progress in grain yield, plant height and test weight in selected advanced lines from the Instituto de Investigaciones Agropecuarias (INIA)-Chile breeding program. The experiments were conducted in a high-yielding Mediterranean environment between 2006 and 2015.

2. Materials and Methods

2.1. Experimental Site, Plant Material and Growing Conditions

The experiments were conducted at the Santa Rosa experimental field station (36°32′ S, 71°55′ W; 220 m.a.s.l.) of the Centro Regional de Investigación (CRI)-Quilamapu, Instituto de Investigaciones Agropecuarias (INIA). The climate corresponds to a humid Mediterranean type. During the experimental period (2006–2015), the monthly minimum average temperature was 3.1 °C (July) and the maximum 29.6 °C (January), and the average annual precipitation was 903 mm (Supplemental Table S1). The soil was a sandy loam, humic haploxerands (Andisol). Soil chemical characteristics of the top 10 cm were: pH 6.0, 8.87 mg kg^{-1} of N-$N0_3$; 17.05 mg kg^{-1} of P (Olsen), 0.45% of N-total, 4.5% of C and 0.33, 5.75, 0.65 and 0.48, cmol kg^{-1} of available K, Ca, Mg and Na, respectively [23].

Two different experiments were conducted. In the first experiment, ten cultivars released by the INIA breeding wheat program from 1964 to 2010 (Table 1) were evaluated during three consecutive seasons (2010 to 2012). The INIA cultivars derive from germplasm introduced from the International Maize and Wheat Improvement Center (CIMMYT) and probably all of them have the *Rht-B1* gene.

Table 1. Cultivars of durum wheat released by the wheat breeding program of the Chilean INIA between 1964 and 2010.

Cultivar	Year [1]	Cross/Pedigree
Alifén	1964	CAPELLI//ST 464
Quilafén	1970	YT54/NI08//LD 357/2 *TC
Chagual INIA	1986	2156 3/AA" S"//PG" S"
Chonta INIA	1990	FRIGATTE"S"//RUFF/FLAMINGO"S"
Licán INIA	1990	RUFF "S"/FG"S"//MEX/3/SHWA"S"
Llareta INIA	1997	D67.54.4.9A//JORI'S'/ROSNER DURUM 119-200-4Y/3/ SAHEL77
Guayacán INIA	1997	ALTAR84/STINT"S"//SILVER
Corcolén INIA	2002	ALGA"S"/3/CANDEALFENS5/FLAMINGO"S"//PETREL"S"/ 4/CHURRILLA"S"/5/AUK"S"/6/RUFF"S"/FLAMINGO"S"// FLAMINGO"S"/CRANE"S"/3/YAVOROS 79/HUITLES"S"
Lleuque INIA	2009	YEL"S"/BAR"S"/3/GR"S"/AFN//CR"S"/5/DOM"S"//CR"S"*2/ GS"S"/3/SCO"S"/4/HORA/6/LAP76/GULL"S"/7/LICAN
QUC 3104–2005 [2]	2010	ALTAR84/ALD"S"//STN"S"/CHEN"S"/ALTAR84/4/ATES1D

[1] Year of cultivar release; [2] experimental line.

The experimental design was a complete block with four replications. Each plot consisted of five rows of 2.5 m length and 0.2 m apart. Sowing dates were in August of each year and the sowing rate was 220 kg ha^{-1}. Fertilization consisted of 1.5 t ha^{-1} of lime (88%–90% $CaCO_3$) before sowing, 260 kg ha^{-1} of diammonium phosphate (46% P_2O_5, 18% N), 200 kg ha^{-1} of potassium magnesium sulfate (22% K_2O, 18% MgO, 22% S), 90 kg ha^{-1} of potassium chloride (60% K_2O), 10 kg ha^{-1} of boronatrocalcite and 3 kg ha^{-1} of zinc sulfate (35% Zn) at sowing. After sowing, an extra 133 kg ha^{-1} of urea (46% of N) was applied at tillering initiation (Zadoks 20; [24]) and 201 kg ha^{-1} at the first node (Zadoks 31). Plots were furrow irrigated according to the needs of the crop (3–4 irrigations of ~50 mm each, per season). Weeds were controlled using the pre-emergence herbicide Bacara Forte 360SC, Bayer Crop Science (800 mL ha^{-1}; 12:12:12% w/v a.i. of flufenacet/flurtamone/diflufenican) and the post-emergence Ajax, Anasac, Chile (10 g ha^{-1}; 50% w/w a.i. of metsulfuron-methyl) and MCPA 750 SL, Anasac, Chile (800 mL ha^{-1}; 95% w/v a.i. of 2-methyl-4-chlorophenoxyacetic acid)). Since the oldest cultivars showed susceptibility to rust (*Puccinia striiformis* and *Puccinia triticina*), two applications were made of the foliar fungicide Juwel-Top, Basf (100 mL ha^{-1}; 12.5:12.5:15% a.i. of kresoxim-methyl/epoxiconazole/phenopropimorph). These applications were made before symptoms appeared, to avoid any interference of these diseases in the development of the plants.

In the second experiment, a selection of 46 advanced lines (F6–F8) of durum wheat from the breeding program (Durum Yield Nursery) and four check cultivars (Llareta-INIA, Corcolén-INIA, Lleuque-INIA and Queule-INIA) were tested each year from 2006 to 2015. Two trials of 25 genotypes each, including check cultivars, were established each year in an α-lattice design with five incomplete blocks per replicate, each block containing five genotypes. There were four replicates per genotype. The plots consisted of five rows of 2 m length and 0.20 m between rows. The seed rate was the equivalent of 220 kg ha^{-1}. The sowing date was August of each year. Crop fertilization and weed control were as recommended for each year. Plots were furrow irrigated according to crop need (3–4 irrigations of ~50 mm each, per season). These trials were regularly conducted by the breeding program in order to test the most promising advance lines in comparison with the commercial cultivars (check cultivars); those advance lines with outstanding performance were evaluated for more than one year, and the rest were replaced by new ones. As a consequence, the set of advance lines evaluated in each year was composed of different elite genotypes.

2.2. Agronomic Traits

In Experiment 1 the following traits were evaluated: (a) Days from emergence to heading (DH) through periodic observations (twice per week), when approximately half of the spikes in the plot had already extruded; (b) the number of spikes per m^2 (SM2) by counting the spikes in a 1.0 m length of a row; (c) the harvest index (HI), determined from a sample from the 1.0 m row at maturity and calculated as the ratio of grain dry weight to total above ground dry weight; (d) the number of kernels per spike (KS) and thousand kernel weight (TKW) from 25 spikes taken at random from each plot and (e) the number of kernels per m^2 (KM2) calculated as SM2 × KS. In Experiments 1 and 2, plant height (PH) from the ground to the top of the spike, excluding awns, was measured at maturity, and GY was assessed by harvesting 2 m^2 (five rows, 2 m long).

2.3. Grain Quality

The test weight was evaluated in Experiments 1 (2010) and 2 (2006 to 2015), in samples of wheat free of impurities (obtained from each genotype and replicate) using a 250 cc Schopper scale (Louis Schopper, Germany). In addition, grain samples obtained from the genotypes and replicates evaluated in Experiment 1 (in 2010) were ground in mill for wet gluten and protein content determination. Wet gluten content was determined according to the International Approved Methods of Analysis (AACCI Method 38–12.02) in 10 g of pure flour mixed with 5.5 mL of a 2% saline solution, which was homogenized and then placed in a gluten washer (Glutomatic® 2200, Perten Instruments, USA) for 5 min; then the wet gluten was weighed and expressed as a percentage of the amount of pure flour. Protein content (%) was also determined in ground grain samples placed in a quartz cuvette and the reflectance spectrum between 800 and 2500 nm was determined using near infrared reflectance spectroscopy (NIRS), Bruker, USA. Yellow berry incidence was assessed on 100 g of kernels, separating and weighing the affected grains and then expressed in percentage.

2.4. Total N Content and C and N Isotope Analyses

Measurements were performed in mature grains harvested in 2011 (Experiment 1). The total N content was analyzed using an elemental analyzer (Flash 1112 EA; ThermoFinnigan, Bremen, Germany). The stable carbon ($^{13}C/^{12}C$) and nitrogen ($^{15}N/^{14}N$) isotope ratios of the same mature grains were determined in the same elemental analyzer coupled with an isotope ratio mass spectrometer (Delta C IRMS, ThermoFinnigan, Bremen, Germany). Nitrogen was expressed as a concentration (g N per g of dry weight) and atropine was used as a system check in the elemental analyses of nitrogen. The $^{13}C/^{12}C$ ratios of plant material were expressed in δ notation: $\delta^{13}C = (^{13}C/^{12}C)$ sample/$(^{13}C/^{12}C)$ standard − 1, where 'sample' refers to plant material and 'standard' of known $^{13}C/^{12}C$ ratios. The $^{15}N/^{14}N$ ratios were also expressed in δ notation ($\delta^{15}N$) using international secondary standards of known $^{15}N/^{14}N$ ratios. More details are described in del Pozo et al. [8]. Measurements were performed at the Scientific Facilities of the University of Barcelona.

2.5. Data Analysis

Complete block analysis of variance (ANOVA) were performed for the set of cultivars evaluated in Experiment 1 using IBM SPSS Statistics software (SPSS Inc, USA). In addition, correlation analyses were performed between the year of cultivar release and agronomic, grain quality and isotope composition traits, and among the different traits. Trends for grain yield, plant height and test weight of 46 advanced lines and cultivars evaluated from 2005 to 2015 in Experiment 2 are also presented.

3. Results

3.1. Agronomic Traits in Cultivars Released During the Past Six Decades

Days to heading differed significantly among cultivars and also the year × cultivar interaction was significant (Table 2); it reduced in the 1990s, but then increased in the 2000s (Figure 1A). Plant height was significantly ($p < 0.001$) reduced from 108 cm in the 1960s to 90 cm in the 1970s, with a slight increase in 2010 (Table 2; Figure 1B).

Table 2. Mean sum of squares of the analysis of variance (ANOVA) for agronomic traits of ten durum wheat cultivars cultivated during three growing seasons (2010–2012).

Source of Variation	d.f.	DH	GY	PH	SM2	KS	KM2	TKW	HI
Year	2	238.9	187.9	418.1	413,645	420.6	345.5×10^6	63.5	0.017
Cultivar	9	84.2	20.3	456.7	36,213	287.0	50.4×10^6	292.5	0.019
Block	3	0.7	**9.0**	6.9	5219	16.3	19.5×10^6	0.2	0.001
Year × Cultivar	18	*2.7*	*3.0*	12.1	16,296	*29.6*	20.1×10^6	*7.1*	0.001
Residual	87	0.4	1.7	10.3	5040	11.5	12.8×10^6	2.9	0.001
Total	120								

Level of significance is indicated in bold ($p < 0.01$) and cursive ($p < 0.05$). DH: Days to heading; GY: Grain yield; PH: Plant height; SM2: Number of spikes per m^2; KS: Kernels per spike; KM2: Kernel number per m^2; TKW: Thousand kernel weight; HI: Harvest index.

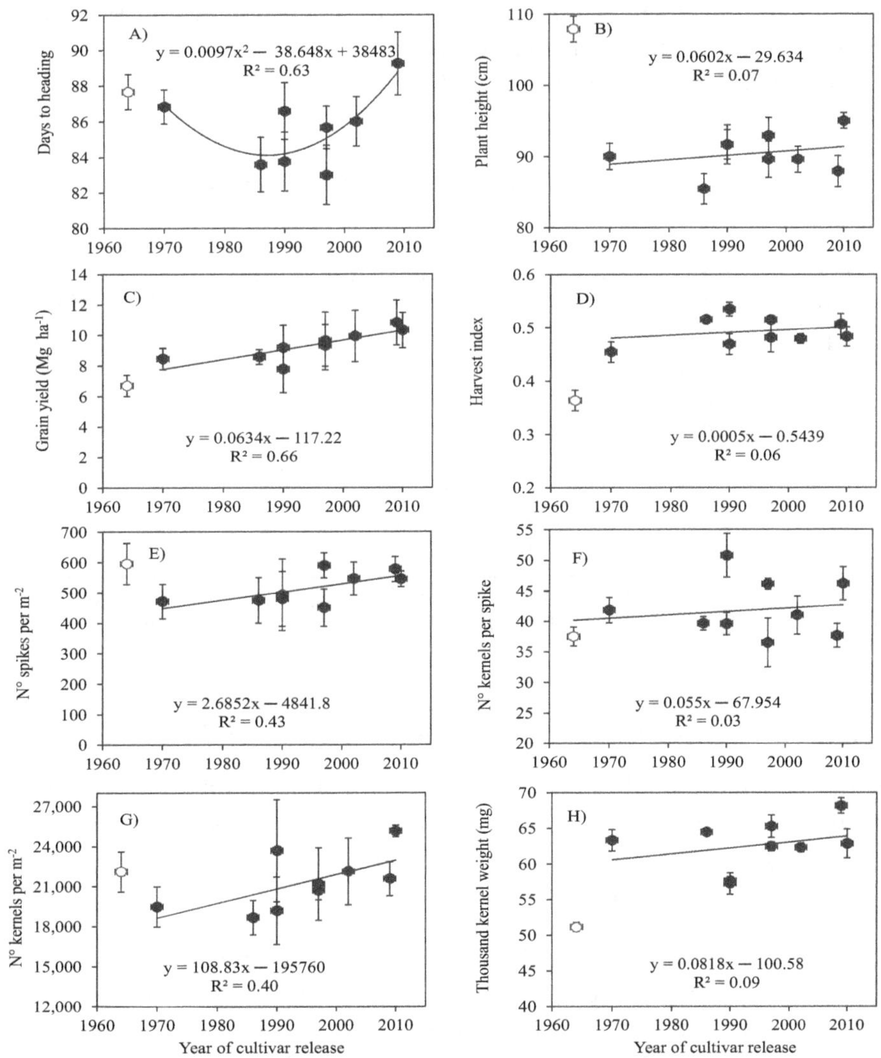

Figure 1. Relationships between the year of release of ten durum wheat cultivars and: Day to heading (**A**), plant height (**B**), grain yield (**C**), harvest index (**D**), number of spikes per square meter (**E**), number of kernels per spike (**F**), number of kernels per square meter (**G**) and thousand kernel weight (**H**). Values correspond to the average (±SE) of three growing seasons (2010–2012) except for HI, which was determined in 2010 and 2011. The oldest (1964) cultivar (open circle) was not considered in the regressions. Mean values of cultivars for each year of evaluation are shown in Supplemental Table S2.

GY exhibited a positive and linear relationship with the year of cultivar release ($R^2 = 0.66$; $p < 0.001$), this analysis excluding the oldest cultivar (1964; Figure 1C). The rate of increase in GY after 1960 was 72.8 kg ha^{-1} per year, and excluding the oldest cultivar it was 63.4 kg ha^{-1} per year. The SM2 of the ten cultivars ranged between 450 and 595 and increased significantly ($R^2 = 0.43$; $p < 0.05$) with the year of release (Figure 1E). The HI was 0.36 in the 1960s and increased to 0.45–0.53 in the 1970s and onwards, whereas TKW was 51.1 g in the 1960s and rose to 57–68 g after the 1970s, but neither trait was correlated with the year of cultivar release (Figure 1D,H). Similarly, KS was not correlated with the year of release (Figure 1F), but KM2 increased significantly ($R^2 = 0.40$; $p < 0.05$) with the year of cultivar release (Figure 1G).

The correlation matrix among the agronomic traits of the 10 cultivars evaluated during three growing seasons indicated that days to heading exhibited a positive and significant correlation with GY ($p < 0.05$) and KS ($p < 0.01$), and GY showed a positive and significant correlation with KM2 ($p < 0.05$) and TKW ($p < 0.001$; Table 3). Plant height was not correlated with GY. However, plant height showed a negative and highly significant ($p < 0.001$) correlation with TKW and HI. SM2 had a positive correlation with KM2 but a negative correlation with KS.

Table 3. Correlation matrix among agronomic traits evaluated in ten cultivars during three growing seasons (2010–2012).

	DH	GY	PH	SM2	KS	KM2	TKW	HI
DH	1.00							
GY	0.44 *	1.00						
PH	0.36	0.08	1.00					
SM2	−0.30	0.08	0.20	1.00				
KS	0.50 **	0.30	0.09	−0.61 ***	1.00			
KM2	0.06	0.39 *	0.33	0.76 ***	0.03	1.00		
TKW	0.14	0.59 ***	−0.55 ***	−0.03	0.01	0.01	1.00	
HI	−0.21	0.02	−0.75 ***	−0.32	0.34	−0.11	0.51 *	1.00

*: $p < 0.05$; **: $p < 0.01$; ***: $p < 0.001$ DH: Days to heading; GY: Grain yield; PH: Plant height; SM2: Spike number per m^2; KS: Kernels per spike; KM2: Kernel number per m^2; TKW: Thousand kernel weight; HI: Harvest index.

3.2. *Grain Quality and Kernel Isotope Composition in Cultivars Released During the Past Six Decades*

The test weight increased curvilinearly with the year of cultivar release (Figure 2A). The gluten and protein content tended to decrease with the year of cultivar release, although the correlations were not significant (Figure 2B,C). Yellow berry was higher in two cultivars, but there was no clear pattern with the year of cultivar release (Figure 2D).

The relationships between the year of cultivar release and N concentration or $\delta^{15}N$ in kernels were not significant (Figure 3A,B). The $\delta^{13}C$ of kernels tended to decrease with the year of cultivar release, although the correlation was not significant (Figure 3C). In addition, $\delta^{13}C$ was negatively correlated ($r = -0.62$; $p < 0.05$) with GY, but $\delta^{15}N$ was not correlated ($r = 0.03$; $p > 0.05$).

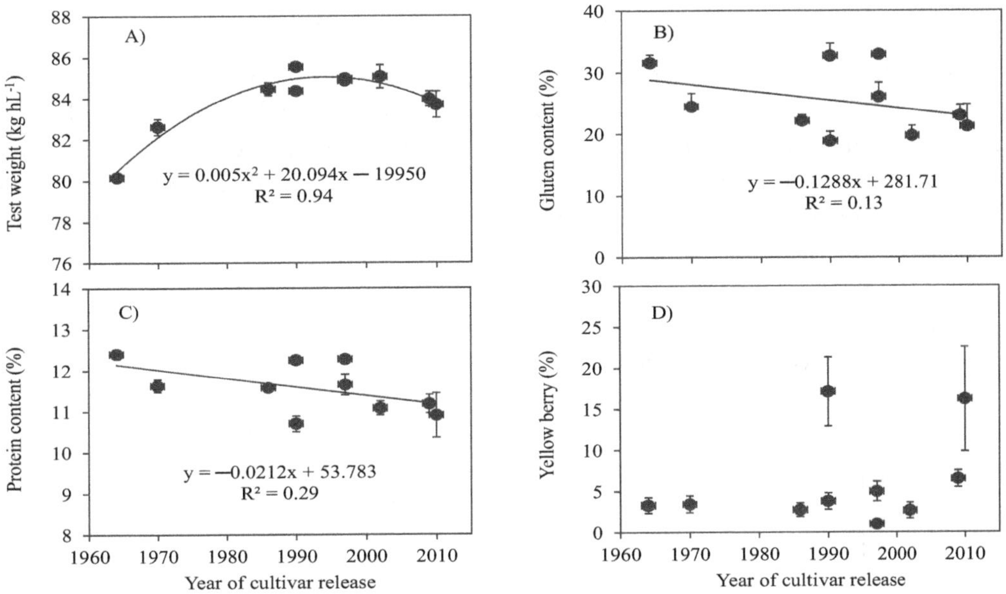

Figure 2. Relationships between the year of release of 10 durum wheat cultivars and kernel test weight (**A**), wet gluten content (**B**), protein content (**C**) and yellow berry (**D**), determined in 2010. Values correspond to the average (±SE) of four replicates.

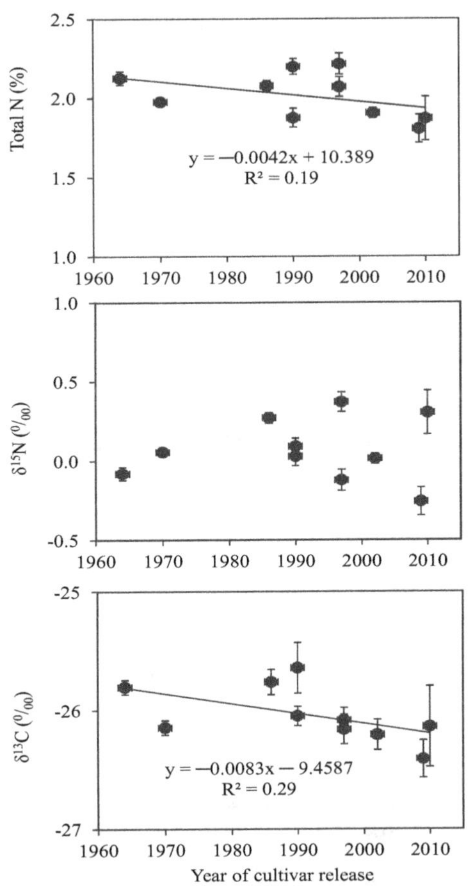

Figure 3. Relationships between the year of release of 10 durum wheat cultivars and the (**A**) total nitrogen, (**B**) natural abundance of ^{15}N (δ^{15}N) and (**C**) carbon isotope composition (δ^{13}C) in kernels, determined in 2011. Values correspond to the average (±SE) of four replicates.

3.3. *Agronomic and Grain Quality Traits in Advanced Lines During the Last Decade*

The GY and plant height of advanced lines increased from 2006 to 2015, reaching a maximum in 2011 with averages of 12.7 ± 0.8 Mg ha^{-1} and 96 ± 4.2 cm, respectively (Figure 4A,B). GY was highly correlated with plant height ($r = 0.85$; $p < 0.001$). The check cv. 'Corcolén' followed a similar trend to the advanced lines. The average GY of advanced lines and cultivars had a positive and significant ($R^2 = 0.50$; $p < 0.001$) relationship with the year of evaluation; the regression analysis indicated that the rate of increase in GY between 2006 and 2015 was 569 kg ha^{-1} per year. The test weight did not increase during this period (Figure 4C). No significant ($p < 0.05$) correlation was found between GY of advanced lines and the average temperature (maximum, minimum or mean) for the wheat growing season (August–January) from 2006 to 2015.

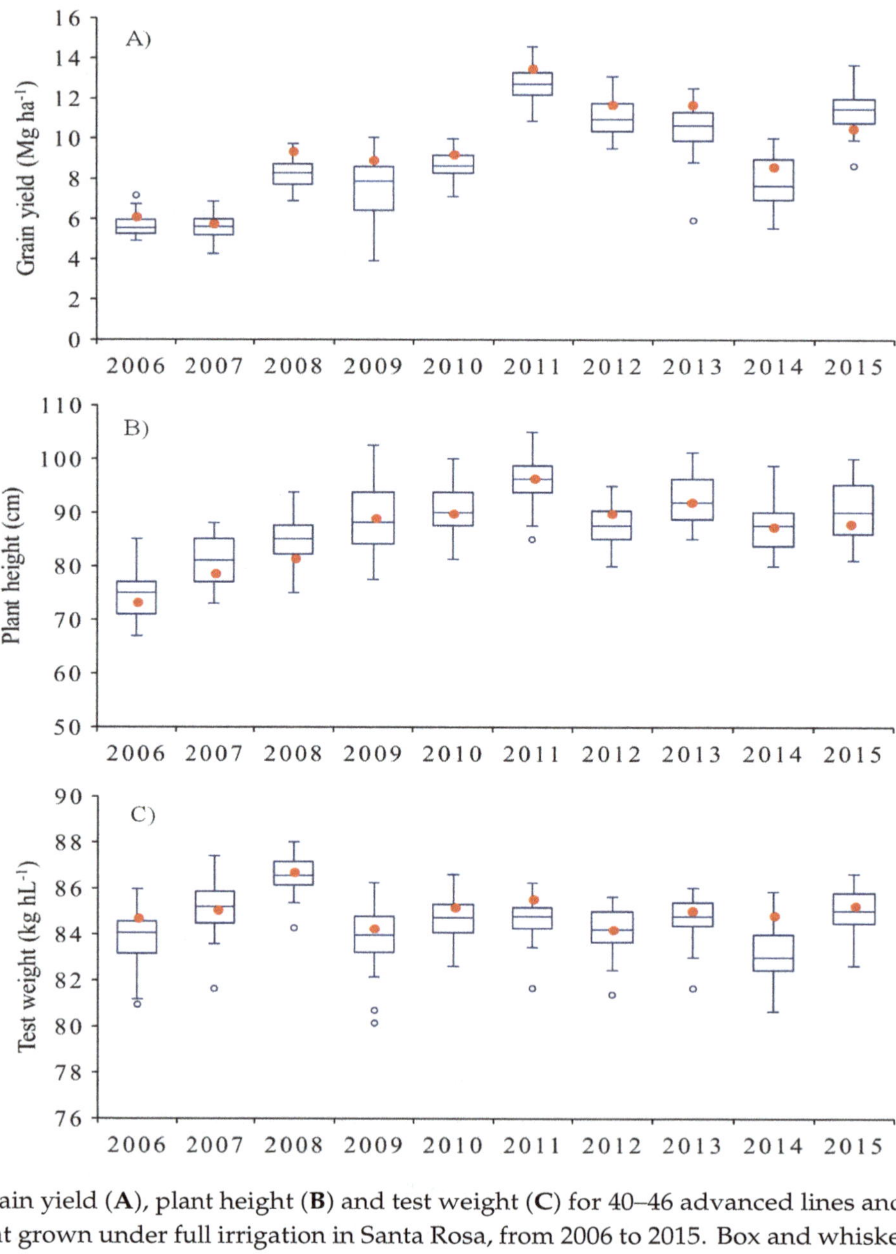

Figure 4. Grain yield (**A**), plant height (**B**) and test weight (**C**) for 40–46 advanced lines and cultivars of durum wheat grown under full irrigation in Santa Rosa, from 2006 to 2015. Box and whisker plots show the population minimum, 25th percentile/median/75th percentile and maximum. The open symbols indicate outlier data and the closed symbols indicate the check cultivar 'Corcolén'.

4. Discussion

4.1. Agronomic Traits

Modern cultivars of spring durum wheat from Chile have a very high yield potential (~13 Mg ha^{-1}) in a Mediterranean environment, under fully irrigated conditions. The yield potential achieved in Chile is clearly higher than values recorded in the Mediterranean basin. For example, high-yielding conditions in Spain usually do not surpass 8 Mg ha^{-1} [25,26], which is clearly lower than those achieved in the Mediterranean conditions of Chile. The high-yielding conditions in Spain usually imply several irrigations per season, particularly during the critical period from stem elongation to the middle grain filling, which alongside natural rainfall aims to balance the water lost due to accumulated evapotranspiration. Even so, the potential yields achieved in Spain are lower than in Chile due to a number of factors, such as Spain's shorter crop cycle duration, its higher night temperatures and the higher temperatures during the reproductive stage. The genetic advance in GY of spring durum wheat in the high-yielding environment of central Chile was 72.8 kg ha^{-1} per year (0.73% per year) for the period 1964–2010, and 63.4 kg ha^{-1} per year when the cultivar released in 1964 was excluded from the analysis (Figure 1). This is higher than the findings for spring bread wheat (43.5 kg ha^{-1} per year or 0.51% per year) for a similar period (1964–2008) and in the same Mediterranean environment [8]. It is also clearly higher than the increase reported for durum wheat in Spain (24 kg ha^{-1} y^{-1}; 0.44% y^{-1}) from 1980 to 2003, with no clear additional improvements occurring thereafter [26]. In northwest Mexico, under fully irrigated conditions, the genetic progress of spring durum and bread wheat varieties developed by CIMMYT was 0.49% and 0.41% per year, respectively, between 1966 and 2003 [27], and 0.88% per year when comparing eight bread wheat cultivars released between 1962 and 1988 [28]. A more recent study conducted at the same site in Mexico indicated that the GY progress was 30 kg ha^{-1} per year (0.59%) for spring bread wheat cultivars developed from 1966 to 2019 [29]. In Spain, under moderately irrigated conditions, the genetic progress of GY was 0.36% and 0.44% for Italian and Spanish cultivars of durum wheat, respectively, for cultivars released between 1920 and 2000 [9]. In South Australia, under rain fed conditions, the annual rate of increase in GY was 25 kg ha^{-1} for 13 cultivars released between 1958 and 2007 [30]. In North China, the annual genetic progress of spring bread wheat ranged from 0.48% (32.0 kg ha^{-1}) for cultivars released between the 1960s and the 1990s [31], and in Henan Province values of 51.3 kg ha^{-1} per year have been reported for the last three decades [32].

The yield progress observed in advanced lines of the INIA-Chile breeding program (Experiment 2), which includes genetic and agronomic progress, has been much higher (569 kg ha^{-1} per year) than in all the studies discussed above. This large increase in GY is explained partly by the genetic progress, but overall the improvements have derived from better agronomic management of durum wheat in the central-south of Chile, and this has included modifications to irrigation and particularly adjustments in fertilization practices conducted during the first three years of the program. In winter bread wheat, the yield progress was 246 kg ha^{-1} per year (2.6%) between 1976 and 1998 in central Chile under fully irrigated conditions [14]. Clearly, fine tuning of crop management can have large impacts on GY in high-yielding environments when lines or cultivars of high yield potential are available.

Plant height was reduced from 107 cm in 1964 to an average of 90 cm for the period 1970–2010 (Figure 1A), and this was the consequence of the introduction of semi-dwarfing genes in Chile in the late 1950s [33]. Plant height was negatively correlated with the year of release in Australia, in cultivars developed between 1958 and 1973, but not in cultivars released after 1973 [30], and in China, in cultivars released between 1960 and 2000 [34]. A negative correlation between plant height and GY was also reported in the study of Zhou et al. [34]. However, the comparison of advanced lines produced during the last decade (Figure 2) showed a positive correlation between plant height and GY. These results suggest that plant height of semi-dwarf wheat below 70–80 cm may limit light interception and thus canopy photosynthesis and yield potential in high-yielding environments.

The HI increased between 1964 and 1970, but after that there were no changes. Furthermore, the correlation between GY and HI was not significant. The maximum values of harvest index (0.53) found in the current work were higher than those reported by Royo et al. [9] in a set of Italian and Spanish cultivars of durum wheat released between 1920 and 2000 and tested in Spain. In studies where cultivars released before and after the green revolution were evaluated, HI and the year of cultivar release were positively correlated (e.g., [9] for durum; [6,8,31] for bread wheat), but there was no correlation in cultivars released after 1970 (Figure 1C; see also [29]).

The increase in GY was positively associated with days to heading and KM2 and TKW (Table 3). The increase in the crop cycle in an irrigated Mediterranean environment contrasts with the breeding trend observed in rain fed Mediterranean areas, where early flowering, shorter duration cultivars are selected to escape post anthesis drought [9,35,36].

TKW increased significantly from 1964 to 1970, but the correlation with the year of cultivar release was not significant for the period 1970–2000 (Figure 1F). Genetic progress in TKW can be positive, negative or null depending on whether kernel weight has been a selection target for breeders and whether there have been changes in the number of grains per year (the trade-off between seed size and number in crops; [37]). For instance, in durum wheat growing in Mediterranean environments, kernel weight was superior in modern cultivars in Turkey [38], but remained unchanged in Italian and Spanish cultivars from the 20th century [9]. In bread wheat, kernel weight has been reduced [8,14,39] or has not changed [31] with genetic improvement.

4.2. *Grain Quality and Kernel Isotope Composition*

The test weight increased in modern cultivars and was positively correlated with TKW ($r = 0.44$; $p < 0.05$). The values of test weight obtained in this work are higher than those found in durum wheat genotypes grown under rain fed conditions in different zones of Spain [40,41]. Unfortunately, gluten and protein content did not improve between 1964 and 2010. Other studies comparing older cultivars or landraces to modern cultivars of durum wheat from Mediterranean countries have revealed lower grain nitrogen or protein content in the more modern cultivars [15,16,20,21,42]. In addition, the presence of *Rht* dwarfing genes in bread and durum wheat seems to reduce the concentration of Zn, Fe, Mn and Mg in kernels [43].

A number of studies have reported a negative correlation between grain protein concentration and GY in durum wheat [44] and in bread wheat [45,46]. It is probable that the lack of genetic progress in protein content is related to the strong increase in GY of the Chilean cultivars. However, this negative relationship should not be a limitation for genetic improvement in quality traits in grains of durum wheat because protein composition seems to be more important than the concentration [16,47].

The relationship between kernel $\delta^{13}C$ and GY was negative, suggesting that genotypes exhibiting higher water use are the most productive [21,48]. In bread wheat under fully irrigated conditions, modern and more productive cultivars showed lower $\delta^{18}O$ and $\delta^{13}C$, and higher stomatal conductance [8,18,21,25]. This negative relationship between $\delta^{13}C$ and GY (or positive relationship between $\Delta^{13}C$ and GY) has also been found in rain fed Mediterranean conditions ([21] for durum wheat; [49,50] for bread wheat), suggesting that the most productive lines are those able to maintain higher stomatal conductance and use more water [51]. In addition, the stomatal conductance of post green revolution wheat cultivars in Australia seem to show a lower sensitivity to vapor pressure deficit above 2 kPa compared to older cultivars [52], and this can be associated with lower (more negative) $\delta^{13}C$ values.

In summary, changes in a number of traits have occurred in durum wheat cultivars selected for high-yielding environments in Chile. The large genetic progress in grain yield was associated with increases in days to heading, KM2 and TKW. The test weight has also increased with the year of cultivar release, but the gluten and protein content have not improved between 1964 and 2010. Interestingly, the increase in yield potential seems related to longer duration and somewhat taller plants that are able to use more water.

5. Conclusions

This study provided evidence that a high genetic advance in GY for durum wheat is feasible under high yielding conditions. The increase in GY was a consequence of a greater number of kernels per m^2 and higher kernel weight in the more modern cultivars. The test weight was lower in the 1960s and increased curvilinearly with year of cultivar release. The gluten and protein content did not improve between 1964 and 2010. GY was negatively correlated with kernel $\delta^{13}C$, suggesting that genotypes exhibiting higher water use are the most productive. The yield progress of a set of advanced lines evaluated between 2006 and 2015 was very high, due to genetic progress, but this was also due to management improvements, particularly adjustment of fertilization practices conducted during the first three years. Unlike other Mediterranean agro-environments, a longer growing cycle together with taller plants seems to be related to the increase in the GY of Chilean durum wheat during recent decades.

Author Contributions: I.M. designed the experiment and selected the germplasm. D.C. and K.R. were in charge of the management of the experiment and evaluation of agronomic traits. M.D.S. contributed to the isotope analysis. A.M.M.-E. contributed to data analysis. A.d.P. performed the data analysis and was in charge of writing the text but all the authors contributed to the manuscript.

Acknowledgments: We thank Alejandro Castro for technical assistance with the field experiments.

References

1. Elias, E.M.; Manthey, F.A. End products: Present and future uses. In *Durum Wheat Breeding Current Approaches and Future Strategies*; Royo, C., Nachit, M.M., di Fonzo, N., Araus, J.L., Pfeiffer, W.H., Slafer, G.A., Eds.; Food Products Press: New York, NY, USA, 2005; pp. 63–86.
2. Acevedo, E.; Silva, P. Trigo candeal: Calidad, mercado y zonas de cultivo. In *Serie Ciencias Agronómicas N°12*; Universidad de Chile, Facultad de Ciencias Agronómicas: Santiago, Chile, 2007. (In Spanish)
3. ODEPA. Estadísticas Productivas. 2017. Available online: http://www.odepa.gob.cl/estadisticas/productivas/ (accessed on 23 November 2018). (In Spanish).
4. Miralles, D.J.; Slafer, G.A. Sink limitations to yield in wheat: How could it be reduced? *J. Agric. Sci.* **2007**, *145*, 139–149. [CrossRef]
5. Hedden, P. The genes of the Green Revolution. *Trends Genet.* **2003**, *19*, 5–9. [CrossRef]
6. Shearman, V.J.; Sylvester-Bradley, R.; Scott, R.K.; Foulkes, M.J. Physiological processes associated with wheat yield progress in the UK. *Crop Sci.* **2005**, *45*, 175–185.
7. Del Pozo, A.; Matus, I.; Araus, J.L.; Serret, D. Agronomic and physiological traits associated with breeding advances of wheat under high-productive Mediterranean conditions. The case of Chile. *Environ. Exp. Bot.* **2014**, *103*, 180–189. [CrossRef]
8. Trethowan, R.M.; Reynolds, M.P.; Ortiz-Monasterio, J.I.; Ortiz, R. The genetic basis of the green revolution in wheat production. *Plant Breed. Rev.* **2007**, *8*, 39–58.
9. Royo, C.; Álvaro, F.; Martos, V.; Ramdani, A.; Isidro, J.; Villegas, D.; García del Moral, L.F. Genetic changes in durum wheat yield components and associated traits in Italian and Spanish varieties during the 20th century. *Euphytica* **2007**, *155*, 259–270. [CrossRef]
10. De Vita, P.; Li Destri Nicosia, O.; Nigro, F.; Platani, C.; Riefolo, C.; di Fonzo, N.; Cattivelli, L. Breeding progress in morpho-physiological, agronomical and qualitative traits of durum wheat cultivars released in Italy during the 20th century. *Eur. J. Agron.* **2007**, *26*, 39–53. [CrossRef]
11. Lo Valvo, P.J.; Miralles, D.J.; Serrago, R.A. Genetic progress in Argentine bread wheat varieties released between 1918 and 2011: Changes in physiological and numerical yield components. *Field Crop Res.* **2018**, *221*, 314–321. [CrossRef]

12. Ortiz-Monasterio, J.I.; Peña, R.J.; Sayre, K.D.; Rajaram, S. CIMMYT's Genetic Progress in Wheat grain quality under four nitrogen rates. *Crop Sci.* **1997**, *37*, 892–898. [CrossRef]
13. Dencic, S.; Kobiljski, B.; Mladenov, N.; Hristov, N.; Pavlovic, M. Long-term breeding for bread making quality in wheat. In *Wheat Production in Stressed Environments*; Buck, H.T., Nisi, J.E., Salomon, N., Eds.; Springer: New York, NY, USA, 2005; pp. 495–501.
14. Matus, I.; Mellado, M.; Pinares, M.; Madariaga, R.; del Pozo, A. Genetic progress in winter wheat cultivars released in Chile from 1920 and 2000. *Child. J. Agric. Res.* **2012**, *72*, 303–308. [CrossRef]
15. Motzo, R.; Fois, S.; Giunta, F. Relationship between grain yield and quality of durum wheats from different eras of breeding. *Euphytica* **2004**, *140*, 147–154. [CrossRef]
16. Subira, J.; Peña, R.J.; Álvaro, F.; Ammar, K.; Ramdani, A.; Royo, C. Breeding progress in the pasta-making quality of durum wheat cultivars released in Italy and Spain during the 20th Century. *Crop Pasture Sci.* **2014**, *65*, 16–26. [CrossRef]
17. Fischer, R.A.D.; Rees, K.D.; Sayre, Z.M.; Lu, A.G.; Condon, A. Larqué-Saavedra, Wheat yield progress associated with higher stomatal conductance and photosynthetic rate, and cooler canopies. *Crop Sci.* **1998**, *38*, 1467–1475. [CrossRef]
18. Barbour, M.M.; Fischer, R.A.; Sayre, K.D.; Farquhar, G.D. Oxygen isotope ratio of leaf and grain material correlates with stomatal conductance and grain yield in irrigated wheat. *Aust. J. Plant Physiol.* **2000**, *27*, 625–637. [CrossRef]
19. Foulkes, M.J.; de Silva, J.; Gaju, O.; Carvalho, P. Relationships between $\delta^{13}C$, $\delta^{18}O$ and grain yield in bread wheat genotypes under favourable irrigated and rain-fed conditions. *Field Crop Res.* **2016**, *196*, 237–250. [CrossRef]
20. Araus, J.L.; Ferrio, J.P.; Buxó, R.; Voltas, J. The historical perspective of dryland agriculture: Lessons learned from 10,000 years of wheat cultivation. *J. Exp. Bot.* **2007**, *58*, 131–145. [CrossRef] [PubMed]
21. Araus, J.L.; Cabrera-Bosquet, L.; Serret, M.D.; Bort, J.; Nieto-Taladriz, M.T. Comparative performance of $\delta^{13}C$, $\delta^{18}O$ and $\delta^{15}N$ for phenotyping durum wheat adaptation to a dryland environment. *Funct. Plant Biol.* **2013**, *40*, 595–608. [CrossRef]
22. Royo, C.; Martos, V.; Ramdani, A.; Villegas, D.; Rharrabti, Y.; García del Moral, L.F. Changes in yield and carbon isotope discrimination of Italian and Spanish durum wheat during the 20th century. *Agron. J.* **2008**, *100*, 352–360. [CrossRef]
23. Zagal, E.; Rodríguez, N.; Vidal, I.; Quezada, L. Actividad microbiana en un suelo de origen volcánico bajo distinto manejo agronómico. *Agric. Técnica* **2002**, *62*, 297–309. (In Spanish) [CrossRef]
24. Zadoks, J.C.; Chang, T.T.; Konzak, C.F. A decimal code for the growth stages of cereals. *Weed Res.* **1974**, *14*, 415–421. [CrossRef]
25. Araus, J.L.; Villegas, D.; Aparicio, N.; García del Moral, L.F.; El Hani, S.; Rharrabti, Y.; Ferrio, J.P.; Royo, C. Environmental factors determining carbon isotope discrimination and yield in durum wheat under Mediterranean conditions. *Crop Sci.* **2003**, *43*, 170–180. [CrossRef]
26. Chairi, F.; Vergara Diaz, O.; Vatter, T.; Aparicio, N.; Nieto-Taladriz, M.T.; Kefauver, S.C.; Bort, J.; Serret, M.D.; Araus, J.L. Post-Green Revolution genetic advance in durum wheat: The case of Spain. *Field Crop Res.* **2018**, *228*, 158–169. [CrossRef]
27. Fischer, R.A. Understanding the physiological basis of yield potential in wheat. *J. Agric. Sci.* **2007**, *145*, 99–113. [CrossRef]
28. Sayre, K.D.; Rajaram, S.; Fischer, R.A. Yield progress in short bread wheats in northwest Mexico. *Crop Sci.* **1997**, *37*, 36–42. [CrossRef]
29. Aisawi, K.A.B.; Reynolds, M.P.; Singh, R.P.; Foulkes, M.J. The physiological basis of the genetic progress in yield potential of CIMMYT spring wheat cultivars from 1966 to 2009. *Crop Sci.* **2015**, *55*, 1749–1764. [CrossRef]
30. Sadras, V.; Lawson, C. Genetic gain in yield and associated changes in phenotype, trait plasticity and competitive ability of South Australian wheat varieties released between 1958 and 2007. *Crop Pasture Sci.* **2011**, *62*, 533–549. [CrossRef]
31. Zhou, Y.; He, Z.H.; Chen, X.M.; Wang, D.S.; Yan, J.; Xia, X.C.; Zhang, Y. Genetic improvement of wheat yield potential in North China. In *Wheat Production in Stressed Environments*; Buck, H.T., Nisi, J.E., Salomón, N., Eds.; Springer: New York, NY, USA, 2007; pp. 583–589.

32. Zheng, T.C.; Zhang, X.K.; Yin, G.H.; Wang, L.N.; Han, Y.L.; Chen, L.; Huang, F.; Tang, J.W.; Xia, X.C.; He, Z.H. Genetic gains in grain yield, net photosynthesis and stomatal conductance achieved in Henan Province of China between 1981 and 2008. *Field Crop Res.* **2011**, *122*, 225–233. [CrossRef]
33. Zapata, C.; Silva, P.; Acevedo, E. Comportamiento de isolíneas de altura en relación con el rendimiento y distribución de asimilados en trigo. *Agric. Técnica* **2004**, *64*, 139–155. (In Spanish) [CrossRef]
34. Zhou, B.; Elazab, A.; Bort, J.; Sanz-Sáez, A.; Nieto-Taladriz, M.T.; Serret, M.D.; Araus, J.L. Agronomic and physiological responses of Chinese facultative wheat genotypes to high-yielding Mediterranean conditions. *J. Agric. Sci.* **2016**, *154*, 870–889. [CrossRef]
35. Perry MWD'Antuono, M.F. Yield improvement and associated characteristics of some Australian spring wheat cultivars introduced between 1860 and 1982. *Aust. J. Agric. Res.* **1989**, *40*, 457–472.
36. Loss, S.P.; Kirby, E.J.M.; Siddique, K.H.M.; Perry, M.W. Grain growth and development of old and modern Australian wheats. *Field Crop Res.* **1989**, *21*, 131–146. [CrossRef]
37. Sadras, V.O. Evolutionary aspects of the trade-off between seed size and number in crops. *Field Crop Res.* **2007**, *100*, 125–138. [CrossRef]
38. Koç, M.; Celaleddin, B.; Genç, I. Photosynthesis and productivity of old and modern durum wheats in a Mediterranean environment. *Crop Sci.* **2003**, *43*, 2089–2098. [CrossRef]
39. Guarda, G.; Padovan, S.; Delogu, G. Grain yield, nitrogen-use efficiency and baking quality of old and modern Italian bread-wheat cultivars grown at different nitrogen levels. *Eur. J. Agron.* **2004**, *21*, 181–192. [CrossRef]
40. Rharrabti, Y.; Royo, C.; Villegas, D.; Paricio, N.; García del Moral, L.F. Durum wheat quality in Mediterranean environments: I. Quality expression under different zones, latitudes and water regimes across Spain. *Field Crop Res.* **2003**, *80*, 123–131. [CrossRef]
41. Garrido-Lestache, E.; López-Bellido, R.J.; López-Bellido, L. Durum wheat quality under Mediterranean conditions as affected by N rate, timing and splitting, N form and S fertilization. *Eur. J. Agron.* **2005**, *23*, 265–278. [CrossRef]
42. Acreche, M.M.; Slafer, G.A. Variation of grain nitrogen content in relation with grain yield in old and modern Spanish wheats grown under a wide range of agronomic conditions in a Mediterranean region. *J. Agric. Sci.* **2009**, *147*, 657–667. [CrossRef]
43. Velu, G.; Singh, R.P.; Huerta, J.; Guzmán, C. Genetic impact of *Rht* dwarfing genes on grain micronutrients concentration in wheat. *Field Crop Res.* **2017**, *214*, 373–377. [CrossRef]
44. Blanco, A.; Mangini, G.; Giancaspro, A.; Giove, S.; Colasuonno, P.; Simeone, R.; Signorile, A.; De Vita, P.; Mastrangelo, A.M.; Cattivelli, L. Relationships between grain protein content and grain yield components through quantitative trait locus analyses in a recombinant inbred line population derived from two elite durum wheat cultivars. *Mol. Breed.* **2012**, *30*, 79–92. [CrossRef]
45. Oury, F.X.; Godin, C. Yield and grain protein concentration in bread wheat: How to use the negative relationship between the two characters to identify favourable genotypes? *Euphytica* **2007**, *157*, 45–57. [CrossRef]
46. Bogard, M.; Allard, V.; Brancourt-Hulmel, M.; Heumez, E.; Machet, J.M.; Jeuffroy, M.-H.; Gate, P.; Martre, P.; Le Gouis, J. Deviation from the grain protein concentration–grain yield negative relationship is highly correlated to post-anthesis N uptake in winter wheat. *J. Exp. Bot.* **2010**, *61*, 4303–4312. [CrossRef] [PubMed]
47. De Santis, M.A.; Giuliani, M.M.; Giuzio, L.; de Vita, P.; Lovegrove, A.; Shewry, P.R.; Flagella, Z. Differences in gluten protein composition between old and modern durum wheat genotypes in relation to 20th century breeding in Italy. *Eur. J. Agron.* **2017**, *87*, 19–29. [CrossRef] [PubMed]
48. Blum, A. Effective use of water (EUW) and not water-use efficiency (WUE) is the target of crop yield improvement under drought stress. *Field Crop Res.* **2009**, *112*, 119–123. [CrossRef]
49. Rebetzke, G.J.; Condon, A.G.; Farquhar, G.D.; Appels, R.; Richards, R.A. Quantitative trait loci for carbon isotope discrimination are repeatable across environments and wheat mapping populations. *Appl. Genet.* **2008**, *118*, 123–137. [CrossRef] [PubMed]
50. Del Pozo, A.; Yáñez, A.; Matus, I.; Tapia, G.; Castillo, D.; Araus, J.L.; Sanchez-Jardón, L. Physiological traits associated with wheat yield potential and performance under water-stress in a Mediterranean environment. *Front. Plant Sci.* **2016**, *7*, 987. [CrossRef] [PubMed]

51. Roche, D. Stomatal conductance is essential for higher yield potential of C3 crops. *Crit. Rev. Plant Sci.* **2015**, *34*, 429–453. [CrossRef]
52. Schoppach, R.; Fleury, D.; Sinclair, T.R.; Sadok, W. Transpiration sensitivity to evaporative demand across 120 years of breeding of Australian wheat cultivars. *J. Agric. Crop Sci.* **2017**, *203*, 219–226. [CrossRef]

Permissions

All chapters in this book were first published by MDPI; hereby published with permission under the Creative Commons Attribution License or equivalent. Every chapter published in this book has been scrutinized by our experts. Their significance has been extensively debated. The topics covered herein carry significant findings which will fuel the growth of the discipline. They may even be implemented as practical applications or may be referred to as a beginning point for another development.

The contributors of this book come from diverse backgrounds, making this book a truly international effort. This book will bring forth new frontiers with its revolutionizing research information and detailed analysis of the nascent developments around the world.

We would like to thank all the contributing authors for lending their expertise to make the book truly unique. They have played a crucial role in the development of this book. Without their invaluable contributions this book wouldn't have been possible. They have made vital efforts to compile up to date information on the varied aspects of this subject to make this book a valuable addition to the collection of many professionals and students.

This book was conceptualized with the vision of imparting up-to-date information and advanced data in this field. To ensure the same, a matchless editorial board was set up. Every individual on the board went through rigorous rounds of assessment to prove their worth. After which they invested a large part of their time researching and compiling the most relevant data for our readers.

The editorial board has been involved in producing this book since its inception. They have spent rigorous hours researching and exploring the diverse topics which have resulted in the successful publishing of this book. They have passed on their knowledge of decades through this book. To expedite this challenging task, the publisher supported the team at every step. A small team of assistant editors was also appointed to further simplify the editing procedure and attain best results for the readers.

Apart from the editorial board, the designing team has also invested a significant amount of their time in understanding the subject and creating the most relevant covers. They scrutinized every image to scout for the most suitable representation of the subject and create an appropriate cover for the book.

The publishing team has been an ardent support to the editorial, designing and production team. Their endless efforts to recruit the best for this project, has resulted in the accomplishment of this book. They are a veteran in the field of academics and their pool of knowledge is as vast as their experience in printing. Their expertise and guidance has proved useful at every step. Their uncompromising quality standards have made this book an exceptional effort. Their encouragement from time to time has been an inspiration for everyone.

The publisher and the editorial board hope that this book will prove to be a valuable piece of knowledge for researchers, students, practitioners and scholars across the globe.

List of Contributors

Hiroki Saito
Graduate School of Agriculture, Kyoto University, Kyoto, Kyoto 606-8502, Japan
Tropical Agriculture Research Front, Japan International Research Center of Agricultural Science, Ishigaki, Okinawa 907-0002, Japan

Yutaka Okumoto and Masayoshi Teraishi
Graduate School of Agriculture, Kyoto University, Kyoto, Kyoto 606-8502, Japan

Takuji Tsukiyama
Graduate School of Agriculture, Kyoto University, Kyoto, Kyoto 606-8502, Japan
Faculty of Agriculture, Kindai University, Nara, Nara 631-8505, Japan

Chong Xu and Takatoshi Tanisaka
Graduate School of Agriculture, Kyoto University, Kyoto, Kyoto 606-8502, Japan
School of Agriculture, Kibi International University, Minami-Awaji 656-0484, Japan

Anna Gagliardi, Federica Carucci, Zina Flagella, Giuseppe Gatta and Marcella Michela Giuliani
Department of Agricultural Food and Environmental Science, University of Foggia, 71122 Foggia FG, Italy

Stefania Masci
Department of Agricultural and Forest Sciences, University of Tuscia, 01100 Viterbo, Italy

Daiki Toyomoto
United Graduate School of Agricultural Sciences, Kagoshima University, 1-21-24 Korimoto, Kagoshima, Kagoshima 890-0065, Japan

Masato Uemura
Faculty of Agriculture, Kagoshima University, 1-21-24 Korimoto, Kagoshima, Kagoshima 890-0065, Japan

Katsuyuki Ichitani
United Graduate School of Agricultural Sciences, Kagoshima University, 1-21-24 Korimoto, Kagoshima, Kagoshima 890-0065, Japan
Faculty of Agriculture, Kagoshima University, 1-21-24 Korimoto, Kagoshima, Kagoshima 890-0065, Japan

Satoru Taura
Institute of Gene Research, Kagoshima University, 1-21-24 Korimoto, Kagoshima, Kagoshima 890-0065, Japan

Tadashi Sato
Graduate School of Agricultural Science, Tohoku University, Sendai, Miyagi 980-8577, Japan

Amadou Tidiane Sall
Institut Sénégalais de Recherches Agricoles (ISRA), Saint-Louis 46024, Senegal
International Center for Agricultural Research in the Dry Areas (ICARDA), Rabat 10000, Morocco

Kemal Seid-Ahmed and Filippo Maria Bassi
International Center for Agricultural Research in the Dry Areas (ICARDA), Rabat 10000, Morocco

Tiberio Chiari
Italian Agency for Development Cooperation (AICS), Addis Ababa 1000, Ethiopia

Wasihun Legesse
Ethiopian Institute Agricultural Research (EIAR), Agricultural Research Center, Debre Zeit 1000, Ethiopia

Rodomiro Ortiz
Sveriges lantbruksuniversitet (SLU), Institutionen för Växtförädling (VF), 23053 Alnarp, Sweden

Maarten van Ginkel
International Center for Agricultural Research in the Dry Areas (ICARDA), Amman 1118, Jordan

Kumpei Shiragaki, Toshinobu Morikawa, Masayuki Oda and Takahiro Iizuka
Graduate School of Life and Environmental Sciences, Osaka Prefecture University, 1-1 Gakuen-cho, Nakaku, Sakai, Osaka 599-8531, Japan

Tsutomu Kuboyama
College of Agriculture, Ibaraki University, 3-21-1 Chuo, Ami, Ibaraki 300-0393, Japan

Takahiro Tezuka
Graduate School of Life and Environmental Sciences, Osaka Prefecture University, 1-1 Gakuen-cho, Nakaku, Sakai, Osaka 599-8531, Japan
Education and Research Field, College of Life, Environment, and Advanced Sciences, Osaka Prefecture University, 1-1 Gakuen-cho, Nakaku, Sakai, Osaka 599-8531, Japan

Filippo M. Bassi
International Center for the Agricultural Research in Dry Area (ICARDA), Rabat 10000, Morocco

Miloudi Nachit and Khaoula El Hassouni
Faculty of sciences, University of Mohammed V, Rabat 10000, Morocco
International Center for the Agricultural Research in Dry Area (ICARDA), Rabat 10000, Morocco

Bouchra Belkadi and Abdelkarim Filali-Maltouf
Faculty of sciences, University of Mohammed V, Rabat 10000, Morocco

Amadou Tidiane-Sall
International Center for the Agricultural Research in Dry Area (ICARDA), Rabat 10000, Morocco
Institut Sénégalais de Recherches Agricoles (ISRA), Saint-Louis 46024, Senegal

Ayed Al-Abdallat
Faculty of Agriculture, The University of Jordan, Amman 11942, Jordan

Hayba Badro, Agnelo Furtado and Robert Henry
Queensland Alliance for Agriculture and Food Innovation, University of Queensland, Brisbane, QLD 4072, Australia

Abdelhamid Mekhlouf
Laboratoire d'Amélioration et de Développement de la Production Végétale et Animale (LADPVA), University of Ferhat ABBAS (UFAS-Sétif1), Setif 19000, Algeria

Ridha Boudiar
Laboratoire d'Amélioration et de Développement de la Production Végétale et Animale (LADPVA), University of Ferhat ABBAS (UFAS-Sétif1), Setif 19000, Algeria
Biotechnology Research Center (CRBt), UV 03 BP E73, Nouvelle Ville Ali Mendjli, Constantine 25016, Algeria

Juan M. González
Department of Biomedicine and Biotechnology, University of Alcalá, 28805 Alcalá de Henares, Spain

Ana M. Casas and Ernesto Igartua
Estación Experimental de Aula Dei, EEAD-CSIC, Avenida Montañana 1005, 50059 Zaragoza, Spain

Cuong D. Nguyen
United Graduate School of Agricultural Sciences, Kagoshima University, Kagoshima 890-0065, Japan
College of Food Industry, 101B Le Huu Trac Street, Son Tra District, Da Nang City 550000, Vietnam

Holden Verdeprado
International Rice Research Institute, Metro Manila 1301, Philippines

Demeter Zita
Faculty of Agriculture, Saga University, Saga 840-8502, Japan

Sachiyo Sanada-Morimura and Masaya Matsumura
NARO Kyushu Okinawa Agricultural Research Center, 2421 Suya, Koshi, Kumamoto 861–1192, Japan

Parminder S. Virk
International Rice Research Institute, Metro Manila 1301, Philippines
International Center for Tropical Agriculture, A.A, 6713 Cali, Colombia

Darshan S. Brar
International Rice Research Institute, Metro Manila 1301, Philippines
School of Agricultural Biotechnology, Punjab Agricultural University, Ludhiana 141027, India

Finbarr G. Horgan
EcoLaVerna Integral Restoration Ecology, Bridestown, Kildinan, Co. Cork, T56 CD39, Ireland

Hideshi Yasui
Plant Breeding Laboratory, Graduate School, Kyushu University, Fukuoka 812-8581, Japan
Plant Breeding Laboratory, Faculty of Agriculture, Kyushu University. 744, Motooka, Nishi-ku, Fukuoka 819-0395, Japan

Daisuke Fujita
United Graduate School of Agricultural Sciences, Kagoshima University, Kagoshima 890-0065, Japan
International Rice Research Institute, Metro Manila 1301, Philippines
Faculty of Agriculture, Saga University, Saga 840-8502, Japan
Plant Breeding Laboratory, Graduate School, Kyushu University, Fukuoka 812-8581, Japan

Edossa Fikiru Wayima
Department of Plant Sciences, College of Agriculture and Natural Resources, Madda Walabu University, Bale-Robe, Ethiopia

Ayalew Ligaba-Osena
Laboratory of Molecular Biology and Biotechnology, The University of North Carolina at Greensboro, 321 McIver St., 308 Sullivan Science Building, Greensboro, NC 27412, USA

Kifle Dagne and Kassahun Tesfaye
Cellular and Molecular Biology Department, College of Natural Sciences, Addis Ababa University, Addis Ababa, Ethiopia

Eunice Magoma Machuka and Samuel Kilonzo Mutiga
Biosciences Eastern and Central Africa-International Livestock Research Institute (BecA-ILRI) Hub, Nairobi, Kenya

Emmanuel Delhaize
CSIRO Agriculture and Food, Canberra 2601, Australia

Takayuki Ogami, Atsushi Yoshimura and Yoshiyuki Yamagata
Plant Breeding Laboratory, Faculty of Agriculture, Kyushu University. 744, Motooka, Nishi-ku, Fukuoka 819-0395, Japan

Ioannis N. Xynias
School of Agricultural Sciences, University of Western Macedonia, Terma Kontopoulou, 53100 Florina, Greece

Ioannis Mylonas, Evangelos G. Korpetis and Aphrodite Tsaballa
Institute of Plant Breeding and Genetic Resources, Hellenic Agricultural Organization–"Demeter", 57001 Thessaloniki, Greece

Elissavet Ninou
Institute of Industrial and Fodder Crops, Hellenic Agricultural Organization Demeter–"Demeter", 41335 Larissa, Greece

Ilias D. Avdikos and Athanasios G. Mavromatis
Laboratory of Genetics and Plant Breeding, School of Agriculture, Aristotle University of Thessaloniki, 54124 Thessaloniki, Greece

Dinh Thi Lam
United Graduate School of Agricultural Sciences, Iwate University, Morioka, Iwate 020-8550, Japan
Institute of Agricultural Science for Southern Vietnam, District 1, Ho Chi Minh City 121, Vietnam

Robert J. Henry
Queensland Alliance for Agriculture and Food Innovation, University of Queensland, Brisbane QLD 4072, Australia

Ryuji Ishikawa
Faculty of Agriculture and Life Science, Hirosaki University, 3 Bunkyo-cho, Hirosaki, Aomori 036-8561, Japan

Motonori Tomita
Research Institute of Green Science and Technology, Shizuoka University, 836 Ohya, Suruga-ku, Shizuoka City, Shizuoka 422-8529, Japan

Keiichiro Ishimoto
Faculty of Agriculture, Tottori University, 4-101 Koyama Minami, Tottori 680-8550, Japan

Ana María Méndez-Espinoza and Alejandro del Pozo
Centro de Mejoramiento Genético y Fenómica Vegetal, Facultad de Ciencias Agrarias, Universidad de Talca, 3460000 Talca, Chile

Iván Matus and Dalma Castillo
CRI-Quilamapu, Instituto de Investigaciones Agropecuarias, 3800062 Chillán, Chile

Kurt Ruf
Centro de Mejoramiento Genético y Fenómica Vegetal, Facultad de Ciencias Agrarias, Universidad de Talca, 3460000 Talca, Chile
CRI-Quilamapu, Instituto de Investigaciones Agropecuarias, 3800062 Chillán, Chile

María Dolores Serret
Centre de Recerca en Agrotecnologia (AGROTECNIO), 2519825198 Lleida, Spain

Index

www.ingramcontent.com/pod-product-compliance
Lightning Source LLC
LaVergne TN
LVHW082123200926
848484LV00008B/1999
* 9 7 9 8 8 8 8 3 6 0 4 6 0 *